Cloud Computing and Digital Media

Fundamentals, Techniques, and Applications

Cloud Computing and Digital Media

Fundamentals, Techniques, and Applications

Edited by
Kuan-Ching Li
Qing Li
Timothy K. Shih

CRC Press
Taylor & Francis Group
Boca Raton London New York

CRC Press is an imprint of the
Taylor & Francis Group, an **informa** business
A CHAPMAN & HALL BOOK

CRC Press
Taylor & Francis Group
6000 Broken Sound Parkway NW, Suite 300
Boca Raton, FL 33487-2742

First issued in paperback 2019

© 2014 by Taylor & Francis Group, LLC
CRC Press is an imprint of Taylor & Francis Group, an Informa business

No claim to original U.S. Government works

ISBN-13: 978-1-4665-6917-1 (hbk)
ISBN-13: 978-0-367-37882-0 (pbk)

Library of Congress Cataloging-in-Publication Data

Cloud computing and digital media : fundamentals, techniques, and applications / edited by Kuan-Ching Li, Qing Li, and Timothy K. Shih.
     pages cm
Includes bibliographical references and index.
ISBN 978-1-4665-6917-1 (hardback)
  1. Cloud computing. 2. Digital communications. 3. Computer storage devices. I. Li, Kuan-Ching.

QA76.585.C5726 2014
004.67'82--dc23

2013041050

Visit the Taylor & Francis Web site at
http://www.taylorandfrancis.com

and the CRC Press Web site at
http://www.crcpress.com

Contents

Foreword

CLOUD COMPUTING WAS INITIALLY ONLY AIMING AT PROVIDING on-demand computing via shared pools of computational infrastructures. In just a few years, cloud computing has dramatically expanded its horizon to offer on-demand services to a broad range of configurable resources-sharing scenarios in networking, servers, storage, software, and applications. Such a thriving development of cloud computing is largely credited to an array of attractive benefits that include on-demand self-service provision, Internet-wide and device-independent access, rapid response to dynamic service requests, and usage-based pricing. The expansion from computational infrastructure sharing to a broader range of common resource sharing propels cloud computing into many new application domains that were not considered possible even when cloud computing was originally introduced.

Although we have witnessed an unprecedented boom in the development of various cloud computing–related technologies, commercially viable cloud computing services are still considered to be at an early stage of market adoption. However, according to many marketing analysts, cloud computing service revenues have been, and continue to be, growing strongly. Based on recent forecasts by leading market analysis firms, the compound growth rate for cloud computing services should remain at 20% or even higher for the next few years. Such a strong revenue growth should in turn fuel more comprehensive innovations in both technology advancement and application development. We have every reason to anticipate a profound penetration of cloud computing technology into all walks of digital life in the years to come.

Among various technical disciplines that have been vigorously impacted by cloud computing, digital media is probably the most prominent beneficiary from the recent advances in cloud computing. One reason for this pronounced impact can be attributed to the unique characteristics of digital media in its enormous data volume and real-time requirement throughout the entire application life cycle from generation, encoding, storage, processing, transmission, reception, and consumption of digital media. Cloud computing services, with their on-demand provision in nature, have been able to offer an extremely flexible platform for hosting a wide variety of digital media applications to take full advantage of virtually unlimited resources for the deployment, management, retrieval, and delivery of digital media services.

Many digital media applications are indeed demanding high computation at the cloud data center for an efficient management of media contents so as to release the burden of computational requirements for media terminals. Such applications are very much suited for the most acknowledged cloud computing service class known as infrastructure as a service (IaaS). The demands for intensive computation typically involve processing volumetric media data with massive parallel machines at cloud centers. More recently, two new types of cloud computing services, known as software as a service (SaaS) and platform as a service (PaaS), have also been recognized as having the potential to substantially change the way digital media content is accessed by consumers distributed over scattered geographical locations worldwide. Among this diverse set of digital media applications, some can be captured as software applications running on an underlying cloud computing infrastructure as SaaS for services that are readily accessible via Web browsers from any terminal at any location. More emerging digital media applications can also be deployed at cloud computing infrastructure using programming languages and toolsets as PaaS to host a variety of digital media toolsets for both enterprise and individual consumers.

The contemporary necessity of a ubiquitous access requirement for ever-increasing mobile device users has boosted the adoption of cloud computing for digital media enterprises and executives. Most cloud centers can be considered as geographically neutral because they can be accessed by mobile devices from locations worldwide. It is this characteristic of cloud services that enables digital media companies to develop new and better ways to quickly and efficiently deliver media content to fine-grained targeted consumers. Using cloud computing services, digital media enterprises shall be able to capture the greatest opportunity of efficient delivery because cloud centers allow content storage, media processing, and media distribution to be colocated and seamlessly coordinated. The cloud-based strategy can also improve media companies' competitive advantage through a faster and universal infiltration of multichannel (both wired and wireless networks) and multiscreen (fixed, tablet, laptop, and smartphones) markets with potentially reduced operation costs.

However, mobile media also poses significant challenges in the evolving new paradigm of cloud computing. At the center of these challenges is the significant unbalance in computational and storage capabilities between the cloud centers and mobile devices that triggers the necessary shift of intensive media operations from thin client mobile devices to cloud centers. Resource optimization becomes the major challenge for cloud-based digital media applications, especially for new media services that involve multichannels and multiscreens. To meet the dynamic demands from various media flows, novel solutions are needed to shift computational and storage loads from mobile devices to the cloud, to perform load balancing within a cloud, and to allocate resources across multiple clouds.

Emerging applications of cloud computing have outspread to a much broader range beyond digital media services. Two noticeable areas of such emerging applications are in health care and education. In health care, one central issue is the migration of current locally hosted electronic health record databases to the cloud-based service infrastructure to achieve reduced health-care integration costs, optimized resource management, and innovative multimedia-based electronic health-care records. In education, the ubiquity of cloud computing service centers facilitates a pervasive learning environment for both continuing education of common people and asynchronous tutoring of personalized learners.

As the landscape of new technologies for cloud computing and its applications changes at a steadfast pace, it is very much desired to have a comprehensive collection of articles in various topics in cloud computing as well as their applications in digital media. This excellent book coedited by Kuan-Ching Li, Qing Li, and Timothy K. Shih covers the fundamentals of cloud and media infrastructure, emerging technologies that integrate digital media with cloud computing, and real-world applications that exemplify the potential of cloud computing for next-generation digital media. Specifically, this book covers resource optimization for multimedia cloud computing, a key technical challenge in adopting cloud computing for various digital media applications. It also contains several important new technologies in cloud computing and digital media such as query processing, semantic classification, music retrieval, mobile multimedia, and video transcoding. In addition, this book also includes several chapters to illustrate emerging health-care and educational applications of cloud computing that shall have a profound impact on the welfare of mass populations in terms of their physical well-being and intellectual life. This book is indeed a must read not only for the researchers, engineers, and graduate students who are working in the related research and development topics but also for technology company executives, especially media company executives, to keep pace with the innovations that may impact their business models and market trends. I expect that the timely contributions from these distinguished colleagues shall have prominent influences on the continued flourishing of research and development in cloud computing and digital media.

Chang Wen Chen
State University of New York
Buffalo, New York

Preface

CLOUD COMPUTING HAS APPEARED AS A NEW TREND FOR BOTH computing and storage. It is a computing paradigm where hardware and network details are abstracted and hidden from the users who no longer need to have expertise in or control over the technology because the infrastructure "in the cloud" should support them. Cloud computing describes a new supplement, consumption, and delivery model based on the Internet, where shared resources, software, and information are provided on demand to computers and other devices, similar to an electricity grid. It has even been said that cloud computing may have a greater effect on our lives than the personal computer and dot-com revolutions combined due to scalability, reliability, and cost benefits that this technology can bring forth.

Digital media is a term that widely covers a large number of topics including entertainment, gaming, digital content, streaming, and authoring. Encompassed with the advancements of microprocessor and networking technologies, digital media is considered as a niche in the market as "the creative convergence of digital arts, science, technology and business for human expression, communication, social interaction and education."

The purpose of this book is to bridge the gap between digital media and cloud computing and to bring together technologies for media/data communication, elastic media/data storage, security, authentication, cross-network media/data fusion, interdevice media interaction/reaction, data centers, PaaS, SaaS, and so on. This book also aims at interesting applications involving digital media in the cloud. In addition, this book points out new research issues for the community to discuss in conferences, seminars, and lectures.

The book contains 15 chapters centered on digital media and cloud computing, covering various topics that can be roughly categorized into three levels: infrastructure where fundamental technologies need to be developed, middleware where integration of technologies and software systems need to be defined, and applications cases from the real world. The book is thus suitable as a timely handbook for senior and graduate students who major in computer science, computer engineering, management information system (MIS), or digital media technologies, as well as professional instructors and product developers. In addition, it can also be used as a textbook in senior research seminars and graduate lectures.

The development and production of this book would not have been possible without the support and assistance of Randi Cohen, computer science acquisitions editor at Chapman & Hall/CRC Press. Cohen brought this project from concept to production and has been a wonderful colleague and friend throughout the process. She deserves the credit for all the tedious work that made our work as editors appear easy. Her warm personality made this project fun, and her advice significantly improved the quality of this book. Kate Gallo, Samantha White, and Ed Curtis worked intensively with us and provided the necessary support to make this book ready.

With the continued and increasingly attracted attention on digital media in cloud computing, we foresee that this fast growing field will flourish just as successfully as the Web has done over the past two decades. We believe that readers can benefit from this book in searching for state-of-the-art research topics as well as in the understanding of techniques and applications in cloud computing, interaction/reaction of mobile devices, and digital media/data processing and communication. Of course, we also hope that readers will like this book and enjoy the journey of studying the fundamental technologies and possible research focuses of digital media and cloud computing.

Kuan-Ching Li
Providence University

Qing Li
City University of Hong Kong

Timothy K. Shih
National Central University

MATLAB® is a registered trademark of the MathWorks, Inc. For product information, please contact:

The MathWorks, Inc.
3 Apple Hill Drive
Natick, MA 01760-2098 USA
Tel: +1 508 647 7000
Fax: +1 508 647 7001
E-mail: info@mathworks.com
Web: www.mathworks.com

Editors

Kuan-Ching Li is a professor in the Department of Computer Science and Information Engineering and the special assistant to the university president at Providence University, Taiwan. He earned his PhD in 2001 from the University of São Paulo, Brazil. He has received awards from NVIDIA, investigator of several National Science Council (NSC) awards, and also has held visiting professorships at universities in China and Brazil. He serves or has served as the chair of several conferences and workshops, and he has organized numerous conferences related to high-performance computing and computational science and engineering. Dr. Li is the editor-in-chief of the technical publications *International Journal of Computational Science and Engineering (IJCSE)* and *International Journal of Embedded Systems (IJES)*, both published by Inderscience, and he also serves on editorial boards and as guest editor for a number of journals. He is a fellow of the Institution of Engineering and Technology (IET), a senior member of the Institute of Electrical and Electronics Engineers (IEEE), and a member of the Taiwan Association for Cloud Computing (TACC). He has coauthored over 100 articles in peer-reviewed journals and conferences on topics that include networked computing, graphics processing unit (GPU) computing, parallel software design, virtualization technologies, and performance evaluation and benchmarking.

Qing Li is a professor in the Department of Computer Science, City University of Hong Kong, where he has been a faculty member since September 1998. He earned his BEng at Hunan University and MSc and PhD at the University of Southern California, Los Angeles, all in computer science. His research interests include database modeling, Web services, multimedia retrieval and management, and e-learning systems. He has been actively involved in the research community and is serving or has served as an editor of several leading technical journals, such as *IEEE Transactions on Knowledge and Data Engineering (TKDE)*, *ACM Transactions on Internet Technology (TOIT)*, *World Wide Web (WWW)*, and *Journal of Web Engineering*, in addition to serving as conference and program chair/co-chair of numerous major international conferences, including ER, CoopIS, and ACM RecSys. Professor Li is a fellow of the Institution of Engineering and Technology (IET, UK) and a senior member of the Institute of Electrical and Electronics Engineers (IEEE, USA) and China Computer Federation (CCF, China). He is also a steering committee member of Database Systems for Advanced Applications (DASFAA), International Conference on Web-based Learning (ICWL), and U-Media.

Timothy K. Shih is a professor at National Central University, Taiwan. He was the dean of the College of Computer Science, Asia University, Taiwan, and the chair of the Department of Computer Science and Information Engineering (CSIE) at Tamkang University, Taiwan. Dr. Shih is a fellow of the Institution of Engineering and Technology (IET). He is also the founding chairman of the IET Taipei Interim Local Network. In addition, he is a senior member of Association for Computing Machinery (ACM) and a senior member of the Institute of Electrical and Electronics Engineers (IEEE). Dr. Shih also joined the Educational Activities Board of the Computer Society. His research interests include multimedia computing and distance learning. He has edited many books and published over 490 papers and book chapters as well as participated in many international academic activities, including the organization of more than 60 international conferences. He was the founder and co-editor-in-chief of the *International Journal of Distance Education Technologies*, published by the Idea Group Publishing, Hershey, Pennsylvania. Dr. Shih is an associate editor of the *IEEE Transactions on Learning Technologies*. He was an associate editor of the *ACM Transactions on Internet Technology* and an associate editor of the *IEEE Transactions on Multimedia*. He has received research awards from the National Science Council (NSC) of Taiwan, the International Institute for Advanced Studies (IIAS) research award from Germany, the Brandon Hall award from the United States, and several best paper awards from international conferences. Dr. Shih has been invited to give more than 40 keynote speeches and plenary talks at international conferences as well as tutorials at IEEE International Conference on Multimedia and Expo (ICME) 2001 and 2006 and ACM Multimedia 2002 and 2007.

Contributors

Paolo Balboni
ICT Legal Consulting
Milan, Italy

Subhash Bhalla
Database Systems Laboratory
University of Aizu
Fukushima, Japan

Gaëlle Calvary
LIG Laboratory
University of Grenoble
Grenoble, France

Lei Chen
Department of Computer Science
and Engineering
Hong Kong University of Science
and Technology
Hong Kong, China

Lung-Pin Chen
Department of Computer
Science
Tunghai University
Taichung, Taiwan

Shu-Ching Chen
School of Computing and
Information Sciences
Florida International University
Miami, Florida

Yung-Hui Chen
Department of Computer
Information and Network
Engineering
Lunghwa University of Science
and Technology
Taoyuan, Taiwan

Ling Feng
Department of Computer Science
and Technology
Tsinghua University
Beijing, China

Fausto C. Fleites
School of Computing and
Information Sciences
Florida International University
Miami, Florida

Ling Guan
Department of Electrical and
  Computer Engineering
Ryerson University
Toronto, Ontario, Canada

Hsin-Yu Ha
School of Computing and
  Information Sciences
Florida International University
Miami, Florida

Yifeng He
Department of Electrical and
  Computer Engineering
Ryerson University
Toronto, Ontario, Canada

Yao-Min Huang
Department of Computer
  Science
National Tsing Hua University
Hsinchu, Taiwan

Jyh-Shing Roger Jang
Department of Computer
  Science
National Tsing Hua University
Hsinchu, Taiwan

Li Jin
Department of Computer Science
  and Technology
Tsinghua University
Beijing, China

Wei-Tsa Kao
Department of Computer Science
National Tsing Hua University
Hsinchu, Taiwan

Ralf Klamma
Informatik 5
RWTH Aachen University
Aachen, Germany

Dejan Kovachev
Informatik 5
RWTH Aachen University
Aachen, Germany

Yann Laurillau
LIG Laboratory
University of Grenoble
Grenoble, France

Wen-Shan Liou
Department of Computer
  Science
National Tsing Hua University
Hsinchu, Taiwan

Laure Martins-Baltar
LIG Laboratory
University of Grenoble
Grenoble, France

Rory McGreal
Technology Enhanced
  Knowledge Research
  Institute
Athabasca University
Athabasca, Alberta, Canada

Pulkit Mehndiratta
Department of Computer Science
  and Technology
Jaypee Institute of Information
  Technology
Noida, India

Xiaoming Nan
Department of Electrical
 and Computer Engineering
Ryerson University
Toronto, Ontario, Canada

Hemjyotasna Parashar
Department of Computer Science
 and Technology
Jaypee Institute of Information
 Technology
Noida, India

Claudio Partesotti
ICT Legal Consulting
Milan, Italy

Mika Rautiainen
Center for Internet Excellence (CIE)
University of Oulu
Oulu, Finland

Griff Richards
Technology Enhanced Knowledge
 Research Institute
Athabasca University
Athabasca, Alberta, Canada

Shelly Sachdeva
Department of Computer Science
 and Technology
Jaypee Institute of Information
 Technology
Noida, India

Brian Stewart
Technology Enhanced Knowledge
 Research Institute
Athabasca University
Athabasca, Alberta, Canada

Matthias Sturm
AlphaPlus
Toronto, Ontario, Canada

Chung-Che Wang
Department of Computer
 Science
National Tsing Hua University
Hsinchu, Taiwan

Hao Wang
Department of Computer Science
 and Technology
Tsinghua University
Beijing, China

Martin M. Weng
Department of Computer
 Science and Information
 Engineering
Tamkang University
New Taipei, Taiwan

Bin Wu
Software School
Xiamen University
Xiamen, China

I-Chen Wu
Department of Computer
 Science
National Chiao Tung University
Hsinchu, Taiwan

Yimin Yang
School of Computing and
 Information Sciences
Florida International University
Miami, Florida

Junfeng Yao
Software School
Xiamen University
Xiamen, China

Tzu-Chun Yeh
Department of Computer Science
National Tsing Hua University
Hsinchu, Taiwan

Neil Y. Yen
School of Computer Science
and Engineering
University of Aizu
Aizuwakamatsu, Japan

Mika Ylianttila
Center for Internet Excellence (CIE)
University of Oulu
Oulu, Finland

Xiaofei Zhang
Hong Kong University of Science
and Technology
Hong Kong, China

Jiehan Zhou
Nipissing University
North Bay, Ontario, Canada

Haibin Zhu
Nipissing University
North Bay, Ontario, Canada

Roger Zimmermann
Department of Computer
Science
National University of Singapore
Singapore

CHAPTER 1

Mobile Multimedia Cloud Computing

An Overview

Jiehan Zhou and Haibin Zhu

Nipissing University
North Bay, Ontario, Canada

Mika Ylianttila and Mika Rautiainen

University of Oulu
Oulu, Finland

CONTENTS

1.1 INTRODUCTION

Mobile multimedia cloud computing provides access to data-intensive services (multimedia services) and data stored in the cloud via power-constrained mobile devices. With the development of multimedia computing, mobile devices, mobile multimedia services, and cloud computing, mobile multimedia cloud computing attracts growing attention from researchers and practitioners [1–3].

Mobile devices refer to miniaturized personal computers (PCs) [4] in the form of pocket PCs, tablet PCs, and smart phones. They provide optional and portable ways for users to experience the computing world. Mobile devices [5] are also becoming the most frequently used terminal to access information through the Internet and social networks. A mobile application (mobile app) [4,6] is a software application designed to run on mobile devices. Mobile apps such as Apple App Store (http://store.apple.com/us), Google Play (https://play.google.com/store?hl = en), Windows Phone Store (http://www.windowsphone.com/en-us/store), and BlackBerry App World (http://appworld.blackberry.com/webstore/?) are usually operated by the owner of the mobile operating system. Original mobile apps were for general purposes, including e-mail, calendars, contacts, stock market information, and weather information. However, the number and variety of apps are quickly increasing to other categories, such as mobile games, factory automation, global positioning system (GPS) and location-based services, banking, ticket purchases, and multimedia applications. Mobile multimedia applications are concerned with intelligent multimedia techniques to facilitate effort-free multimedia experiences on mobile devices, including media acquisition, editing, sharing, browsing, management, search, advertising, and related user interface [7]. However, mobile multimedia service still needs to meet bandwidth requirements and stringent timing constraints [8].

Cloud computing creates a new way of designing, developing, testing, deploying, running, and maintaining applications on the Internet [9]. The cloud center distributes processing power, applications, and large systems among a group of machines. A cloud computing platform consists of a variety of services for developing, testing, running, deploying, and maintaining applications in the cloud. Cloud computing services are grouped into three types: (1) application as a service is generally accessed through a Web browser and uses the cloud for processing power and data storage, such as Gmail (http://gmail.com); (2) platform as a service (PaaS) offers the infrastructure on which such applications are built and run, along with the computing power to deliver them, such as Google App Engine (http://code.google.com/appengine/); and (3) infrastructure as a service (IaaS) offers sheer computing resources without a development platform layer, such as Amazon's Elastic Compute Cloud (Amazon EC2; http://aws.amazon.com/ec2/). Cloud computing makes it possible for almost anyone to deploy tools that can scale up and down to serve as many users as desired. The cloud does have certain drawbacks, such as service availability and data security. However, economical cloud computing is being increasingly adopted by a growing number of Internet users without investing much capital in physical machines that need to be maintained and upgraded on-site.

With the integration of mobile devices, mobile multimedia applications, and cloud computing, mobile multimedia cloud computing presents a noteworthy technology to provide cloud multimedia services for generating, editing, processing, and searching multimedia contents, such as images, video, audio, and graphics via the cloud and mobile devices. Zhu et al. [3] addressed multimedia cloud computing from multimedia-aware cloud (media cloud) and cloud-aware multimedia (cloud media) perspectives. Multimedia cloud computing eliminates full installation of multimedia applications on a user's computer or device. Thus it alleviates the burden of multimedia software maintenance and upgrades as well as saving the battery of mobile phones. Kovachev et al. [10] proposed the i5CLoud, a hybrid cloud architecture, that serves as a substrate for scalable and fast time-to-market mobile multimedia services and demonstrates the applicability of emerging mobile multimedia cloud computing. SeViAnno [11] is an MPEG-7-based interactive semantic video annotation Web platform with the main objective of finding a well-balanced trade-off between a simple user interface and video semantization complexity. It allows standard-based video annotation with multigranular community-aware tagging functionalities. Virtual Campfire [12] embraces a set of advanced applications for communities of practice. It is a framework for mobile multimedia management concerned with mobile multimedia semantics, multimedia metadata, multimedia content management, ontology models, and multimedia uncertainty management.

However, mobile multimedia cloud computing is still at the infant stage of the integration of cloud computing, mobile multimedia, and the Web. More research is needed to have a comprehensive review of the current state of the art and practices of mobile multimedia cloud computing techniques. This chapter presents the state of the art and practices of mobile multimedia cloud computing. The rest of the chapter is organized as follows: Section 1.2 reviews the scenario examples of mobile multimedia cloud computing examined in recent studies. Section 1.3 explains the requirements for multimedia cloud computing architecture. Section 1.4 describes the architecture for mobile multimedia cloud computing designed in recent studies. Section 1.5 discusses existing and potential multimedia cloud services. And Section 1.6 draws a conclusion.

1.2 OVERVIEW OF MOBILE MULTIMEDIA CLOUD COMPUTING SCENARIOS

In this section, we review the scenarios examined in the existing literature for identifying the challenges imposed by mobile multimedia cloud computing, which need to be addressed to make mobile multimedia applications feasible. Table 1.1 presents the scenarios examined in recent studies with the application name, its description, and its focused cloud services.

In the cloud mobile gaming (CMG) scenario, Wang et al. [1] presumed to employ cloud computing techniques to host a gaming server, which is responsible for executing the appropriate gaming engine and streaming the resulting gaming video to the client device. This is termed CMG and enables rich multiplayer Internet games on mobile devices, where computation-intensive tasks such as graphic rendering are executed on cloud servers in response to gaming commands on a mobile device, with the resulting video being streamed back to the mobile device in near real time. CMG eliminates the need for mobile devices to download and execute computation-intensive video processing.

In the Virtual Campfire scenario, Cao et al. [12] examined the following three services enabling communities to share knowledge about multimedia contents. (1) In multimedia creation and sharing, the user creates and enriches multimedia content with respective metadata on various mobile devices, such as the Apple iPhone. Technical and contextual semantic metadata on the mobile device (device type, media file size, video codec, etc.) are automatically merged with manual annotations by the user. (2) In the multimedia search and retrieval, the user uses various multimedia search and retrieval methods such as plain keyword tags and semantic context-aware queries based on SPARQL [9,13]. The multimedia search results are presented as a thumbnail gallery. (3) In the recontextualization in complex collaboration, there are three services for the recontextualization of media. The first service facilitates the user to record the condition of the destroyed Buddha figures in the Bamiyan Valley during a campaign. All contents with additional stored GPS coordinates can be requested.

TABLE 1.1   Scenarios Examined in the Existing Literature toward Mobile Multimedia Cloud Computing

Name	Brief Description	Cloud Services
CMG [1]	One of the most compute- and mobile bandwidth-intensive multimedia cloud applications	Graphic rendering
Virtual Campfire [12]	Established in the German excellence cluster UMIC,[a] intending to facilitate intergenerational knowledge exchange by means of a virtual gathering for multimedia contents	Multimedia content creation and sharing, search and retrieval, recontextualization in collaboration
Collaborative metadata management and multimedia sharing [2]	Provide a set of services for mobile clients to perform acquisition of multimedia, to annotate multimedia collaboratively in real time, and to share the multimedia, while exploiting rich mobile context information	Metadata management
Mobile and Web video integration [2]	Platform-independent video sharing through an Android application for context-aware mobile video acquisition and semantic annotation	Multimedia annotation and rendering
Mobile video streaming and processing [2]	Android-based video sharing application for context-aware mobile video acquisition and semantic annotation	Video streaming
MEC-based Photosynth [3]	Cloud-based parallel synthing with a load balancer, for reducing the computation time when dealing with a large number of users	Image conversion Feature execution Image matching Reconstruction

[a] UMIC, Ultra High-Speed Mobile Information and Communication.

The user can collaboratively tag contents by using recombination or embedding techniques. The second service is a mobile media viewer. The third service is a collaborative storytelling service. In the end, Cao et al. illustrate two future scenarios including (1) the 3D video scenario and (2) the remote sensing scenario. In the first scenario, further integration of device functionalities (GPS, digital documentation) and 3D information can be realized by using cost-efficient standard video documentation hardware or even advanced mobile phone cameras. Thus, computational efforts can be incorporated into new 3D environment for storytelling or game-like 3D worlds, such as Second Life (http://secondlife.com/). In the second scenario, the remote sensing data from high-resolution satellites can be incorporated into complex and collaborative planning processes for urban or regional planners, for example, in cultural site management.

In the collaborative metadata management and multimedia sharing scenario, Koavchev et al. [2] depicted the workflow as follows: Imagine that a research team, consisting of experts on archeology, architecture, history, and so on, is documenting an archeological site. (1) The documentation expert takes pictures or videos of a discovered artifact on-site. (2) He tags the content with basic metadata. (3) He stores the tagged content to the cloud. (4) The architecture expert annotates the multimedia content and (5) pushes the updates to the local workforce or historian. (6) The historian corrects the annotation, stores the corrections in the cloud, and (7) pushes the updates to all the subscribed team members. Zhou et al. [14] examined a similar scenario to address how multimedia experiences are extended and enhanced by consuming content and multimedia-intensive services within a community.

In the mobile and Web video integration scenario, Koavchev et al. [2] demonstrated an Android-based video sharing application for context-aware mobile video acquisition and semantic annotation. In this application, videos are recorded with a phone camera and can be previewed and annotated. The annotation is based on the MPEG-7 metadata standard. The basic types include agent, concept, event, object, place, and time. After the video content is recorded and annotated, the users can upload the video content and metadata to the cloud repository. In the cloud, the transcoding service transcodes the video into streamable formats and stores the different versions of the video. At the same time, the semantic metadata services handle the metadata content and store it in the MPEG-7 metadata store.

In the video streaming and processing scenario, Koavchev et al. [2] stated that the cloud computing paradigm is ideal to improve the user multimedia experience via mobile devices. The scenario shows the user records some events using the phone camera. The video is live streamed to the cloud. The cloud provides various services for video processing, such as transcoding and intelligent video processing services with feature extraction, automatic annotation, and personalization of videos. This annotated video is further streamed and available for watching by other users. Figure 1.1 illustrates the user experience of cloud-enhanced video browsing.

To reduce the computation time when dealing with a large number of users, Zhu et al. [3] demonstrated a cloud computing environment to achieve the major computation tasks of Photosynth, that is, image conversion, feature extraction, image matching, and reconstruction. Each

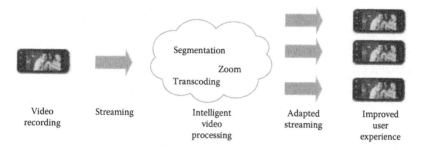

FIGURE 1.1 Improving user experience for mobile video by video processing cloud services.

computation task of Photosynth is conducted in a media-edge cloud (MEC). The proposed parallel synthing consists of user- and task-level parallelization. In the former, all tasks of synthing from one user are allocated to one server to compute, but the tasks from all users can be done simultaneously in parallel in the MEC. In the latter, all tasks of synthing from one user are allocated to N servers to compute in parallel.

In sum, the mobile multimedia cloud computing shares the same scenario with the traditional multimedia applications for distributedly and collaboratively creating, annotating, and sharing the content to enhance and extend the user multimedia experience. However, in the mobile multimedia cloud service, it is a big problem for mobile devices to provide intelligent video processing, such as processing of videos, because they need a lot of resources and are very central processing unit (CPU) intensive. The integration of multimedia applications into cloud computing is investigated as an efficient alternative approach that has been gaining growing attention. The efficient use of scalable computational resources in cloud computing enables a great number of users to concurrently enhance and extend the user multimedia experience on mobile devices.

1.3 OVERVIEW OF ARCHITECTURAL REQUIREMENTS FOR MOBILE MULTIMEDIA CLOUD COMPUTING

Multimedia processing is energy consuming and computing power intensive, and it has a critical demand for quality of multimedia experience as well. Due to the limited hardware resources of mobile devices, it may be promising to investigate the paradigm of multimedia cloud computing using cloud computing techniques to enhance and extend the user multimedia experience. On the one hand, cloud computing efficiently consolidates and shares computing resources and distributes processing power and applications in the units of utility services. On the other hand, multimedia cloud computing needs to address the challenges of reducing the cost of using mobile network and making cloud multimedia services scalable in the context of concurrent users and communication costs due to limited battery life and computing power as well as the narrow wireless bandwidth presented by mobile device [1,2,15]. This section presents the architectural requirements for mobile multimedia cloud computing indicated by recent studies (Table 1.2).

Concerning the design of the cloud multimedia streaming on limited mobile hardware resources, Chang et al. [16] presented the following three key challenges for system developers.

1. Data dependence for dynamic adjustable video encoding. Multimedia encoding and decoding often depends on the information on mobile devices. A suitable dynamic adjustable video encoding through a cloud needs to be designed to prevent failure of decoding.

2. Power-efficient content delivery. Mobile devices usually have limited power supplies; therefore, it is necessary for mass data computing to develop power-efficient mechanisms to reduce energy consumption and achieve user experience quality.

3. Bandwidth-aware multimedia delivery. If network bandwidth is not sufficient, it may easily cause download waiting time during play. Therefore, an adjustable multimedia encoding algorithm is required to dynamically adjust the suitable encoding for the multimedia file playing in the mobile device.

TABLE 1.2 Requirements for Mobile Multimedia Cloud Computing Indicated by the Recent Studies

Requirement	Description
Cloud multimedia stream [16]	Data dependence for adjustable video encoding, power-efficient content delivery, and bandwidth-aware multimedia delivery
Multimedia cloud computing [3]	Heterogeneities of the multimedia service, the QoS, the network, and the device
Cloud mobile media [1]	Response time, user experience, cloud computing cost, mobile network bandwidth, and scalability
Mobile multimedia cloud computing [2]	Three crucial perspectives: technology, mobile multimedia, and user and community

Zhu et al. [3] stated that multimedia processing in the cloud imposes great challenges. They highlighted several fundamental challenges for multimedia cloud computing. (1) Multimedia and service heterogeneity. Because there are so many different types of multimedia services, such as photo sharing and editing, image-based rendering, multimedia streaming, video transcoding, and multimedia content delivery, the cloud has to support all of them simultaneously for a large base of users. (2) Quality-of-service (QoS) heterogeneity. Different multimedia services have different QoS requirements; the cloud has to support different QoS requirements for various multimedia services. (3) Network heterogeneity. Because there are different networks, such as the Internet, wireless local area networks (WLANs), and third-generation wireless networks, with different network characteristics, such as bandwidth, delay, and jitter, the cloud has to adapt multimedia content for optimal delivery to various types of devices with different network bandwidths and latencies. (4) Device heterogeneity. Because there are so many different types of devices, such as televisions, PCs, and mobile phones, with different capacities for multimedia processing, the cloud has to adjust multimedia processing to fit the different types of devices, including CPU, graphics processing unit (GPU), display, memory, storage, and power.

Wang et al. [1] analyzed the requirements imposed by mobile multimedia cloud computing, including response time, user experience, cloud computing cost, mobile network bandwidth, and scalability to a large number of users; other important requirements are energy consumption, privacy, and security.

Koavchev et al. [2] investigated the requirements for mobile multimedia cloud architecture from three crucial perspectives: technology, mobile multimedia, and user and community. The technology perspective establishes a basic technical support to facilitate mobile cloud computing. The mobile multimedia perspective concerns the capabilities of multimedia processing. The last perspective is related to users' experiences in multimedia delivery and sharing. Table 1.3 details the three perspectives.

1.4 OVERVIEW OF THE ARCHITECTURE DESIGN TOWARD MOBILE MULTIMEDIA CLOUD COMPUTING

This section reviews the architecture toward mobile multimedia cloud computing designed in recent studies (Table 1.4).

In order to improve the current development practices, combining with a mobile cloud computing delivery model, Koavchev et al. [2] proposed a four-layered i5 multimedia cloud architecture (Figure 1.2). The infrastructure and platform layer focus on requirements from the

TABLE 1.3   Three Perspectives Addressing Requirements of Mobile Multimedia Cloud Computing

	Perspective	Description
	Data management	Cloud storage is well suitable for content management, but is inferior to metadata management.
	Communication	Broadband Internet connection is needed to meet the required QoE.[a] XMPP (http://xmpp.org) and SIP[b] [17] together with their extensions are powerful for cloud services.
	Computation	The huge cloud processing power is not fully accessible to mobile devices.
Mobile multimedia	Multimedia formats and transcoding	Different mobile device media platforms are based on different formats and coding.
	Multimedia semantics	Multimedia semantic analysis is needed for discovering complex relations, which are serving as input for reasoning in the media interpretation.
	Multimedia modeling	Modeling multimedia content sensed by mobile devices provides valuable context information for indexing and querying of the multimedia content.
User and community	Sharing and collaboration	XMPP-based communication is needed to enhance real-time multimedia collaboration on multimedia metadata, adaptation, and sharing.
	Ubiquitous multimedia services	Users expect to have ubiquitous access to their multimedia content by switching from one device to another.
	Privacy and security	Ensure that the data and processing is secure and remains private, and the data transmission between the cloud and the mobile device is secured.

[a] QoE, quality of experience.
[b] SIP, session initiation protocol.

technology perspective and use virtualization technology, which separates the software from the underlying hardware resources. The virtual machines are grouped into three realms: processing realm for parallel processing, streaming realm for scalable handling of streaming requests, and general realm for running other servers such as extensible messaging and presence protocol (XMPP) or Web server. DeltaCloud application programming interface (API) layer enables cross-cloud interoperability on infrastructure level with other cloud providers, for example, Amazon

TABLE 1.4   Architecture Designed in Recent Studies toward Mobile Multimedia Cloud Computing

Architecture	Brief Description
i5CLoud architecture [2]	It consists of four layers of infrastructure, platform, multimedia services, and application.
Cloud mobile media architecture [1]	It is capable of dynamically rendering multimedia in the cloud servers, depending on the mobile network and cloud computing constraints.
Multimedia streaming service architecture over cloud computing [16]	It provides dynamic adjustable streaming services while considering mobile device resources, multimedia codec characteristics, and the current network environment.
Multimedia cloud computing [3]	It provides multimedia applications and services over the Internet with desired QoS.

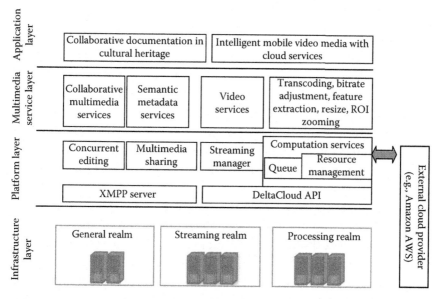

FIGURE 1.2   i5CLoud architecture for multimedia applications. ROI, region-of-interest.

EC2. The DeltaCloud core framework assists in creating intermediary drivers that interpret the DeltaCloud representational state transfer (REST) API on the front while communicating with cloud providers using their own native APIs on the back. The concurrent editing and multimedia sharing components are the engine for the collaborative multimedia and semantic metadata services. MPEG-7 metadata standards are employed to realize semantic metadata services. Video processing

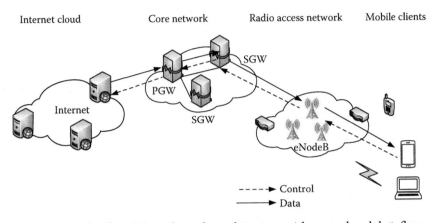

Internet cloud   Core network   Radio access network   Mobile clients

FIGURE 1.3   Cloud mobile multimedia architecture with control and data flows. PGW, packet data gateway; SGW, service gateway. (From Wang, S. and Dey, S., *IEEE Transactions on Multimedia*, 99, 1, 2013. With permission.)

services improve mobile users' experience. The application layer provides a set of services for mobile users to create multimedia to collaboratively annotate and share.

Wang et al. [1] described a typical architecture for cloud mobile multimedia applications, including end-to-end flow of control and data between the mobile devices and the Internet cloud servers (Figure 1.3). A typical cloud multimedia application primarily relies on cloud computing IaaS and PaaS resources in public, private, or hybrid clouds. A multimedia application has a thin client on mobile devices, which provide the appropriate user interfaces (gesture, touch screen, voice, and text based) to enable users to interact with the application. The resulting control commands are transmitted uplink through cellular radio access network (RAN) or WiFi access points to appropriate gateways in an operator core network (CN) and finally to the cloud. Consequently, the multimedia content produced by the multimedia cloud service is transmitted downlink through the CN and RAN back to the mobile device. Then the client decodes and displays the content on the mobile device display.

To address the restricted bandwidth and improve the quality of multimedia video playback, Chang et al. [16] proposed a novel cloud multimedia streaming architecture for providing dynamic adjustable streaming services (Figure 1.4), which consist of two parts: the service function of the cloud equipment (i.e., cloud service) and the information modules provided by the mobile device (i.e., mobile device service). Table 1.5 describes the architecture modules.

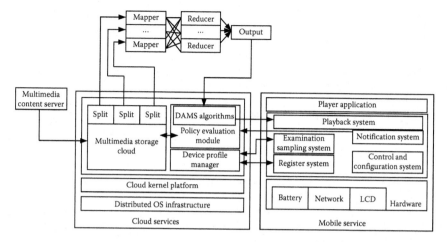

FIGURE 1.4 Multimedia streaming service architecture over cloud computing. OS, operating system. DAMS, dynamic adjustable streaming; LCD, liquid-crystal display. (From Chang, S., Lai, C., and Huang, Y., *Computer Communications*, 35, 1798–1808, 2012. With permission.)

TABLE 1.5 Module Description for the Multimedia Streaming Service Architecture

Module	Sub-Module	Description
Cloud service	Device profile management	Records the features of mobile devices such as the maximum power of the processor, the codec type, the available highest upload speed of the network, and the highest available download speed of the network
	Policy evaluation module	Determines the multimedia coding parameters, in terms of mobile device parameters
	Multimedia storage cloud	Provided by some multimedia storage devices
Mobile device service	Register system	Registers a device profile manager over the cloud
	Notification system	Has the hardware monitor and notification component, which is used to monitor the real-time information of battery and network bandwidth
	Examination sampling system	Measures system efficiency, including the parameters in the DPM[a]
	Playback system	Parses metadata to obtain film coding characteristics and relevant information
	Control and configuration system	Offers user–machine interface interaction settings and controls hardware module functions

[a] DPM, device profile manager.

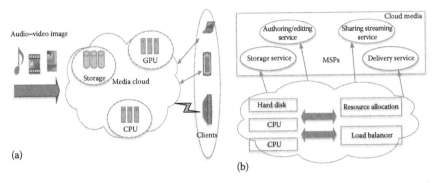

FIGURE 1.5 Architecture of multimedia cloud computing: Media cloud (a) and cloud media (b) services. (From Zhu, W., Luo, C., Wang, J., and Li, S., "Multimedia Cloud Computing," *IEEE Signal Processing Magazine*, 28, 59–69, 2011. ©2011 IEEE. With permission.)

It is foreseen that cloud computing could become a disruptive technology for mobile multimedia applications and services [18]. In order to meet multimedia's QoS requirements in cloud computing for multimedia services over the Internet and mobile wireless networks, Zhu et al. [3] proposed a multimedia cloud computing framework that leverages cloud computing to provide multimedia applications and services over the Internet. The principal conceptual architecture is shown in Figure 1.5. Zhu et al. addressed multimedia cloud computing from multimedia-aware cloud (media cloud) and cloud-aware multimedia (cloud media) perspectives. The media cloud (Figure 1.5a) focuses on how a cloud can perform distributed multimedia processing and storage and QoS provisioning for multimedia services. In a media cloud, the storage, CPU, and GPU are presented at the edge (i.e., MEC) to provide distributed parallel processing and QoS adaptation for various types of devices. The MEC stores, processes, and transmits media data at the edge, thus achieving a shorter delay. In this way, the media cloud, composed of MECs, can be managed in a centralized or peer-to-peer (P2P) manner. The cloud media (Figure 1.5b) focuses on how multimedia services and applications, such as storage and sharing, authoring and mashup, adaptation and delivery, and rendering and retrieval, can optimally utilize cloud computing resources to achieve better quality of experience (QoE). As depicted in Figure 1.5b, the media cloud provides raw resources, such as hard disk, CPU, and GPU, rented by the media service providers (MSPs) to serve users. MSPs use media cloud resources to develop their multimedia applications and services, for example, storage, editing, streaming, and delivery.

1.5 OVERVIEW OF MULTIMEDIA CLOUD SERVICES

Mobile multimedia cloud computing presents a significant technology to provide multimedia services to generate, edit, process, and search multimedia contents via the cloud and mobile devices. Traditionally, there exist different types of multimedia services, such as photo sharing and editing, multimedia streaming, image searching and rendering, and multimedia content delivery. A typical multimedia life cycle is composed of acquisition, storage, processing, dissemination, and presentation [3]. Theoretically, the cloud should support these types of multimedia services. This section presents multimedia cloud services examined by the recent studies (Table 1.6).

Cloud multimedia authoring as a service [3] is the process of editing multimedia contents, whereas a mashup deals with combining multiple segments from different multimedia sources. A cloud can make online authoring and mashup very effective, providing more functions to clients, since it has powerful computation and storage resources that are widely distributed geographically. Cloud multimedia authoring can avoid the preinstallation of an authoring software in clients. With the use of the cloud multimedia authoring service, users conduct editing in the media cloud. One of the key challenges in cloud multimedia authoring is the computing and communication costs in processing multiple segments from single or multiple sources. Zhu et al. [3] pointed out that future research needs to tackle distributed storage and processing in the cloud, online previewing on mobile devices.

Cloud multimedia storage as a service is a model of networked online storage where multimedia content is stored in virtualized pools of storage. The

TABLE 1.6   Multimedia Cloud Services Examined by the Recent Studies

Service	Description
Cloud multimedia authoring	The process of editing segments of multimedia contents in the media cloud
Cloud storage	The advantage of being "always-on," higher level of reliability than local storage
Cloud multimedia rendering	Conducting multimedia rendering in the cloud, instead of on the mobile device
Cloud multimedia streaming	Potentially achieving much a lower latency and providing much a higher bandwidth due to a large number of servers deployed in the cloud
Cloud multimedia adaptation	Conducting both offline and online media adaptation to different types of terminals
Cloud multimedia retrieval	Achieving a higher search quality with acceptable computation time, resulting in better performance

cloud multimedia storage service can be categorized into consumer- and developer-oriented services [3]. Consumer-oriented cloud storage service holds the storage service on its own servers. Amazon Simple Storage Service (S3) [19] and Openomy [20] are developer-oriented cloud storage services, which go with the typical cloud provisioning "pay only for what you use."

Cloud multimedia rendering as a service [1] is a promising category that has the potential of significantly enhancing the user multimedia experience. Despite the growing capacities of mobile devices, there is a broadening gap with the increasing requirements for 3D and multiview rendering techniques. Cloud multimedia rendering can bridge this gap by conducting rendering in the cloud instead of on the mobile device. Therefore, it potentially allows mobile users to experience multimedia with the same quality available to high-end PC users [21]. To address the challenges of low cloud cost and network bandwidth and high scalability, Wang et al. [1] proposed a rendering adaptation technique, which can dynamically vary the richness and complexity of graphic rendering depending on the network and server constraints, thereby impacting both the bit rate of the rendered video that needs to be streamed back from the cloud server to the mobile device and the computation load on the cloud servers. Zhu et al. [3] emphasized that the cloud equipped with GPU can perform rendering due to its strong computing capability. They categorized two types of cloud-based rendering: (1) to conduct all the rendering in the cloud and (2) to conduct only computation-intensive part of the rendering in the cloud while the rest would be performed on the client. More specifically, an MEC with a proxy can serve mobile clients with high QoE since rendering (e.g., view interpolation) can be done in the proxy. Research challenges include how to efficiently and dynamically allocate the rendering resources and design a proxy for assisting mobile phones on rendering computation.

Cloud multimedia streaming as a service utilizes cloud computing resources to perform computation-intensive tasks of encoding and transcoding in order to adapt to different devices and networks. Cloud multimedia streaming services utilize the elasticity provided by cloud computing to cost-effectively handle peak demands. Cloud-based streaming can potentially achieve a much lower latency and provide a much higher bandwidth due to the large number of servers deployed in the cloud. Cloud multimedia sharing services also increase media QoS because cloud–client connections almost always provide a higher bandwidth and a shorter delay than client–client connections. The complexities of cloud multimedia sharing mainly reside in naming, addressing, and accessing control [3].

Cloud multimedia adaptation as a service [3] transforms input multimedia contents into an output video in a form that meets the needs of heterogeneous devices. It plays an important role in multimedia delivery. In general, video adaptation needs a large amount of computing, especially when there are a vast number of simultaneous consumer requests. Because of the strong computing and storage power of the cloud, cloud multimedia adaptation can conduct both offline and online media adaptation to different types of terminals. CloudCoder is a good example of a cloud-based video adaptation service that was built on the Microsoft Azure platform [22]. CloudCoder is integrated into the origin digital central management platform while offloading much of the processing to the cloud. The number of transcoder instances automatically scale to handle the increased or decreased volume. Zhu et al. [3] presented a cloud-based video adaptation framework in which the cloud video adaptation in a media cloud is responsible for collecting customized parameters, such as screen size, bandwidth, and generating various distributions either offline or on the fly. One of the future research topics is how to perform video adaptation on the fly.

Cloud multimedia retrieval as a service is a good application example of cloud computing used to search digital images in a large database based on the image content. Zhu et al. [3] discussed how content-based image retrieval (CBIR) [23] can be integrated into cloud computing. CBIR includes multimedia feature extraction, similarity measurement, and relevance feedback. The key challenges in CBIR are how to improve the search quality and how to reduce computation time. Searching in a database such as the Internet is becoming computation intensive. With the use of the strong computing capacity of a media cloud, one can achieve a higher search quality with acceptable computation time, resulting in better performance.

1.6 CONCLUSION

Multimedia computing needs powerful computing and storage capacity for handling multimedia content while achieving the desired QoE, such as response time, computing cost, network bandwidth, and concurrent user numbers. Mobile devices are constrained in resources of memory, computing power, and battery lifetime in the handling of multimedia content. Cloud computing has the ability to develop on-demand computing and storage capacities by networking computer server resources. Integrating cloud computing into mobile multimedia applications has a profound impact on the entire life cycle of multimedia contents, such as authoring, storing, rendering, streaming and sharing, and retrieving. With the use of cloud multimedia services, potential mobile cloud multimedia applications include storing documents, photos, music, and videos in the cloud; streaming audio and video in the cloud; coding/decoding audio and video in the cloud; interactive cloud advertisements; and mobile cloud gaming.

In this chapter, we presented the state of the art and practices of emerging mobile multimedia cloud computing with perspectives of scenario examination, requirement analysis, architecture design, and cloud multimedia services. Research in mobile multimedia cloud computing is still in its infancy, and many issues in cloud multimedia services remain open, for example, how to design a proxy in a media cloud for manipulating 3D content on demand to favor both network bandwidth usage and graphical rendering process, how to optimize and simplify 3D content to reduce the energy consumption of a mobile device, how to accelerate mobile multimedia cloud computing utilizing P2P technology (i.e., P2P-enabled mobile multimedia cloud computing), and so on.

ACKNOWLEDGMENTS

This work was carried out through the adaptive content delivery cluster (ACDC) project, which was funded by Tekes, the Finnish Funding Agency for Technology and Innovation. We also thank Associate Professor Chung-Horng Lung for his hosting while the first author was a visiting research fellow in Carleton University, Ottawa, Ontario.

REFERENCES

1. Wang, S. and Dey, S., "Adaptive mobile cloud computing to enable rich mobile multimedia applications," *IEEE Transactions on Multimedia*, 99: 1, 2013.
2. Koavchev, D., Cao, Y., and Klamma, R., "Mobile multimedia cloud computing and the web," in *Workshop on Multimedia on the Web*, September 8, pp. 21–26, Graz, Austria, IEEE Press, 2008.
3. Zhu, W., Luo, C., Wang, J., and Li, S., "Multimedia cloud computing," *IEEE Signal Processing Magazine*, 28: 59–69, 2011.
4. Lee, V., Schneider, H., and Schell, R., *Mobile Applications: Architecture, Design, and Development*, Upper Saddle River, NJ: Prentice Hall, 2004.
5. Hua, X. S., Mei, T., and Hanjalic, A., *Online Multimedia Advertising: Techniques and Technologies*, Hershey, PA: IGI Global, 2010.
6. Tracy, K. W., "Mobile Application Development Experiences on Apple's iOS and Android OS," *IEEE Potentials*, 31(4): 30, 34, 2012.
7. Steinmetz, R. and Nahrstedt, K., (Eds.) *Multimedia Applications*, Berlin: Springer, 2004.
8. Luo, H., Egbert, A., and Stahlhut, T., "QoS architecture for cloud-based media computing," in *IEEE 3rd International Conference on Software Engineering and Service Science*, pp. 769–772, June 22–24, Beijing, IEEE Press, 2012.

9. Wang, S. and Dey, S., "Modeling and characterizing user experience in a cloud server based mobile gaming approach," in *IEEE Conference on Global Telecommunications*, pp. 1–7, 2009.

10. Kovachev, D., Cao, Y., and Klamma, R., "Building mobile multimedia services: A hybrid cloud computing approach," *Multimedia Tools Application*, 5: 1–29, 2012.

11. Cao, Y., Renzel, D., Jarke, M., Klamma, R., Lottko, M., Toubekis, G., and Jansen, M., "Well-balanced usability and annotation complexity in interactive video semantization," in *4th International Conference on Multimedia and Ubiquitous Engineering*, pp. 1–8, 2010.

12. Cao, Y., Klamma, R., and Jarke, M., "Mobile multimedia management for virtual campfire: The German excellence research cluster UMIC," *International Journal on Computer Systems, Science and Engineering*, 25(3): 251–265, 2010.

13. Cao, Y., Klamma, R., and Khodaei, M., "A multimedia service with MPEG-7 metadata and context semantics," in *Proceedings of the 9th Workshop on Multimedia Metadata*, March 19–20, Toulouse, 2009.

14. Zhou, J., Rautiainen, M., and Ylianttila, M., "Community coordinated multimedia: Converging content-driven and service-driven models," in *IEEE International Conference on Multimedia and Expo*, pp. 365–368, 2008.

15. Li, L., Li, X., Youxia, S., and Wen, L., "Research on mobile multimedia broadcasting service integration based on cloud computing," in *International Conference on Multimedia Technology*, pp. 1–4, 2010.

16. Chang, S., Lai, C., and Huang, Y., "Dynamic adjustable multimedia streaming service architecture over cloud computing," *Computer Communications*, 35: 1798–1808, 2012.

17. Schulzrinne, H. and Wedlund, E., "Application-layer mobility using SIP," in *IEEE Globecom Workshop on Service Portability and Virtual Customer Environments*, December 1, IEEE, San Francisco, CA, pp. 29–36, 2000.

18. ABI Research. "Mobile cloud computing," Available at http://www.abiresearch. com/research/1003385-Mobile+Cloud+Computing, accessed on February 27, 2013.

19. Amazon S3. Available at https://s3.amazonaws.com/, accessed on February 27, 2013.

20. Openomy. Available at http://www.killerstartups.com/web-app-tools/ openomy-com-more-free-storage/, accessed on February 27, 2013.

21. Wang, S. and Dey, S., "Cloud mobile gaming: Modeling and measuring user experience in mobile wireless networks," *SIGMOBILE Mobile Computing and Communications Review*, 16: 10–21, 2012.

22. Origin Digital. "Video services provider to reduce transcoding costs up to half." Available at http://www.Microsoft.Com/casestudies/Case_Study_Detail. aspx?CaseStudyID=4000005952, accessed on February 27, 2013.

23. Smeulders, A. W. M., Worring, M., Santini, S., Gupta, A., and Jain, R., "Content-based image retrieval at the end of the early years," *IEEE Transactions on Pattern Analysis and Machine Intelligence*, 22: 1349–1380, 2000.

Resource Optimization for Multimedia Cloud Computing

Xiaoming Nan, Yifeng He, and Ling Guan

Ryerson University
Toronto, Ontario, Canada

CONTENTS

2.1 INTRODUCTION

In the recent years, we have witnessed the fast development of cloud computing from a dream into a commercially viable technology. According to the forecast from International Data Corporation (IDC) [1], the world-wide public cloud computing services will edge toward $100 billion by 2016 and enjoy an annual growth rate of 26.4%, which is 5 times the traditional information technology (IT) industry. As the emerging computing paradigm, cloud computing manages a shared pool of physical servers in data centers to provide on-demand computation, communication, and storage resources as services in a scalable and virtualized manner. In the traditional computing, both data and software are operated at user side, whereas in cloud computing, the tasks that require an intensive computation or a large storage are processed in the powerful data center. By using cloud-based applications, users are free from application installation and software maintenance.

The emergence of cloud computing has greatly facilitated online multimedia applications. The elastic and on-demand resource provisioning in cloud can effectively satisfy intensive resource demands of multimedia processing. In particular, the scalability of cloud can handle frequent surges of requests, which is demonstrated with substantial advantages over traditional server clusters [2]. In cloud-based multimedia applications, computationally intensive tasks are processed in data centers, thus greatly reducing hardware requirements on the user side. Moreover, users are able to access remote multimedia applications from anywhere at any time, even through resource-constrained devices. Therefore, cloud-based multimedia applications, such as online photo editing [3], cloud-based video retrieval [4], and various social media applications, have been increasingly adopted in daily life.

Multimedia applications also bring new challenges to current general-purpose cloud computing. Nowadays, the general-purpose cloud employs a utility-based resource management to allocate computation resources (e.g., CPU, memory, storage, etc.). In utility-based resource allocations, cloud resources are packaged into virtual machines (VMs) as a metered service. By using VMs, cloud resources can be provisioned or released with minimal efforts. As cloud providers, such as Amazon Elastic Compute Cloud (Amazon EC2) [5], the only guaranteed parameter in the service-level agreement (SLA) is the resource availability, that is, users can access rented resources at any time. However, for multimedia applications, in addition to the computation resources, another important factor is the stringent quality-of-service (QoS) requirements in terms of service response time, jitter, and packet loss. If the general-purpose cloud is used to deal with multimedia applications without considering the QoS requirements, the media experience may become unacceptable to users. Hence, an effective resource allocation scheme is strongly needed for multimedia cloud computing. Recently, a lot of research efforts have been made on the resource allocation and QoS provisioning for cloud-based multimedia applications [6–26].

From the perspective of multimedia service providers, there are two fundamental concerns: the QoS and the resource cost. Multimedia applications are typically delay sensitive. Therefore, the service response time is widely adopted as the major QoS factor to measure the performance of cloud-based multimedia applications. The service response time is defined as the duration from the time when the application request arrives at the data center to the time when the service result completely departs from the data center. A lower service response time will lead to a higher QoS. However, multimedia applications have different service response time requirements and dynamic resource demands. It is challenging to optimally allocate resources to meet different service response time requirements for all applications. Besides the service response time, the second concern is the resource cost. Cloud computing involves various computing resources, and applications have different resource demands. As multimedia service providers, they are concerned about how to optimally allocate different resources to satisfy all applications at the minimal cost. The service process in multimedia cloud can generally be divided into three consecutive phases: the scheduling phase, the computation phase, and the transmission phase. Inappropriate resource allocation among the three phases will lead to resource waste and QoS degradation. For example, with excessive computing resource and inadequate bandwidth resource allocated in multimedia cloud, requests will be processed fast, but the service results cannot be transmitted efficiently due to the limited bandwidth capacity. Therefore, it is challenging for multimedia service providers to optimally configure cloud resources among the three phases to minimize the resource cost while guaranteeing service response time requirements.

In this chapter, we investigate the optimal resource allocation problem for multimedia cloud computing. We first provide a review of recent advances on cloud-based multimedia services and resource allocation in cloud computing. We then propose queuing model-based resource optimization schemes for multimedia cloud computing, in which we present the proposed system models, the problem formulations, and the simulation results. Finally, the future research directions in the area of multimedia cloud computing will be discussed.

The remainder of this chapter is organized as follows: Section 2.2 discusses the state of the art of optimal resource allocation and QoS provisioning in cloud computing. Section 2.3 presents the proposed queuing model-based resource optimization schemes, including system models, problem formulations, and simulation results. Section 2.4 discusses the future research directions. Finally, Section 2.5 provides the summary.

2.2 RELATED WORK

In this section, we present the recent advances on cloud-based multimedia services and resource allocation in cloud computing.

2.2.1 Cloud-Based Multimedia Services

The development of cloud computing has a profound effect on the development of multimedia services. Conventionally, the multimedia storage, streaming, processing, and retrieval services are provided by private clusters, which are too expensive for small businesses. The "pay-as-you-go" model of public cloud would greatly facilitate multimedia service providers, especially the start-up businesses. For multimedia service providers, they just pay for the computing and storage resources they used, rather than maintaining costly private clusters. Hence, cloud-based multimedia services have been increasingly adopted in the recent years.

Cloud-based live video streaming provides an "always-on" video streaming platform so that users can access preferred channels at arbitrary time. Huang et al. [6] present *CloudStream*, a cloud-based peer-to-peer (P2P) live video streaming platform that utilizes public cloud servers to construct a scalable video delivery platform with Scalable Video Coding (SVC). In the CloudStream, a large number of users can receive live video stream at the same time by dynamically arranging the available resources based on the streaming quality requested by the users. Inspired by the similar idea, Pan et al. [7] present a framework of adaptive mobile video streaming and user behavior-oriented video prefetching in clouds, which is named as *adaptive mobile video prefetching cloud (AMVP-Cloud)*. In the AMVP-Cloud, a private agent is generated for each mobile user to provide "nonterminating" video streaming and to adapt to the fluctuation of link quality based on the SVC technique and feedback from users. With background prefetching among video agents and local storage of mobile side, the AMVP-Cloud realizes "nonwaiting" experiences for mobile users.

Besides live video streaming, cloud-based media storage is another hot research topic. Liu et al. [8] propose a real-world video-on-demand (VoD) system, which is called *Novasky*. The Novasky is based on a P2P storage cloud that can store and refresh video stream in a decentralized way. The storage cloud is a large-scale pool of storage space, which can be accessed conveniently by users who need storage resources. In contrast to the traditional P2P VoD with local caching, peers in P2P storage cloud are interconnected with a high-bandwidth network, which supplies a stellar level of performance when streaming on-demand videos to participating users. Based on the P2P storage cloud, the Novasky is able to deliver over 1,000 cinematic-quality video streams to over 10,000 users at bit rates of 1–2 Mbps, which is much higher than the bit rate of 400 Kbps in traditional P2P VoD. In addition to the P2P storage cloud, content delivery cloud (CDC) offers elastic, scalable, and low-cost storage services for users. For multimedia streaming, the latency of content delivery can be minimized by caching the media content onto the edge server of CDC. The work by Bao and Yu [9] studies the caching algorithm for scalable multimedia over CDC, in which the edge server can calculate a truncation ratio for the cached scalable multimedia contents to balance the quality and the resource usage.

Mobile multimedia is becoming popular nowadays. The shipments of smart phones in the United States have exceeded the traditional computer segments since 2010 [10]. However, due to the power limitation, it is still hard to fully operate computationally intensive multimedia applications on mobile devices. Cloud computing can help to address this issue by providing multimedia applications with powerful computation resources and storage spaces. Mobile users, therefore, can easily access the cloud mobile media applications through wireless connectivity. Zhang et al. [11] present an interactive mobile visual search application, which can understand visual queries captured by the built-in camera such that mobile-based social activities can be recommended for users. On the client end, a mobile user can take a photo and indicate an interested object with so-called "O" gesture. On the cloud, a recognition-by-search mechanism is implemented on cloud servers to identify users' visual intent. By incorporating visual search results with sensory context [e.g., global positioning system (GPS) location], the relevant social activities can be recommended for users.

2.2.2 Resource Allocation in Cloud Computing

The resource allocation in cloud environment is an important and challenging research topic. Verma et al. [12] formulate the problem of dynamic placement of applications in virtualized heterogeneous systems as a continuous optimization: The placement of VMs at each time frame is optimized to minimize resource consumption under certain performance requirements. Chaisiri et al. [13] study the trade-off between the advance reservation and the on-demand resource allocation, and propose a VM placement algorithm based on stochastic integer programming. The proposed algorithm minimizes the total cost of resource provision in infrastructure as a service (IaaS) cloud. Wang et al. [14] present a virtual appliance-based automatic resource provisioning framework for large virtualized data centers. Their framework can dynamically allocate resources to applications by adding or removing VMs on physical servers. Verma et al. [12], Chaisiri et al. [13], and Wang et al. [14] study cloud resource allocation from VM placement perspective. Bacigalupo et al. [15] quantitatively compare the effectiveness of different techniques on response time prediction. They study different cloud services with different priorities, including urgent cloud services that demand cloud resource at short notice and dynamic enterprise systems that need to adapt to frequent changes in the workload. Based on these cloud services, the layered queuing network and historical performance model are quantitatively compared in terms of prediction accuracy. Song et al. [16] present a resource allocation approach according to application priorities in multiapplication virtualized cluster. This approach requires machine learning to obtain the utility functions for applications and defines the application priorities in advance. Lin and Qi [17] develop a self-organizing model to manage cloud resources in the absence of centralized management control. Nan et al. [18] present optimal cloud resource allocation in priority service scheme to minimize the resource cost. Appleby et al. [19] present a prototype of infrastructure, which can dynamically allocate cloud resources for an e-business computing utility. Xu et al. [20] propose a two-level resource management system with local controllers at the VM level and a global controller at the server level. However, they focus only on resource allocation among VMs within a cloud server [19,20].

Recently, there is an upsurge of research interests in QoS provisioning for cloud-based multimedia applications. Zhu et al. [21] introduce multimedia cloud computing from the perspectives of the multimedia-aware cloud and the cloud-aware multimedia, and propose a media-edge cloud computing architecture in order to reduce transmission delays between the users and the data centers. Wen et al. [22] present an effective load-balancing technique for a cloud-based multimedia system, which can schedule VM resources for different user requests with a minimal cost. Wu et al. [23] present a system to utilize cloud resources for VoD applications and propose a dynamic cloud resource provisioning algorithm to support VoD streaming at a minimal cloud utilization cost. Wang et al. [24] present a framework for cloud-assisted live media streaming, in which cloud servers can be adaptively adjusted according to dynamic demands.

2.3 QUEUING MODEL-BASED RESOURCE OPTIMIZATION

In this section, we present queuing model-based resource optimization schemes for cloud-based multimedia applications. System models are discussed in Section 2.3.1 to characterize the service process at the cloud data center. Based on the system models, we study the relationship between the service response time and the allocated cloud resources. Moreover, we examine the resource allocation problem in the first-come first-served (FCFS) service case and the priority service case. In each case, we formulate and solve the service response time minimization problem and the resource cost minimization problem.

2.3.1 System Models

2.3.1.1 Data Center Architecture

Currently, most of the clouds are built in the form of data centers [25,26]. The architecture of the multimedia cloud data center is illustrated in Figure 2.1, which consists of a master server, a number of computing servers, and a transmission server. The master server and computing servers are virtual clusters [27] composed of multiple VM instances, whereas the transmission server is used to transmit the service results or media data. The master server serves as a scheduler, receiving all requests and then distributing them to the computing servers. The number of computing servers is denoted by N. The computing servers process the requests using the allocated computation resources. In order to provide efficient services, the required multimedia data are shared by the computing servers, and the master server and computing servers are connected by high-speed communication links. After processing, service results will be transmitted

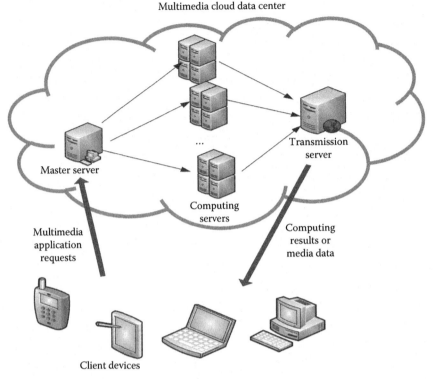

FIGURE 2.1   Data center architecture for multimedia applications.

back to the users by the transmission server. In practical cloud platforms, such as Amazon EC2 [5] or Windows Azure [28], the bandwidth resource is specified by the total transmission amount. Therefore, we use the allocated bandwidth to represent the resource of the transmission server. The computing servers and the transmission server share results in the memory, and thus, there is no delay between the computing servers and the transmission server in our data center architecture. Owing to the time-varying workload, the resources in the cloud have to be dynamically adjusted. Therefore, we divide the time domain into time slots with a fixed length Γ. In our work, the cloud resources will be dynamically allocated in every time slot t.

According to different usages, there are two types of cloud: storage-oriented cloud, such as Amazon Simple Storage Service (Amazon S3) [29], and computation-oriented cloud, such as Amazon EC2 [5]. In this chapter, we study the computation-oriented cloud. The allocated cloud resources in our study include the scheduling resource, the computation resource, and the bandwidth resource. The scheduling resource is represented by the scheduling rate S^t indicating the number of requests scheduled per second; the computation resource at computing server i is represented by the processing rate C_i^t indicating the number of instructions executed per second; and the bandwidth resource is represented by the transmission rate B^t indicating the number of bits transmitted per second. Thus, the resource allocation in the multimedia cloud determines the optimal scheduling resource S^t, computation resource C_i^t ($\forall i = 1,\ldots,N$), and bandwidth resource B^t at time slot t.

Suppose that M classes of applications are provided. For each class of application, there are four parameters: the average request arrival rate λ_j^t, the average task size F_j, the average result size D_j, and the service response time requirement τ_j. The workload of each application is time varying, and thus, the resource demands are dynamically changing. Moreover, the response time requirement affects the resource demands.

2.3.1.2 Queuing Model

The proposed queuing model is shown in Figure 2.2. The model consists of three concatenated queuing systems: the scheduling queue, the computation queue, and the transmission queue. The master server maintains the scheduling queue. Since two consecutive arriving requests may be sent from different users, the inter-arrival time is a random variable, which can be modeled as an exponential random variable [30]. Therefore, the arrivals of requests follow a Poisson process. The average arrival rate is denoted by λ^t. The requests are scheduled to the computing servers at the rate S^t. Each computing server has a corresponding computation queue to process requests. The service results are sent back to the users at the rate B^t by the transmission server. We assume that the service availability is guaranteed

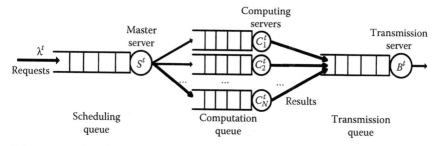

FIGURE 2.2   Queuing model of the data center in multimedia cloud.

by the SLA and no request is dropped during the process. Therefore, the number of results is equal to that of received requests.

2.3.1.3 Cost Model

The allocated cloud resources in our study include the scheduling resource at the master server, the computation resource at the computing servers, and the bandwidth resource at the transmission server. We employ a linear function to model the relationship of the resource cost and the allocated resources. The total resource cost $\zeta^{\text{tot}(t)}$ at time slot t can be formulated as

$$\zeta^{\text{tot}(t)} = \left(\alpha S^t + \beta \sum_{i=1}^{N} C_i^t + \gamma B^t \right) \Gamma \tag{2.1}$$

where:
  Γ is the time slot length
  S^t is the allocated scheduling resource
  C_i^t is the allocated computation resource at the computing server i
  B^t is the allocated bandwidth resource
  α, β, and γ are the cost rates for scheduling, computation, and transmission, respectively

The linear cost model in Equation 2.1 has been justified by the numerical analysis in Reference 5.

2.3.2 Problem Formulation

Based on the proposed system models, we study the resource optimization problems in the FCFS service case and the priority service case. In each case, we optimize cloud resources to minimize the service response time and the resource cost, respectively.

2.3.2.1 FCFS Service Case

In the FCFS service case, the requests of users are served in the order that they arrived at the data center. All requests are processed with the same priority. Suppose that there are M classes of applications provided in multimedia cloud. Applications have different processing procedures, task sizes, and result sizes, as well as different requirements on service response time. The mean arrival rates of the requests at time slot t are denoted as $\lambda_1^t, \lambda_2^t, \ldots, \lambda_M^t$, respectively. According to the composition property of Poisson process [31], the total arrivals of requests follow the Poisson process with an average arrival rate $\lambda^t = \sum_{i=1}^{M} \lambda_i^t$. Thus, the scheduling queue can be modeled as an $M/M/1$ queuing system [30] with the mean service rate S^t at the master server. To maintain a stable queue, $\lambda^t < S^t$ is required. The response time of the scheduling queue is given by

$$T^{\text{sch}(t)}_{\text{FCFS}} = \frac{1/S^t}{1 - \lambda^t / S^t}$$

Each application requires a different service procedure. M computing servers are employed to process M classes of applications, in which computing server i is dedicated to serve requests for class i application. Since the VMs can be dynamically provisioned and released in the cloud, the number of computing servers can be dynamically changed according to the number of applications. For class-i application, the average task size is denoted by F_i, and the execution time is a random variable, which is assumed to follow exponential distribution with an average of F_i/C_i^t [32,33]. To maintain a stable queue, the constraint $\lambda_i^t < C_i^t/F_i$ should be satisfied. The response time at computing server i is given by

$$T^{\text{com}(i)(t)}_{\text{FCFS}} = \frac{F_i/C_i^t}{1 - \lambda_i^t F_i / C_i^t}$$

Therefore, the average response time in computation phase can be formulated as

$$T^{\text{com}(t)}_{\text{FCFS}} = \sum_{i=1}^{M} (\lambda_i^t/\lambda^t) T^{\text{com}(i)(t)}_{\text{FCFS}} = \sum_{i=1}^{M} \frac{\lambda_i^t F_i / C_i^t}{\lambda^t (1 - \lambda_i^t F_i / C_i^t)}$$

After processing, all service results are sent to the transmission queue. Since a service result is generated for each request and the system is a closed system, the average arrival rate of the results at the transmission queue is also λ^t. For class-i application, the average result size is denoted by D_i. Different applications have different result sizes, leading to different transmission time. The transmission time for the class-i service result is exponentially distributed with mean service time D_i/B, where B is the bandwidth resource at the transmission server. Therefore, the transmission queue can be viewed as a queuing system in which service results are grouped into a single arrival stream and the service distribution is a mixture of M exponential distributions. In fact, the service time follows hyperexponential distribution [30]. The transmission queue is actually an $M/H_M/1$ queuing system, where H_M represents a hyperexponential M distribution. The response time of the $M/H_M/1$ queuing system can be derived from $M/G/1$ queuing system [30]. The response time of the transmission queue is formulated as

$$T_{\text{FCFS}}^{\text{tra}(t)} = \frac{\sum_{i=1}^{M} \lambda_i^t D_i^2}{(B^t)^2 - B^t \sum_{i=1}^{M} \lambda_i^t D_i} + \frac{\sum_{i=1}^{M} \lambda_i^t D_i}{\lambda^t B^t}$$

To ensure a stable queue, $\sum_{i=1}^{M} \lambda_i^t D_i < B^t$ is required.

Based on the above derivations, we can get the total service response time in the FCFS service case as follows:

$$T_{\text{FCFS}}^{\text{tot}(t)} = T_{\text{FCFS}}^{\text{sch}(t)} + T_{\text{FCFS}}^{\text{com}(t)} + T_{\text{FCFS}}^{\text{tra}(t)}$$

$$= \frac{1/S^t}{1 - \lambda^t / S^t} + \sum_{i=1}^{M} \frac{\lambda_i^t F_i / C_i^t}{\lambda^t (1 - \lambda_i^t F_i / C_i^t)}$$

$$+ \frac{\sum_{i=1}^{M} \lambda_i^t D_i^2}{(B^t)^2 - B^t \sum_{i=1}^{M} \lambda_i^t D_i} + \frac{\sum_{i=1}^{M} \lambda_i^t D_i}{\lambda^t B^t} \tag{2.2}$$

Moreover, the mean service response time for class-i service is formulated as

$$T_{\text{FCFS}}^{\text{tot}(t)} = T_{\text{FCFS}}^{\text{sch}(t)} + T_{\text{FCFS}}^{\text{com}(i)(t)} + T_{\text{FCFS}}^{\text{tra}(t)}. \tag{2.3}$$

The total resource cost for the multiple-class service can be formulated as

$$\zeta_{\text{FCFS}}^{\text{tot}(t)} = \left(\alpha S^t + \beta \sum_{i=1}^{N} C_i^t + \gamma B^t \right) \Gamma \tag{2.4}$$

1. *Service response time minimization problem.* Since there are different types of multimedia applications, such as video conference, cloud television, and 3D rendering, multimedia service providers should supply different types of multimedia services to users simultaneously. However, it is challenging to provide various multimedia services with a minimal total service response time under a certain budget constraint. Therefore, we formulate the service response time minimization problem, which can be stated as follows: to minimize the total service response time in the FCFS service case by jointly optimizing the allocated scheduling resource at the master server, the computation resource at each computing

server, and the bandwidth resource at the transmission server, subject to the queuing stability constraint in each queuing system and the resource cost constraint. Mathematically, the problem can be formulated as follows:

$$\underset{\{S^t, C_1^t, ..., C_M^t, B^t\}}{\text{Minimize}} \quad T_{\text{FCFS}}^{\text{tot}(t)} \tag{2.5}$$

subject to

$$\lambda^t < S^t,$$

$$\lambda_i^t F_i < C_i^t, \forall i = 1, ..., M$$

$$\sum_{i=1}^{M} \lambda_i^t D_i < B^t,$$

$$\left(\alpha S^t + \beta \sum_{i=1}^{M} C_i^t + \gamma B^t \right) \Gamma \leq \zeta_{\max}$$

where:

$T_{\text{FCFS}}^{\text{tot}(t)}$, given by Equation 2.2, is the mean service response time for all applications

$\zeta_{\max}$ is the upper bound of the resource cost

The service response time minimization problem (Equation 2.5) is a convex optimization problem [34]. We apply the primal–dual interior-point methods [34] to solve the optimization problem (Equation 2.5).

2. *Resource cost minimization problem.* Since applications have different requirements on service response time, multimedia service providers have to guarantee QoS provisioning for all applications. However, it is challenging to configure cloud resources to provide satisfactory services at the minimal resource cost. Thus, we formulate the resource cost minimization problem, which can be stated as: to minimize the total resource cost in the FCFS service case by jointly optimizing the allocated scheduling resource, the computation resource, and the bandwidth resource, subject to the queuing stability constraint in each queuing system and the requirement on service response time for each application. Mathematically, the problem can be formulated as follows.

$$\underset{\{S^t, C_1^t, ..., C_M^t, B^t\}}{\text{Minimize}} \quad \zeta_{\text{FCFS}}^{\text{tot}(t)} = \left(\alpha S^t + \beta \sum_{i=1}^{M} C_i^t + \gamma B^t \right) \Gamma \tag{2.6}$$

subject to

$$\lambda^t < S^t$$

$$\lambda_i^t F_i < C_i^t, \forall i = 1,\ldots,M$$

$$\sum_{i=1}^{M} \lambda_i^t D_i < B^t$$

$$T_{\text{FCFS}}^{\text{tot}(i)(t)} \leq \tau_i, \forall i = 1,\ldots,M$$

where:

$T_{\text{FCFS}}^{\text{tot}(i)(t)}$ is the service response time for class-i application, which is given by Equation 2.3

τ_i is the upper bound of the service response time for class-i application

The resource cost minimization problem (Equation 2.6) is a convex optimization problem, which can be solved efficiently using the primal–dual interior-point methods [34].

2.3.2.2 Priority Service Case

The FCFS service scheme is not suitable for the multimedia applications that require differentiated services. For example, the urgent multimedia applications, such as the real-time health monitoring, need to be processed as soon as possible; thus, such requests should have a higher priority than the other requests. We extend our resource optimization to the priority service case, in which multiple applications with different priorities are provided. The requests for the higher priority applications should be processed ahead of those for the lower priority applications. Specifically, we study the preemptive priority queuing discipline, in which the requests with a higher priority obtain the service immediately even if other requests with a lower priority are being served, and the preempted requests will later be resumed from the last preemption point.

Suppose that there are M classes of applications with different priorities, which are denoted as class-1, 2, …, M, respectively. A smaller class number corresponds to a higher priority. The mean arrival rate of class-j requests is denoted by λ_j^t. According to the composition property, the total request arrivals follow the Poisson process with an average $\lambda^t = \sum_{j=1}^{M} \lambda_j^t$. When requests arrive at the data center, the master server always schedules the request with the highest priority first. The lower priority requests can be scheduled only after all higher priority requests have left the schedule queue. Therefore, the scheduling queue is modeled as an $M/M/1$ queuing system with a preemptive priority service. The response time for scheduling class-j requests is given by

$$T_{\text{prio}}^{\text{sch}(j)(t)} = \frac{1/S^t}{1-\sigma_{\text{sch}}^{j-1}} + \frac{\sum_{k=1}^{j} \lambda_k^t / (s^t)^2}{(1-\sigma_{\text{sch}}^{j-1})(1-\sigma_{\text{sch}}^{j})}$$

where:

$$\sigma_{\text{sch}}^{j} = \sum_{k=1}^{j} \lambda_k^t / S^t$$

To make the scheduling queue stable, $\sigma_{\text{sch}}^{M} = \sum_{k=1}^{M} \lambda_k^t / S^t < 1$ should be satisfied. Since the scheduling rates are the same for all classes of requests, the mean response time at the master server is given by

$$T_{\text{prio}}^{\text{sch}(t)} = \frac{1/S^t}{1-\lambda^t/S^t}$$

In the computation phase, M computation queues are used to store requests with the corresponding priority. Moreover, the total computation resources are aggregated to provide service. The requests with the highest priority have preemptive right to obtain service immediately. The total computation resource is denoted by C^t, and the average task size of class-j service is F_j. The service time for computing class-j requests is assumed to be exponentially distributed with mean time of F_j / C^t. According to Reference 30, the response time for processing class-j requests is given by

$$T_{\text{prio}}^{\text{com}(j)(t)} = \frac{F_j / C^t}{1-\sigma_{\text{com}}^{j-1}} + \frac{\sum_{k=1}^{j} \lambda_k^t F_k^2 / (C^t)^2}{(1-\sigma_{\text{com}}^{j-1})(1-\sigma_{\text{com}}^{j})}$$

where:

$$\sigma_{\text{com}}^{j} = \sum_{k=1}^{j} \lambda_k^t F_k / C^t$$

Moreover, $\sigma_{\text{com}}^{M} < 1$ is required to maintain the queue stable. Since the service rates are different for different applications, the mean response time at computing server is given by [30]

$$T_{\text{prio}}^{\text{com}(t)} = \sum_{j=1}^{M} \frac{\lambda_j^t}{\lambda^t} T_{\text{prio}}^{\text{com}(j)(t)}$$

After processing, all service results are sent to the transmission queue. The results in the higher priority classes are transmitted prior to those in the lower priority classes. The allocated bandwidth resource is denoted by B^t, and the average result size of class-j application is denoted by D_j. Thus, the mean transmission time is given by D_j/B^t. The response time for transmitting class-j results is given by

$$T_{\text{prio}}^{\text{tra}(j)(t)} = \frac{D_j/B^t}{1-\sigma_{\text{tra}}^{j-1}} + \frac{\sum_{k=1}^{j}\lambda_k^t D_k^2/(B^t)^2}{(1-\sigma_{\text{tra}}^{j-1})(1-\sigma_{\text{tra}}^{j})}$$

where:

$$\sigma_{\text{tra}}^j = \sum_{k=1}^{j}\lambda_k^t D_k/B^t$$

To ensure the transmission queue is stable, $\sigma_{\text{tra}}^M < 1$ should be satisfied. Thus, the mean response time at the transmission server can be formulated as

$$T_{\text{prio}}^{\text{tra}(t)} = \sum_{j=1}^{M}(\lambda_j^t/\lambda^t)\, T_{\text{prio}}^{\text{tra}(j)(t)}$$

Based on the above derivations, the total service response time in the priority service case is the summation of response time in the three phases, which can be given by

$$T_{\text{prio}}^{\text{tot}(t)} = T_{\text{prio}}^{\text{sch}(t)} + T_{\text{prio}}^{\text{com}(t)} + T_{\text{prio}}^{\text{tra}(t)} \tag{2.7}$$

Furthermore, we can get the service response time for the class-j application as follows:

$$T_{\text{prio}}^{\text{tot}(j)(t)} = T_{\text{prio}}^{\text{sch}(j)(t)} + T_{\text{prio}}^{\text{com}(j)(t)} + T_{\text{prio}}^{\text{tra}(j)(t)} \tag{2.8}$$

The total resource cost in the priority service case is formulated as

$$\zeta_{\text{prio}}^{\text{tot}(t)} = (\alpha S^t + \beta C^t + \gamma B^t)\Gamma \tag{2.9}$$

1. *Service response time minimization problem.* In multimedia cloud, priority service discipline has been used in many applications. The applications with more stringent delay requirement should receive a higher priority service, whereas the less delay-sensitive multimedia applications can be served in a lower priority. The multimedia service providers have to support different priority services and minimize the mean service response time. We formulate the service response time minimization problem in the priority service

case, which can be stated as follows: to minimize the mean service response time for all applications by jointly optimizing the allocated scheduling resource, the computation resource, and the bandwidth resource, subject to the queuing stability constraint in each queuing system and the resource cost constraint. Mathematically, the service response time minimization problem can be formulated as

$$\underset{\{S^t, C^t, B^t\}}{\text{Minimize}} \quad T_{\text{prio}}^{\text{tot}(t)} \tag{2.10}$$

subject to

$$\lambda^t < S^t$$

$$\sum_{j=1}^{M} \lambda_j^t F_j < C^t$$

$$\sum_{j=1}^{M} \lambda_j^t D_j < B^t$$

$$(\alpha S^t + \beta C^t + \gamma B^t)\Gamma \le \zeta_{\max}$$

where:

$T_{\text{prio}}^{\text{tot}(t)}$ is given in Equation 2.7

$\zeta_{\max}$ is the upper bound of the resource cost

The optimization problem (Equation 2.10) is a convex optimization problem, which can be efficiently solved using the primal–dual interior-point methods [34].

In the priority service case, the requests with the highest priority have the preemptive right to receive the service. Therefore, the imposition of priorities decreases the mean delay of higher priority requests and increases the mean delay of lower priority requests. However, the effect of imposing priorities for the overall service response time of all requests is not determined. To address the issue of how the imposition of priorities affects the overall service response time, Schrage and Miller [35] propose the shortest processing time (SPT) rule, which is described as follows: (1) If the objective of a queue is to reduce the overall mean delay, a higher priority should be given to the class of requests that has a faster service rate. (2) If the overall objective in multimedia cloud is to reduce the service response time for one specific application, this application should be given the highest priority.

2. *Resource cost minimization problem.* Different priority applications have different requirements on service response time. It is challenging for multimedia cloud providers to support multiple QoS provisioning at the minimal resource cost. Therefore, we formulate the resource cost minimization problem in the priority service case, which can be stated as follows: to minimize the total resource cost in the priority service case by jointly optimizing the allocated scheduling resource, the computation resource, and the bandwidth resource, subject to the queuing stability constraints and the service response time constraint for each application. Mathematically, the resource cost minimization problem can be formulated as follows:

$$\underset{\{S^t, C^t, B^t\}}{\text{Minimize}} \quad \zeta_{\text{prio}}^{\text{tot}(t)} = (\alpha S^t + \beta C^t + \gamma B^t)\Gamma \tag{2.11}$$

subject to

$$\lambda^t < S^t$$

$$\sum_{j=1}^{M} \lambda_j^t F_j < C^t$$

$$\sum_{j=1}^{M} \lambda_j^t D_j < B^t$$

$$T_{\text{prio}}^{\text{tot}(j)(t)} < \tau_j, \forall j = 1, ..., M$$

where:
  $T_{\text{prio}}^{\text{tot}(j)(t)}$, given in Equation 2.8, is the service response time for class-j application
  τ_j is the upper bound of the service response time for class-j application

The resource cost minimization problem (Equation 2.11) can also be solved efficiently using the primal–dual interior-point methods [34].

TABLE 2.1 Parameter Settings for Cloud-Based Multimedia Applications

Application class	1	2	3	4	5
Percentage of request arrival rate (%)	10	15	20	25	30
Task size (MIs)	300	350	400	450	500
Result size (megabit)	7	8	10	12	12
Upper bound of service response time (seconds)	0.1	0.15	0.2	0.25	0.2

2.3.3 Simulations

We perform simulations to evaluate the proposed resource allocation schemes. In our simulations, we set the simulation parameters based on Windows Azure [28], which provides on-demand computation, storage, and networking resources as utilities through Microsoft data centers. The cloud resources of the master server, the computing server, and the transmission server are charged by the scheduling cost rate $\alpha = 5 \times 10^{-4}$ dollars per request, the computation cost rate $\beta = 6 \times 10^{-6}$ dollars per million instructions (MIs), and the transmission cost rate $\gamma = 0.08$ dollars per gigabit, respectively. The length of the time slot is set to 1 hour, which is the same as the resource allocation time unit in Azure. Five classes of multimedia applications are provided in the data center. Each class of applications has a different arrival rate, a different task size, a different result size, and a different requirement on the service response time. Table 2.1 shows the parameter settings for cloud-based multimedia applications.

2.3.3.1 Simulation Results for FCFS Service Case

We first compare the performance between the proposed optimal allocation scheme, in which cloud resources are allocated optimally by solving the optimization problem (Equation 2.5 or 2.6), and the equal allocation scheme, in which the resource cost for scheduling, computation, and transmission is allocated equally. Figure 2.3a shows the comparison of service response time. From Figure 2.3a, we can see that the proposed optimal allocation scheme takes a smaller service response time than the equal allocation scheme under the same resource constraint. Figure 2.3b shows the detailed service rates when the request arrival rate is 150 requests per second. Too many resources are allocated to the master server in the equal allocation scheme, which results in the less resources in the computing servers and the transmission server.

The comparison of resource cost in the multiple-class service case is shown in Figure 2.4a, from which we can see that the proposed optimal allocation scheme achieves a much lower resource cost than the equal

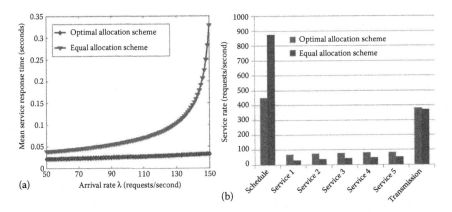

FIGURE 2.3 Simulation results between the proposed optimal allocation scheme and the equal allocation scheme in the FCFS service case: (a) Comparison of mean service response time and (b) comparison of service rates.

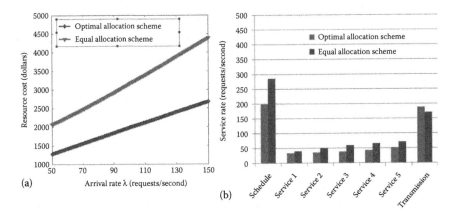

FIGURE 2.4 Simulation results between the proposed optimal allocation scheme and the equal allocation scheme in the FCFS service case: (a) Comparison of resource cost and (b) comparison of service rates.

allocation scheme. The detailed service rate of each server is shown in Figure 2.4b, when the request arrival rate is 150 requests per second. In contrast to the proposed scheme, the equal allocation scheme assigns less resources to the transmission server, which causes longer waiting time in transmission queue and degrades the system performance.

2.3.3.2 Simulation Results for Priority Service Case
In the priority service case, five classes of services with different priorities are provided in the data center. The parameters of each class are the same

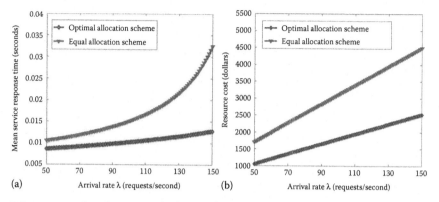

(a)

(b)

FIGURE 2.5   Simulation results between the proposed optimal allocation scheme and the equal allocation scheme in the priority service case: (a) Comparison of mean service response time and (b) comparison of resource cost.

as the parameter settings in Table 2.1. Moreover, a smaller class number corresponds to a higher priority.

We compare the mean service response time and the resource cost in the priority service case between the proposed optimal resource allocation scheme, in which cloud resources are allocated by solving the optimization problem (Equation 2.10 or 2.11), and the equal allocation scheme, in which the resource cost for scheduling, computation, and transmission is allocated equally. Figure 2.5a shows the comparison of the mean service response time. In the simulation, the resource cost constraint is set to 5000 dollars. From Figure 2.5a, we can find that the proposed resource allocation scheme can take a smaller service response time than the equal allocation scheme under the same resource constraint. The comparison of the resource cost in the priority service case is shown in Figure 2.5b, from which we can see that the proposed optimal resource allocation scheme achieves a much lower resource cost than the equal allocation scheme.

2.4 FUTURE RESEARCH DIRECTIONS

In spite of the progress made in cloud computing, there remain a number of important open issues as follows:

1. *Resource demand prediction.* Since the process of cloud resource allocation takes time to complete, it will be too late to prevent the QoS degradation if the resource reallocation is only carried out when

resources become insufficient. Therefore, an accurate resource demand prediction model is required to forecast the resource demand in the near-term future based on the previous statistics.

2. *Workload monitoring.* The workload in cloud is changing in real time. To allocate resources to satisfy the dynamic workload, especially the burst of requests, a live workload monitoring is needed for cloud providers. In addition, it is a challenge to dynamically allocate the cloud resources to handle the time-varying workload.

3. *Workload scheduling.* There are two levels of scheduling in cloud computing. The first level is the user-level scheduling, in which the requests for one application are distributed to different VMs according to the current workload. By balancing workload among the VMs, the user-level scheduling can effectively avoid episodic congestions in the cloud. Compared to the user-level scheduling, the task-level scheduling performs in a finer granularity. An application can be decomposed into a set of tasks, each of which requires different resources. The task-level scheduling is to assign different tasks to different VMs so that the performance can be maximized.

4. *Resource migration.* With current techniques, VMs and application migrations have been implemented in the local area network (LAN) environment. In the future, cloud should be able to migrate VMs and services to other clouds, which can greatly improve the robustness of cloud data center.

5. *Joint resource optimization.* Currently, most of the resource optimization methods focus only on the cloud side, while ignoring the transmission path and the user side. In fact, it is a challenging task to maximize or minimize an end-to-end QoS metric by jointly optimizing the resources in the cloud, at the client, and along the transmission path between the cloud and the client.

2.5 SUMMARY

Multimedia cloud computing, as a specific cloud computing model, focuses on how cloud can effectively provide multimedia services and guarantee QoS provisioning. Optimal resource allocation in multimedia cloud computing can greatly improve the performance for multimedia applications. In this chapter, we investigate the optimal resource allocation for multimedia cloud computing. We first provide a review of recent advances on cloud-based multimedia services and resource allocation in cloud computing. We then present a queuing model-based resource optimization scheme for multimedia cloud computing. Finally, the future research directions in the area of multimedia cloud computing have been discussed.

REFERENCES

1. F. Gens, M. Adam, M. Ahorlu et al., Worldwide and regional public IT cloud services 2012–2016 forecast. Available at http://www.idc.com/getdoc .jsp?containerId=236552#.UV5TMhxwcdU, accessed on April 2013.
2. J. Peng, X. Zhang, Z. Lei, B. Zhang, W. Zhang, and Q. Li, Comparison of several cloud computing platforms, in *Proceedings of IEEE International Symposium on Information Science and Engineering*, December, pp. 23–27, Shanghai, 2009.
3. Pixlr online photo editor. Available at http://pixlr.com/, accessed on April 2013.
4. H. Li and Y. Zhuang, V2 RMC: Vertical video retrieval system in mobile cloud computing environment, in *Proceedings of IEEE International Conference on Intelligent Computation Technology and Automation*, January, pp. 378–381, Hunan, China, 2012.
5. Amazon Elastic Compute Cloud. Available at http://aws.amazon.com/ec2/, accessed on April 2013.
6. Z. Huang, C. Mei, L. E. Li, and T. Woo, CloudStream: Delivering high-quality streaming videos through a cloud-based SVC proxy, in *Proceedings of IEEE INFOCOM*, April, pp. 201–205, Shanghai, 2011.
7. B. Pan, X. Wang, C. Hong, and S. Kim, AMVP-Cloud: A framework of adaptive mobile video streaming and user behavior oriented video pre-fetching in the clouds, in *Proceedings of IEEE 12th International Conference on Computer and Information Technology*, October, pp. 398–405, Chengdu, 2012.
8. F. Liu, S. Shen, B. Li, B. Li, H. Yin, and S. Li, Novasky: Cinematic-quality VoD in a P2P storage cloud, in *Proceedings of IEEE INFOCOM*, April, pp. 936–944, Shanghai, 2011.
9. X. Bao and R. Yu, Streaming of scalable multimedia over content delivery cloud, in *Proceedings of Asia-Pacific Signal and Information Processing Association Annual Summit and Conference*, December, pp. 1–5, Hollywood, CA, 2012.
10. The Smartphone Age. Available at http://www.magnoliabroadband.com/ index.php?option=com_rsblog&layout=view&cid=9:the-smartphone-age-smartphone-shipments-exceed-pcs&Itemid=50, accessed on October 2013.
11. N. Zhang, T. Mei, X. Hua, L. Guan, and S. Li, Interactive mobile visual search for social activities completion using query image contextual model, in *Proceedings of IEEE International Workshop on Multimedia Signal Processing*, September, pp. 238–243, Banff, AB, 2012.
12. A. Verma, P. Ahuja, and A. Neogi, pMapper: Power and migration cost aware application placement in virtualized systems, *Middleware, Lecture Notes in Computer Science*, 5346: 243–264, 2008.

13. S. Chaisiri, B. Lee, and D. Niyato, Optimal virtual machine placement across multiple cloud providers, in *Proceedings of IEEE Asia-Pacific Services Computing Conference*, December, pp. 103–110, Singapore, 2009.

14. X. Wang, D. Lan, G. Wang, X. Fang, M. Ye, Y. Chen, and Q. Wang, Appliance-based autonomic provisioning framework for virtualized outsourcing data center, in *Proceedings of IEEE International Conference on Autonomic Computing*, June, pp. 29–38, Jacksonville, FL, 2007.

15. D. Bacigalupo, J. van Hemert, A. Usmani, D. Dillenberger, G. Wills, and S. Jarvis, Resource management of enterprise cloud systems using layered queuing and historical performance models, in *Proceedings of IEEE International Symposium on Parallel and Distributed Processing*, April, pp. 1–8, Atlanta, GA, 2010.

16. Y. Song, H. Wang, Y. Li, B. Feng, and Y. Sun, Multi-tiered on-demand resource scheduling for VM-based data center, in *Proceedings of IEEE/ACM International Symposium on Cluster Computing and the Grid*, May, pp. 148–155, Shanghai, 2009.

17. W. Lin and D. Qi, Research on resource self-organizing model for cloud computing, in *Proceedings of IEEE International Conference on Internet Technology and Applications*, August, pp. 1–5, Wuhan, 2010.

18. X. Nan, Y. He, and L. Guan, Optimal resource allocation for multimedia cloud in priority service scheme, in *Proceedings of IEEE International Symposium on Circuits and Systems*, May, pp. 1111–1114, Seoul, Republic of Korea, 2012.

19. K. Appleby, S. Fakhouri, L. Fong, G. Goldszmidt, M. Kalantar, S. Krishnakumar, D. Pazel, J. Pershing, and B. Rochwerger, Oceano-SLA based management of a computing utility, in *Proceedings of IEEE International Symposium on Integrated Network Management*, May, pp. 855–868, Seattle, WA, 2011.

20. J. Xu, M. Zhao, J. Fortes, R. Carpenter, and M. Yousif, On the use of fuzzy modeling in virtualized data center management, in *Proceedings of IEEE International Conference on Autonomic Computing*, June, pp. 25–29, Jacksonville, FL, 2007.

21. W. Zhu, C. Luo, J. Wang, and S. Li, Multimedia cloud computing, *IEEE Signal Processing Magazine*, 28(3): 59–69, 2011.

22. H. Wen, Z. Hai-ying, L. Chuang, and Y. Yang, Effective load balancing for cloud-based multimedia system, in *Proceedings of IEEE International Conference on Electronic and Mechanical Engineering and Information Technology*, August, pp. 165–168, Heilongjiang, 2011.

23. Y. Wu, C. Wu, B. Li, X. Qiu, and F. Lau, Cloudmedia: When cloud on demand meets video on demand, in *Proceedings of IEEE International Conference on Distributed Computing Systems*, June, pp. 268–277, Minneapolis, MN, 2011.

24. F. Wang, J. Liu, and M. Chen, Calms: Cloud-assisted live media streaming for globalized demands with time/region diversities, in *Proceedings of IEEE INFOCOM*, March, pp. 199–207, Orlando, FL, 2012.

25. W. Zhu, C. Luo, J. Wang, and S. Li, Multimedia cloud computing, *IEEE Signal Processing Magazine*, 28(3): 59–69, 2011.

26. Q. Zhang, L. Cheng, and R. Boutaba, Cloud computing: State-of-the-art and research challenges, *Journal of Internet Services and Applications*, 1(1): 7–18, 2010.
27. M. Murphy, B. Kagey, M. Fenn, and S. Goasguen, Dynamic provisioning of virtual organization clusters, in *Proceedings of IEEE/ACM International Symposium on Cluster Computing and the Grid*, May, pp. 364–371, Shanghai, 2009.
28. Microsoft Azure. Available at http://www.microsoft.com/windowsazure/, accessed on April 2013.
29. Amazon Simple Storage Service. Available at http://aws.amazon.com/s3/, accessed on April 2013.
30. D. Gross, *Fundamentals of Queueing Theory*, New York: Wiley, 2008.
31. J. Cooper, The Poisson and exponential distributions, *Mathematical Spectrum*, 37(3): 123–125, 2005.
32. B. Yang, F. Tan, Y. Dai, and S. Guo, Performance evaluation of cloud service considering fault recovery, in *Proceedings of the 1st International Conference on Cloud Computing*, October, pp. 571–576, Munich, 2009.
33. D. Ardagna, S. Casolari, and B. Panicucci, Flexible distributed capacity allocation and load redirect algorithms for cloud systems, in *Proceedings of IEEE International Conference on Cloud Computing*, July, pp. 163–170, Washington, DC, 2011.
34. S. Boyd and L. Vandenberghe, *Convex Optimization*, Cambridge: Cambridge University Press, 2004.
35. L. Schrage and L. Miller, The queue M/G/1 with the shortest remaining processing time discipline, *Operations Research*, 14(4): 670–684, 1966.

CHAPTER 3

Supporting Practices in Professional Communities Using Mobile Cloud Services

Dejan Kovachev and Ralf Klamma

RWTH Aachen University
Aachen, Germany

CONTENTS

3.1 INTRODUCTION

Over the past decades, new media, new technologies and devices, and new ways of communication continuously define new formats of practice of professionals or knowledge workers. For instance, the worldwide access to heterogeneous information over the Internet has created a lot of new means for cooperative work. Social software well known by examples such as the digital photo sharing platform flickr.com, the digital video sharing platform youtube.com, or the social bookmarking platform del.icio.us can be broadly defined as an environment that supports the activities in digital social networks [1]. Professionals change their work styles according to the new possibilities.

Similarly, mobile devices together with *mobile multimedia* are changing practices, just as the computer and the Internet have fundamentally had pushed a similar transformation few decades ago. Mobile phones have been transformed into digital Swiss Army knives equipped with a multitude of applications that can achieve a variety of tasks. Mobile devices make the information fusion of real life and virtual life possible. Individuals get more flexibility and mobility to communicate with other persons. Emerging mobile technologies include enhanced flexibility and mobility for end users, improving productivity and enabling social interactivity via personal mobile devices' availability of context information. Handheld devices such as smartphones and tablets, mobile televisions, digital camcorders, and personal media players become an integral part of the Web, whereas multimedia is one of the core technologies underpinning the mobile growth worldwide.

These concepts are being unified under the emerging *cloud computing* paradigm [2]. The cloud computing model has recently been well explored for enterprise consumers and service providers [3], but little attention is paid to mobile communities. How cloud computing supports mobile multimedia is the key to cost-efficient design, development, and delivery of mobile community services.

We are at the early stage of confluence of cloud computing, mobile multimedia, and the Web. In current development practices and in the literature, we can observe the trends that may lead to unnecessary frictions in the development of professional mobile Web multimedia applications. Cloud computing has great potential to leverage the current issues with mobile production and use of multimedia materials, in general, and with mobile communities, in particular. However, reviewing novel tools and techniques to deal with the resulting complexity is essential.

The contributions of this chapter comprise a design view, platform, and abstraction levels that lower the barrier for mobile multimedia services which leverage the cloud. In this chapter, we describe both conceptual models and software frameworks. The conceptual models capture specific requirements for building efficient mobile multimedia cloud services. The software frameworks enable the realization of such services. Furthermore, we use several use cases to describe the effects of mobile multimedia cloud services on the practices in professional communities.

3.2 BACKGROUND

3.2.1 Professional Communities

Wenger et al. [4] defines "*Communities of Practice* (CoP) as groups of people who share a concern, a set of problems, or a passion about a topic, and who deepen their knowledge and expertise in this area by interacting on an on-going basis." The most important processes in a CoP are collective learning and the production of shared meaning and collective identity. The social practice consists of explicit and tacit knowledge, and competencies. The concept of CoP is helpful to understand and support cooperation, knowledge management and collaborative learning [5]. The CoP can be seen as shared histories of learning [6]. CoP combines the social practice of the community and the identity of its individual members. Moreover, the term *professional communities* refers to CoP in some professional domain, for example, medicine or construction, where the professional inherently learns by social participation in CoP. In organizations, informal CoP want to share knowledge about their profession.

Information systems (ISs) for professional communities face several challenges. Principles such as legitimate peripheral participation, group knowledge, situated learning, informality, and colocation need to be considered in the design of the IS. First, community membership and social status are highly dynamic and vaguely defined. The number of users to support with a community IS can oscillate between tens of users to millions of users in short time frames. Second, the development process of ISs is less stable. Commonly, community members act as stakeholders in the requirements engineering, which results in the need of continuous and collaborative IS adaptation.

For example, the workplace is currently being shifted from centralized offices to mobile on-site places, causing transformation of professional communities into *mobile communities*. With respect to their information technology (IT) needs, mobile communities introduce unique requirements. These on-site communities are characterized by a high degree of collaborative work, mobility, and integration of data coming from many members, devices, and sensors. A mobile community, therefore, needs tools for communication, collaboration, coordination, and sensing as well as for member, community, and event awareness. Community members are often distributed geographically; therefore, their interactions are mediated by digital channels for direct communication and indirect exchange of information objects. Tools for communication are natively supported on mobile devices. However, multiple forms of communication, such as voice, messages, chat, and video streaming, should also be supported.

3.2.2 Media-Centric ISs

Multimedia ISs have played a significant role in supporting professional communities. Members of a mobile community need to collaborate around different multimedia artifacts, such as images or videos. On the data management level, various data need to be captured, created, stored, managed, and prepared for further processing in applications. However, the massive amount of user-generated multimedia content does not necessarily imply the content quality and the social value. Mobile multimedia ISs need to empower individuals and communities with services for adding value to the content easily. For example, techniques from data mining, machine learning, computer vision, and recommender systems can help to further detect, filter, sort, or enhance multimedia content.

As shown in Figure 3.1 [7], the development of ISs for CoP needs a support for digital media and related communication tools between community members and collaboration tools over digital media objects.

Moreover, the software development simplicity was one of the key success factors of the Web. Such level of simplicity, however, is still not achieved

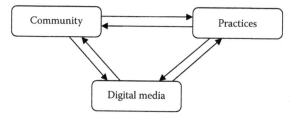

FIGURE 3.1 IS for CoP. (Adapted from de Michelis, G. et al. *Communications of the ACM*, 41, 64–70, 1998.)

for mobile multimedia ISs, although it is the driving force behind mobile community growth. The issue of poor user experience (UX) in many mobile multimedia applications can be associated with high development costs, since the creation and utilization of a multimedia processing and distribution infrastructure is a nontrivial task for small groups of developers. For instance, application developers need constantly to deal with the format and resolution "gap" between Internet videos and mobile devices. Another example of the issue is the provision of up/downstreaming infrastructure. Even more, professional communities are often unable to express their exact needs at the beginning, that is, they develop their requirements as they use the technology. Therefore, flexible, adaptable, and interoperable ISs for professional communities are needed.

For multimedia services, in addition to central processing unit (CPU) and storage requirements, another very important factor is the quality of service (QoS) and the quality of experience (QoE). For a long time, there was no common agreement on the term *user experience*. For some people, it is only seen as being solely user and interface dependent. This understanding of UX changed during the past few years.

The definition of a general UX definition is very hard. First, it is connected with many concepts such as emotional, affective, experiential, hedonistic, and aesthetic variables [8]. Researchers include and exclude these variables arbitrary, depending on their background and interests. Second, the unit of analysis can range from a single-user interaction aspect to all aspects including interaction with the community. Third, UX research is fragmented and complicated by various theoretical models.

In 2009, Law et al. [9] presented a survey on UX to define this term. They designed a questionnaire with three sections: UX Statements, UX Definitions, and Your Background. In "UX Statements," they gave 23 statements and the participants were asked to indicate their level of agreement using a five-point scale. Two hundred and seventy-five researchers and practitioners from academia and industry were asked to fill out the questionnaire. Most researchers agree that UX is influenced by the current internal state of the person, earlier experiences, and the current context. Additionally, most researchers agree on the fact that UX should be assessed while interacting with an artifact. Nevertheless, it also has effects long after the interaction. Finally, in 2010 an ISO definition of UX has been published, which is as follows: "a person's perceptions and responses that result from the use or anticipated use of a product, system or service" [10].

UX in the context of mobile media depends on several aspects. The most obvious one is the small screen size of mobile phones. Even 4-inch displays are small in comparison with today's LCD televisions (often starting at 32 inch). Other device-related issues affecting the UX are the limited battery power, changing bandwidth, WiFi handover, and others. Furthermore, the attention on mobile video is lower than that on the television. For watching a movie, people sit down in front of a television and can spend 2 hours there without a problem. By comparison, mobile users are most often in movement. Taking the train to work or other transportation services is often the common situation of mobile phone usage such as video consumption. This time is limited and the attention is influenced by the surroundings [11].

For long time, high-quality multimedia was reserved to professional organizations equipped with expensive hardware. The distribution of multimedia was limited to hard copies such as video home system (VHS), video CD (VCD), and digital video disk (DVD). The development of Web 2.0, inexpensive digital cameras, and mobile devices has spurred Internet multimedia rapidly. People can generate, edit, process, retrieve, and distribute multimedia content such as images, video, audio, and graphics, much easier than before.

Web 2.0 paved the way of knowledge sharing for CoP. Web 2.0 represents the concepts and tools that put a more social dimension into operation and foster collective intelligence. The more members Web 2.0 services use, the better they become. Such services allow people not only to create, organize, and share knowledge, but also to collaborate, interact virtually, and make new knowledge. However, Web 2.0 models do not solve the issue with continuous changes of IS infrastructure for community services. The ultimate goal of community ISs is to remove the IT engineers from the development loop. The communities should be able to monitor, analyze, and adapt their ISs independently.

Fortunately, the complexity of existing IT systems has resulted in finding new ways for provision and abstraction of systems, networks, and services. Cloud computing can be considered as an innovation driver for IT—its concept imposes new set of requirements on IT systems (scalability, dynamic reconfiguration, multitenancy, etc.), which in turn result in technological breakthroughs. Prior to cloud computing, multimedia storage, processing, and distribution services were provided by different vendors with their proprietary solutions (Table 3.1).

With the rise of Web 2.0, the amounts of data now available to collect, store, manage, analyze, and share are growing continuously. It is becoming common that applications need to scale to datasets of the magnitude of the Web or at least to some fraction of it. Tackling large data problems is not just reserved for large companies, but it is being done by many small communities and individuals. However, it is challenging to handle the large amounts of data within their own organization without having large budget, time, and professional expertise. This is why cloud computing could make techniques for search, mining, and analysis easily accessible to anyone from anywhere. Large-scale data and cloud computing are closely linked.

Cloud computing is fundamental to Web 2.0 development—it enables convenient Web access to data and services. Sharing large amounts of content is one of the salient features of Web 2.0. Clouds further extend these capabilities with externalization of the computing and storage resources. Clouds provide virtually "unlimited" capacities at users' disposal. Cloud computing further extends Web 2.0 with means for flexible composition of arbitrary services, where the user pays only for the consumed resources. This enables communities to be supported with personalized set of services, and not just typical ones, which is important for professional communities with specialized needs. There is no minimum fee and start-up cost. Cloud vendors charge go storage and bandwidth. No installation of media software and upgrade are needed. Expensive and complex licenses are aggregated in the cloud offers. The always-on cloud storage provides worldwide distribution of popular multimedia content. The cloud

TABLE 3.1 Comparison of Key Features between Web 2.0 and Cloud Computing Paradigms

Web 2.0	Cloud Computing
Massive amounts of content	Externalization of computing and storage
Limited practices (sharing, delivery)	Easier to change practices by changing the cloud infrastructure
Predefined multimedia operations and business models	Flexible composition of cloud services based on pay-as-you-go (utility) model and SaaS[a]

[a] SaaS, software as a service.

can alleviate the computation of user devices and save battery energy of mobile phones. These can greatly facilitate small amateur organizations and individuals. Cloud computing boils down multimedia sharing to simple hyperlink sharing. Sharing through a cloud also improves the QoS because cloud–client connections provide higher bandwidth and lower latency than client–client connections.

Cloud computing is a factor for change and innovation, thanks to the increased efficiency of IT infrastructure utilization leading to lower costs. Cloud computing reduces the *cost-effectiveness* for the implementation of the hardware, software, and license for all. Large data centers benefit from *economies of scale*. It refers to reductions in unit cost as the size of a facility and the usage levels of other inputs increase [12]. Large data centers can be run more cost efficiently than private computing infrastructures. They provide resources for a large number of users. They are able to amortize the demand fluctuations per user basis, since the cloud resource providers can aggregate the overall demand in a smooth and predictable manner.

3.2.3 Mobile Clouds

Cloud computing is focused on pooling of resources, whereas mobile technology is focused on pooling and sharing of resources locally enabling alternative use cases for mobile infrastructure, platforms, and service delivery. Mobile cloud computing is envisioned to tackle the limitations of mobile devices by integrating cloud computing into the mobile environment. Resource-demanding applications such as 3D video games are being increasingly demanded on mobile phones. The capabilities of mobile networks and devices craft new ways of ubiquitous interaction over Web 2.0 digital social networks. Consequently, mobile devices, Web 2.0, and social software result in an exponential growth of user-generated mobile multimedia on Web 2.0, which is a driving force for further mobile cloud innovation.

Even if the hardware and mobile networks of the mobile devices continue to evolve and improve, the mobile devices will always be resource-poor and less secure, with unstable connectivity and constrained energy. Resource poverty is a major obstacle for many applications [13]. Therefore, computation on mobile devices will always involve a compromise. For example, on-the-fly editing of video clips on a mobile phone is prohibited by the energy and time consumption. Same performance and functionalities on the mobile devices still cannot be obtained as those on their desktop personal computers (PCs) or even notebooks when dealing with tasks containing complicated or resource-demanding operations.

Mobile devices can be seen as entry points and interface of cloud online services. The combination of cloud computing, wireless communication infrastructure, portable computing devices, location-based services, mobile Web, and so on has laid the foundation for a novel computing model, called *mobile cloud computing*, which allows users an online access to unlimited computing power and storage space. Taking the cloud computing features in the mobile domain, we can define:

> Mobile cloud computing is a model for transparent elastic augmentation of mobile device capabilities via ubiquitous wireless access to cloud storage and computing resources, with context-aware dynamic adjusting of virtual device capacities in respect to change in operating conditions, while preserving available sensing and interactivity capabilities of mobile devices.

To make this vision a reality beyond simple services, mobile cloud computing has many hurdles to overcome. Existing cloud computing tools tackle only specific problems such as parallelized processing on massive data volumes, flexible virtual machine (VM) management, or large data storage. However, these tools provide little support for mobile clouds. The full potential of mobile cloud applications can only be unleashed, if computation and storage are offloaded into the cloud, but without hurting user interactivity, introducing latency, or limiting application possibilities. The applications should benefit from the rich built-in sensors that open new doorways to more smart mobile applications. As the mobile environments change, the application has to shift computation between device and cloud without operation interruptions, considering many external and internal parameters. The mobile cloud computing model needs to address the mobile constraints in success to supporting "unlimited" computing capabilities for applications. Such model should be applicable to different scenarios. The research challenges include how to abstract the complex heterogeneous underlying technology, how to model all the different parameters that influence the performance and interactivity of the application, how to achieve optimal adaptation under different constraints, and how to integrate computation and storage with the cloud while preserving privacy and security. The adoption of cloud computing affects the security in mobile systems. The aspects are related to ensuring that the data and processing controlled by a third party are secure and remain private, and the transmission of data between the cloud and the mobile device is

TABLE 3.2   General Observations and Requirements of the Main Entities in Cloud
Computing Support for Professional Communities

Entity	General Observations
Professional communities	• Vague community membership and social status
	• Dynamic membership
	• Dynamic IS requirements emerging and evolving during the IS usage
Media-centric IS	• Basic operations support (create, store, share, consume, etc.) for increasing number of media formats
	• Massive amounts of content
	• UX
Mobile clouds	• Need to surmount devices' shortcomings
	• Adapted media
	• Privacy and security

secured [14]. Holistic trust models of the devices, applications, communication channels, and cloud service providers are required [15].

A summary of the general observations and requirements of the main entities in cloud computing support for professional communities is given in Table 3.2.

3.3 MOBILE CLOUD MODELS

The clouds have a huge processing power at their disposal, but it is still challenging to make them truly accessible to mobile devices. The traditional client–server model and Web services/applications can be considered as the most widespread cloud application architectures. However, several other approaches to augmenting the computation capabilities of constrained mobile devices have been proposed. Offloading has gained big attention in mobile cloud computing research, because it has similar aims as the emerging cloud computing paradigm, that is, to surmount mobile devices' shortcomings by augmenting their capabilities with external resources. The full potential of mobile cloud applications can only be unleashed, if computation and storage are offloaded into the cloud, but without hurting user interactivity, introducing latency, or limiting application possibilities.

3.3.1 Cloud-Based Mobile Applications

A model where the services are delivered to mobile devices using the traditional client–server model, but cloud computing concepts are applied to the "server" architecture. The bulk of research literature covers this type of applications as examples of mobile cloud computing. This is understandable, since it is the most simple way to augment the capabilities of mobile devices. The client–server model has been well researched and established in practice for distributed systems and Web applications. In this case, the mobile app, regardless of whether it is native or Web based, acts as front end to the services provided in the cloud, which in turn benefits from the interconnected services, huge computational power, and storage capacities.

Multimedia services in the cloud typically experience unpredictable bursts of data access and delivery. The cloud's utility-like allocation mechanism for computing and storing resources is very effective for dynamic requirements such as those of professional communities. However, the generic cloud services provided by major cloud vendors are insufficient to deliver acceptable mobile UX. The fluctuations in mobile wireless networks, limited capacities of mobile devices, and QoE needs for mobile media impose more specific requirements than those for the wired Internet cases. For example, streaming high definition (HD) content to a mobile phone would cause bad UX even if the device could support decoding of such content.

With the growing scale of Web applications and the data associated with them, scalable cloud data management becomes a necessary part of the cloud ecosystem. Some of the popular scalable storage technologies in the moment are Amazon Simple Storage Service (S3), Google BigTable, Hadoop HBase and Hadoop distributed file system (HDFS), and so on. Basically, these distributed blob and key–value storage systems are very suitable for multimedia content, that is, they are scalable and reliable as they use distributed and replicated storage over many virtual servers or network drives. Cloud-based mobile multimedia applications, in general, use cloud services and cloud infrastructure to meet the requirements for mobile multimedia, including content delivery networks (CDNs), peer-to-peer (P2P) multimedia delivery, and parallelized (high-performance) processing of multimedia content. The CDN delivers multimedia content to end users with a better performance, a lower latency, and a higher availability. Amazon Web Services, for example, provide their CloudFront service to application developers for such needs. High-performance multimedia processing, typically, has referred to speeding up the transformation process of multimedia content with the use of powerful server machines. The current trend in practice is to use many commodity hardware machines to perform the same operation with lower operational costs. This approach has become popular under the term "Datacenter as a Computer" [16]. This is why large cloud vendors rely on custom server construction with tens of thousands of cheap computers with conventional multicore processors. They are cheaper and more energy efficient. One powerful machine costs more than two not-so-powerful machines that have the same performance. Having a larger number of small computational units gives an easier way of tackling the fault-tolerant issue. These systems exploit the power of parallelism and at the same time provide reliability at the software level. Instead of using expensive hardware, the system takes advantage of thousands of inexpensive independent components with anticipated hardware failure. The software is responsible for ensuring data replications and computation predictability.

3.3.2 Cloud-Aware Mobile Applications

The relative resource poverty of mobile devices as well as their lower trust and robustness leads to reliance on static servers [17]. But the need to cope with unreliable wireless networks and limited power capacity argues for self-reliance. Mobile cloud computing approaches must balance between these aspects. This balance must dynamically react on changes in the mobile environment. Mobile applications need to be adaptive, that is, the responsibilities of the client and the server need to be adaptively reassigned.

The cloud computing concepts can be considered from a viewpoint of the mobile device. In fact, we want to achieve a virtually more powerful device, but in contrast to the previous model, we want to keep all the application logic and control on the device. There are many reasons why this kind of application model is desired, for example, to retain privacy and security without sharing the code and data with the cloud, or simply reuse the existing desktop applications on the mobile device, or reduce the communication latency introduced by the remote cloud.

Offloading has gained big attention in mobile cloud computing research, because it has similar aims as the emerging cloud computing paradigm, that is, to surmount mobile devices' shortcomings by augmenting their capabilities with external resources. Offloading or augmented execution refers to a technique used to overcome the limitations of mobile phones in terms of computation, memory, and battery.

Such applications, whose code can be partitioned and certain parts are offloaded in a remote cloud [18,19], are called *elastic mobile applications*. Basically, this model of elastic mobile applications provides to the developers an illusion as if he/she is programming virtually much more powerful mobile devices than the actual capacities. Moreover, elastic mobile application can run as stand-alone mobile application but also use external resources adaptively. Which portions of the application are executed remotely is decided at run time based on resource availability. Offloading is a different approach to augment mobile devices' capabilities compared to the traditional client/ server model prevalent on the Web. Offloading enables mobile applications to use external resources adaptively, that is, different portions of the application are executed remotely based on resource availability. For example, in case of unstable wireless Internet connectivity, the mobile applications can still be executed on the device. In contrast, client/server applications have static partitioning of code, data, and business logic between the server and the client, which is done at the development phase.

In order to dynamically shift the computation between a mobile device and a cloud, applications needed to be split in loosely coupled modules interacting with each other. The modules are dynamically instantiated on and shifted between mobile devices and cloud depending on the several metric parameters modeled in a cost model. These parameters can include the module execution time, resource consumption, battery level, monetary costs, security, or network bandwidth. A key aspect is user waiting time, that is, the time a user waits from invoking some actions on the device's interface until a desired output or exception is returned to the user. User wait time is important for deciding whether to do the processing locally or remotely.

Which parts of the mobile application run on the device and which on the cloud can be decided based on cost model. The cost model takes inputs from both device and cloud, and runs optimization algorithms to decide execution configuration of applications (Figure 3.2). Zhang et al. [20]

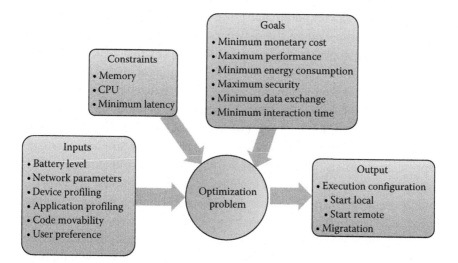

FIGURE 3.2   Cost model parameters for adapting mobile applications.

use naive Bayes classifiers to find the optimal execution configuration from all possible configurations using given CPU, memory and network consumption, user preferences, and log data from the application. Guirgiu et al. [21] model the application behavior using a resource consumption graph. Every bundle or module composing the application has memory consumption, generated input and output traffic, and code size. Application's distribution between the server and the phone is then optimized. The server is assumed to have infinite resources and the client has several resource constraints. The partitioning problem seeks to find an optimal cut in the graph satisfying an objective function and device's constraints. The objective function tries to minimize the interactions between the phone and the server, while taking into account the overhead of acquiring and installing the necessary bundles.

However, optimization involving many interrelated parameters in the cost model can be time or computation consuming, and even can override the cost savings. Therefore, approximate and fast optimization techniques involving prediction are needed. The model could predict costs of different partitioning configurations before running the application and deciding on the best one [22].

3.3.3 Fog/Edge Computing

Reduced latency, media content and processing, and aggregation are pushed at the edge of the network, that is, mobile network base stations or WiFi hot spots. In a nutshell, fog computing offers combined virtualized resources such as computational power, storage capacity, and networking services at the edge of the networks, that is, closer to the end users. Fog computing supports applications and services that require very low latency, location awareness, and mobility. Fog computing complements the cloud services. Fog computing is a highly virtualized platform that provides compute, storage, and networking services between the end devices and the traditional cloud data centers, typically, but not exclusively located at the edge of network [23]. Satyanarayanan et al. [13] define a similar concept called cloudlets, which are software/hardware architectural elements that exist on the convergence of mobile and cloud computing. They are the middle element in the three-tier architecture—mobile device, cloudlet, and cloud. Cloudlets emerge as enabling technology for resource-intensive but also latency-sensitive mobile applications. The critical dependence on a distant cloud is replaced by dependence on a nearby cloudlet and best-effort synchronization with the distant cloud (Figure 3.3).

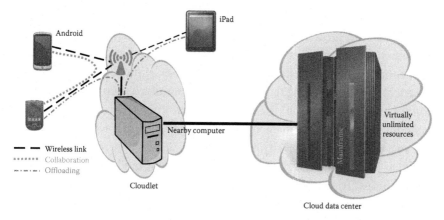

FIGURE 3.3   Fundamental concept of fog computing (i.e., cloudlets).

Cloudlets or fog computing nodes possess sufficient compute power to host resource-intensive tasks from multiple mobile devices. Moreover, they can enable collaboration features with very low latency, aggregation of stream data, local analytics, P2P multimedia streaming, and so on. The end-to-end response times of applications within a cloudlet are fast and predictable. In addition, cloudlets feature good connectivity to large data center-based clouds. Cloudlet resembles a "datacenter-in-a-box" and a self-managing architecture to enable simplistic deployment at any place with Internet connectivity such as a local business office or a coffee shop.

3.4 MOBILE MULTIMEDIA CLOUD SUPPORT OF PROFESSIONAL COMMUNITIES' PRACTICES

To successfully support the practices of any kind of a professional community, independent of the size or domain of interest, understanding the knowledge sharing processes within communities is needed. Supporting professional community practices faces many challenges. There are several reasons that perplex the implementation of successful community IS, for example, processes such as situated learning, shared group knowledge, mobility, and colocation need to be taken into account when designing the IS. Moreover, the needs are community specific. Community members are not able to express precise requirements at the beginning, that is, the requirements emerge along the system use. Therefore, the community needs mechanisms to add, configure, and remove services on the fly. Besides, the advances of multimedia technology require constant support of novel hardware and network capabilities. A full spectrum of multimedia content technologies needs to be supported.

The central process in professional communities is sharing of knowledge about the profession. Organizational knowledge management and professional learning are closely connected. The socialization, externalization, combination, and internationalization (SECI) model by Nonaka and Takeuchi [24] has been widely accepted as a standard model for organizational knowledge creation. It emphasizes that knowledge is continuously embedded, recreated, and reconstructed through interactive, dynamic, and social networking activity. Spaniol et al. [25] further refined the SECI model as a media-centric knowledge management model for professional communities. It combines the types of knowledge of community members, the tacit and explicit knowledge, and the process of digital media discourses within CoP and their media operations.

The media-specific theory [26] distinguishes three basic media operations:

- Transcription: a media-specific operation that makes media collections more readable
- Localization: an operation to transfer global media into local practices, which can be further divided into
  - Formalized localization
  - Practiced localization
- (Re-)addressing: an operation that stabilizes and optimizes the accessibility in global communication

Spaniol et al. [25] integrate these media operations into the learning and knowledge sharing processes of professional communities (Figure 3.4). As seen in the figure, the individuals internalize knowledge from some sources. The knowledge is then communicated with others by (1) the human–human interaction, which is called practiced localization that, in turn, fosters content's socialization within the CoP, thus forming a shared history, and (2) human transcription, which means creating new digital artifacts on an externalized medium. The externalized artifacts are then processed by the IS, which is called formalized localization of the media artifacts. The artifacts are combined and made available for further use. The semiautomatic addressing operation closes the circle that represents a context-aware delivery and presentation of the medial artifacts.

Table 3.3 gives an example mapping between the media operations defined in the theory and possible cloud services. This mapping is not

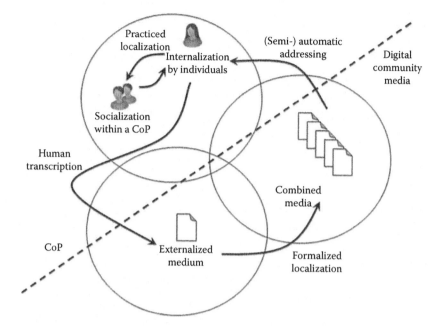

FIGURE 3.4 Media-centric theory of learning in CoP. (Adapted from Spaniol, M. and Klamma, R., *Knowledge Networks: The Social Software Perspective*, 46–60.)

TABLE 3.3 Mapping between the Media-Theoretic Operations and the Cloud Services

Media-Theoretic Operation	Cloud Multimedia Service
Transcription	• Metadata creation
	• Ubiquitous
	• Multimedia acquisition with digital media devices
	• Physical-to-virtual input methods: OCR,[a] object recognition, and voice recognition
Formalized localization	• Real-time audio/video/text communication
	• Multimedia transcoding
	• Multimedia indexing and processing
	• Story creation
Practiced localization	• Content and metadata collaboration
	• Multimedia sharing
	• Tagging
	• Storytelling
Readdressing	• Recommender systems
	• Multimedia retargeting
	• Adaptive streaming
	• MAR[b]

[a] OCR, optical character recognition.
[b] MAR, mobile augmented reality.

complete nor extensive. Many of these services benefit from the cloud infrastructure in terms of compute, storage, and networking resources, for example, multimedia transcoding and recommender systems.

Metadata, or descriptive data about the multimedia content, let us tie different multimedia processes in a life cycle together. Kosch et al. [27] identify two main parts of the metadata space. First, metadata production occurs at or after content production. At this stage, the metadata consist of creation information, automatically extracted information (low-level features such as histograms and segment recognition), and human-generated information (high-level semantics such as scene descriptions and emotional impressions). Second, metadata consumption occurs at the media dissemination and presentation stages of the content's life cycle. For example, metadata facilitate retrieval capabilities for large multimedia databases and guide the adaptation of content to achieve the desired QoS. Metadata are consumed at different stages—at authoring, indexing, and proxy level, and end-device level.

Ubiquitous multimedia acquisition. Mobile devices are becoming an indivisible part of the Web today. Mobile/Web integration is meant not only mobile Web pages only, but also an integration of mobile devices as equal nodes on the Internet, the same as desktop PCs and servers. Actually, the Web nowadays is a common communication channel for multimedia content that is "presumed" on different personal computing devices such as desktops, laptops, tablets, or smartphones. The ever-increasing amounts of user-generated multimedia data require scalable data management in the clouds and ubiquitous delivery through the Web. A seamless multimedia integration with clouds is needed where the clouds heavy-lift the necessary multimedia operations such as transcoding, adaptation, highly available storage, responsive delivery, and scalable processing resources.

Physical-to-virtual world input methods such as optical character recognition (OCR), object recognition on camera's video stream, and voice recognition contribute significantly to the UX especially at field work. For example, novel methods such as OCR help alleviate some of the inherited issues of the small computing devices. OCR refers to the process of acquisition of text and layout information, through analyzing and processing of image files. Compared to the traditional input way of typing, OCR technique has many advantages such as speed and high efficiency for large texts.

Multimedia transcoding becomes more common procedure as the interoperability between different media devices becomes more important. One of the biggest challenges in future multimedia application development is device heterogeneity. Future users are likely to own many types of devices. Users when switching from one device to another would expect to have ubiquitous access to their multimedia content. Cloud computing is one of the promising solutions to offload the tedious multimedia processing on mobile devices and to make the storage and access transparent. Transcoding, generally, is the process of converting one coded format to another. Video transcoding, for example, can adapt the bit rate to meet an available channel capacity or reduce the spatial or temporal resolution to match the constraints of mobile device screens. Video transcoding and processing are data intensive and time and resource consuming. Clouds play a significant role in reducing the costs for upfront investment in infrastructure and in cases of variable demands. *Multimedia indexing* refers to the process of multimedia processing to identify content objects and cues that can later be used for content-based multimedia retrieval. Indexing solutions usually involve resource-expensive computer vision and machine learning algorithms, which can also benefit from a cloud infrastructure.

Content and metadata collaboration. Collaboration plays a significant role among groups of people (e.g., coworkers) who are trying to perform tasks to achieve a common goal or having similar interests. Groupware technology assists a group of people to communicate, manipulate, and modify shared digital objects in a coherent manner [28]. Mobile multimedia applications lack support for real-time collaborative work. Typically, mobile device usage is limited to creating and sharing content, whereas the collaborative operations are performed asynchronously on desktop computers or laptops. Yet real-time collaboration is necessary in many use cases for both on-site professional and amateur communities. These on-site communities are characterized by a high degree of collaborative work, mobility, and integration of data coming from many members. Real-time collaboration provides the ability to iterate quickly by permitting members to work in parallel. In addition, mobile real-time collaboration (MRTC) enables users to benefit from location awareness and work on spatially distributed tasks. Capturing and using the context by multiple collaborators in real time increases productivity.

The user-generated multimedia content changes relatively slowly after its creation. However, the associated metadata are under constant modification. For example, a video creator initially describes and tags a new video. But after sharing the video, many other people contribute to the video with annotations, hyperlinks, comments, ratings, and so on. Therefore, the success of multimedia services highly depends on features for metadata sharing and collaborative metadata editing.

The right set of underlying communication protocols is crucial in MRTC. Google Wave Federation Protocol is an excellent example of an Extensible Messaging and Presence Protocol (XMPP)-based communication and collaboration platform for concurrently editable structured documents and real-time sharing between multiple participants. Novell Vibe Cloud is a Web-based social collaboration platform for enterprises, providing social messaging and online document coediting along with file management, groups, profiles, blogs and wikis, and security and management controls. Both Google Wave Federation Protocol and Novell Vibe are sophisticated collaborative editing software, but their reliance on powerful clients (i.e., desktop Web browsers) limits the usefulness for custom mobile applications.

Tagging is a powerful and flexible approach to organize the content and learning processes in a personalized manner. With the rise of Web 2.0, the word *tag* has been used in almost every Web 2.0 or Web page. Rather than using a standard set of categories defined by experts, everybody creates one's own categories in the form of tags. Tagging helps users collect, find, and organize multimedia effectively. Tags can be available to all online users and user community groups, or only be accessed by the creator privately. Tags are applied to different resources such as images, videos, Web pages, blog entries, and news entries. Various Web resources are organized through tags.

Digital storytelling and story creation. Storytelling intertwines semantic knowledge by linking it with the narrative experiences gained from episodic knowledge. Storytelling is an important aspect for knowledge sharing and learning in professional communities. Telling, sharing, and experiencing stories are common ways to overcome problems by learning from the experiences of other members. One of the major reasons for the limited adoption of digital storytelling in organizational ISs may be that authoring of stories is extremely challenging. Suitable tools and simple methodologies need to be put in place to support the authors in using different media. The development of a shared practice integrates the negotiation of meaning between the members as well as the mutual engagement in joint enterprises and a common repertoire of activities, symbols, and multimedia artifacts. Storytelling and story creation through interactive and effective stories enable joining conceptual and episodic knowledge creation processes with semantically enriched multimedia.

Semantic multimedia retargeting seeks to remedy some of the issues with UX in mobile video applications by making use of cloud services for fast and intelligent video processing. Cloud computing has great potential to leverage the current issues with mobile production and use of multimedia materials, in general, and with mobile UX, in particular. *Multimedia processing* techniques such as automatic video zooming, segmentation, and event/object detection are often proposed techniques for video retargeting to mobile devices. For example, zooming and panning to the regions of interest within the spatial display dimension can be utilized. This kind of zooming displays the cropped region of interest (ROI) at a higher resolution, that is, observing more details. Panning enables watching the same level of zoom (size of ROI), but with other ROI coordinates. For example, in soccer game, this would mean watching how player dribbles with the ball more closely, whereas by panning one can observe other players during the game such as the goalkeeper.

Adaptive streaming. This multimedia streaming technique adjusts the video and audio quality with variable network connection. The video quality decreases when the network connection is not very good. In this way, the quality of video and audio starts degrading gracefully and correspondingly with network bandwidth. In this way, streaming adapts with changing connection leading to high UX with better quality of video and the user is not able to perceive these changes in quality. Examples of this technology are in Apple's HTTP Live Streaming protocol, Microsoft's IIS Smooth Streaming, and 3GPP Adaptive HTTP Streaming. The current streaming technology is dynamic adaptive streaming over HTTP (DASH). Here, the client has control over the delivery. The client is responsible for choosing the alternatives according to the network bandwidth. It is not a system, protocol, presentation, codec, interactivity, or client specification but provides a format to enable efficient and high quality delivery of streaming services over the Internet [29]. In dynamic adaptive streaming, the client has control over the delivery. The additional features of MPEG-DASH are switching and selectable streams, advertisement insertion, segments with variable duration, multiple base URLs, clock drift for live sessions, and scalable video coding support. With the feature of switching and selectable streams, the clients have the option to choose between the audio streams from different languages and between the videos from different cameras. Similarly, the advertisements can be inserted between periods or segments.

Mobile augmented reality (MAR) is a natural complement to mobile computing, since the physical world becomes a part of the user interface (e.g., in video streaming). Accessing and understanding information related to the real world becomes easier. This has led to a widespread commercial adoption of MAR in many domains such as education and instruction, cultural heritage, assisted directions, marketing, shopping, and gaming. For example, Google Sky Map gives a new and intelligent window on the night sky or CarFinder creates a visible marker showing the parked car, its distance away, and the direction in which to head. Furthermore, in order to support diverse digital content, several popular MAR applications have shifted from special-purpose applications into MAR browsers that can display the third-party content. Such content providers use predefined application program interfaces (APIs) that can be used to feed the content to the MAR browser based on context parameters.

Recommender systems have emerged as a kind of information filtering systems that help users deal with information overload. They are applied successfully in many domains, especially in e-business applications such as Amazon.com. The basic idea of recommender systems is to suggest to users items, for example, movies, books, and music that they may be interested in. Since computing is moving toward pervasive and ubiquitous applications, it becomes increasingly important to incorporate contextual aspects into the interaction in order to deliver the right information to the right users, in the right place, and at the right time. Considering the high level of computational efforts needed to generate recommendations, cloud-based recommender systems are many times the only viable solution.

3.5 APPLICATION OF MOBILE MULTIMEDIA CLOUDS

This section describes several use-case prototype systems that have been developed within our group. They demonstrate the application of the mobile cloud approaches in supporting professional community demands.

3.5.1 SeViAnno and AnViAnno: Ubiquitous Multimedia Acquisition and Annotation

Virtual Campfire [30] embraces a set of advanced applications for CoP. It is a framework for mobile multimedia management concerned with mobile multimedia semantics, multimedia metadata, multimedia context management, ontology models, and multimedia uncertainty management. SeViAnno [31] is an MPEG-7-based interactive semantic video annotation Web platform with the main objective to find a well-balanced trade-off between a simple user interface and video semantization complexity. It allows standard-based video annotation with multigranular

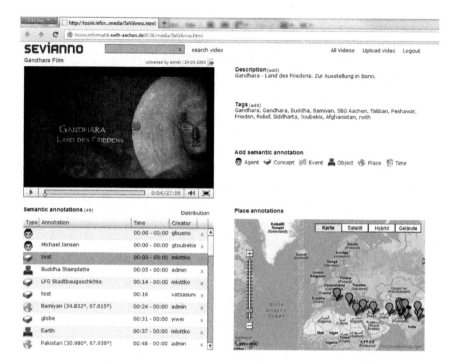

FIGURE 3.5   The SeViAnno user interface with a video player, video information and video list, user-created annotations, and Google map mash-up for place annotations.

community-aware tagging functionalities. Various annotation approaches are integrated and depicted in Figure 3.5.

AnViAnno is an Android application for context-aware mobile video acquisition and semantic annotation (Figure 3.6) [32]. The annotation is based on the MPEG-7 metadata standard. With AnViAnno, users can also semantically annotate videos. The annotation is based on the MPEG-7 metadata standard [33]. MPEG-7 is one of the most complete existing standards for multimedia metadata. It is Extensible Markup Language (XML)-based standard and consists of several components: systems, description definition language, visual, audio, multimedia description schemes, reference software, conformance testing, extraction and use of MPEG-7 descriptions, profiles and levels, schema definition, MPEG-7 profile schemata, and query format. However, several different approaches have been used as metadata formats in multimedia applications. To enable interoperability between systems using these different formats, we have implemented or used mapping services. For example, our MPEG-7 to Resource Description Framework (RDF) converter [34] is able to convert MPEG-7 documents into RDF documents for further reasoning the fact deriving about the multimedia.

FIGURE 3.6 Screen snapshots of AnViAnno—an Android client application for semantic multimedia.

Users are able to capture and annotate the videos with rich semantics in the MPEG-7 standard. The user-generated annotations are further used to navigate within the video content or improve the retrieval from multimedia collections. For example, users can navigate through the video(s) using a seek bar or semantic annotations. The videos and their annotations are exposed to other internal Lightweight Application Server (LAS) MPEG-7 services and external clients.

3.5.2 MVCS: Multimedia Retargeting and the Case of UX Improvement

With the mobile video cloud services (MVCS) prototype [35], we seek to remedy some of the issues with UX in mobile video applications by making use of cloud services for fast and intelligent video processing. The first problem is related to the small viewing size and bit rate limitation. Zooming, ROI enhancement, and bit rate adaptation have been proposed as solutions. The next problem mobile users have to deal with is browsing videos. The aim is to create a system that allows to browse and access the video content in a finer, per-segment basis. Quite different approaches to improve the UX of mobile video using semantic annotation are also existing.

Video processing is a CPU-intensive task. Much research work proposes MapReduce-based [36] cloud solutions, for example, to transform images and videos or to transcode media content into various video formats for different devices and different Internet connections. Basically, the video file is split into multiple parts, processed in parallel, and merged in the correct order. MapReduce can be used not only to speed up transcoding but also for feature detection in videos as you can take the frames as images.

The enhancement of the UX of mobile video in our setup consists of three parts (Figure 3.7). First, thumbnail cue frames are generated at the transitioning scenes (i.e., events in the video). Figure 3.8 shows the concept of video browsing in more detail. The thumbnail seek bar is placed on the

Segment thumbnail enables browsing by scenes

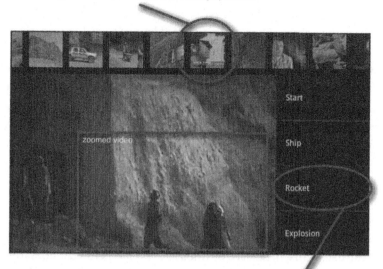

Enables browsing by tags

FIGURE 3.7   Video stream browsing based on video segmentation and automatically generated metadata.

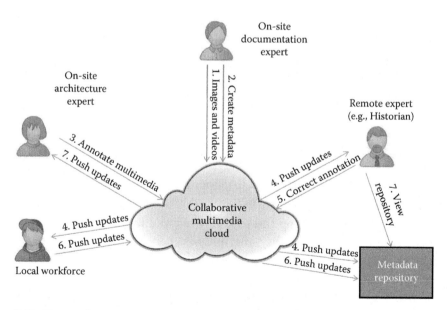

FIGURE 3.8   Semantic annotations created using the mobile collaborative cloud services.

top part. It consists of thumbnails of all the scenes of the video ordered by their occurrence. This makes it easy to browse videos. The user can orientate himself/herself by the thumbnails and does not need to wait until the video is loaded at a certain time point. This works well regarding the low bandwidth. As described earlier, the thumbnails have such a small resolution that they are loaded very fast. Furthermore, a lazy list has been implemented so that it requires even less bandwidth as only currently viewable images are loaded. Clicking on a thumbnail redirects the user directly to the corresponding scene in the video. The user can now search the content much faster than in a traditional video player. This again improves the orientation for the user. If the user clicks on an image, he/she is directly redirected to the corresponding time point in the video. Furthermore, the seek bar focuses on the current scene and scrolls automatically.

Second, the tag list (right) consists of tags that have been added manually by the user himself/herself, by other users, or generated automatically. Like the thumbnails, the tags are ordered by their timely occurrence in the video. If a user clicks on a tag, the video stream goes directly to the corresponding position. Both components, that is, the segment-based seek bar and the tag list, are implemented to overcome the mobile UX problem of video browsing on a mobile device.

Finally, the third part of mobile UX improvement contains the video player itself. As device information including the screen size is sent to the cloud, the necessary zooming ratio can be calculated. Depending on the screen size of the device, a zooming ratio is calculated and depending of the objects a center position of the zooming field is defined. The rectangle overlay on the video player symbolizes the zoomed video. Two persons are recognized by the object recognition service, and therefore, the zooming field is set at this position. The user just sees this part of the video and can better concentrate on the persons.

3.5.3 Mobile Community Field Collaboration: Working with Augmented Reality and Semantic Multimedia

Mobile devices are altering the way we work and live our lives in many ways. For instance, commodity smartphones and tablets have proven to do a better job in many cases than specially developed (and expensive) devices in many professional fields such as disaster management or military. Mobile devices are also ideal tools for ubiquitous production and sharing of multimedia content. Besides that, smartphones equipped with handful sensors provide a platform for context-aware interactions with digital objects. However, mobile multimedia applications lack support for real-time collaborative work.

To illustrate this concept, we consider the use case [37] of MAR browsers, which are typical examples that provide rich multimedia UX, but fail to provide collaborative features. MAR becomes increasingly feasible on inexpensive hardware at mass market effect.

A collaboration environment needs to provide shared workspace and conversational support. In addition, mobile collaboration environments should consider the mobility of users, that is, their dynamical spatial context. In our case, the mobile collaboration services are tailored to the restrictions and features of mobile devices. The workspace facilitates sharing of multimedia content and context, access to multimedia store, and maintaining a consistent state of multimedia content and its semantics. The semantics usually express the context of the multimedia content, for example, location and time where this was captured, creators, low-level semantics (such as histogram features), high-level semantics such as annotations and tags, and so on. The semantics are represented in a metadata format.

Figure 3.8 demonstrates a use case of the mobile collaborative services during a digital documentation fieldwork by different users in a cultural heritage domain. A documentation expert acquires multimedia with the help of the built-in camera and sensors at his/her mobile device. Then the multimedia is shared via the media repository and a point of interest (POI) is also added with spatial context. Collaborators annotate the multimedia by creating and editing annotations in real time. These annotations are propagated to all other collaborators via a collaborative editing infrastructure. Furthermore, collaborators use chat functionality to converse about various reasons such as discussing about content and requesting assistance for annotation. Consumers view the multimedia via MAR as POIs at either camera view or map view. They can also choose a POI to see further details about it. Every multimedia artifact has multiple semantic base types that are used for annotation.

In the use case of MAR browsers, we exploit a device's features to demonstrate the feasibility of a collaborative augmented reality.

For the augmented reality functionality, the multimedia content uses a POI data type. Users can preform spatial range queries over the POIs. Every POI has a reference to a multimedia, longitude, latitude, altitude, and precision. Our multimedia artifacts can be any kind of data that can be rendered on a mobile device, for example, video, images, and 3D objects. Title, description, and keywords form the basic metadata about the multimedia.

Augmented reality information, basic multimedia metadata, and semantic annotations are stored as XML documents. This design choice eases the interoperability and the development of collaboration features. The synchronization is done by keeping a copy of the XML document at every client in the session and ensuring timely updates on the copies in case of edit operations or conflicts.

The collaborative multimedia annotation system is based on the XMPP and MPEG-7 semantics. The work is mainly inspired from cultural heritage scenarios for digital documentation of historical sites, where the described mobile collaboration technology allows professional communities to transform their collaborative work practices in the field. Evaluation indicates that such solutions increase the awareness of community members for activities of coworkers and the productivity in the field in general.

3.5.4 Mobile Augmentation Cloud Services: Cloud-Aware Computation Offloading for Augmenting Devices' Capabilities

Offloading enables mobile applications to use external resources adaptively, that is, different portions of the application are executed remotely based on resource availability. For example, in case of unstable wireless Internet connectivity, the mobile applications can still be executed on the device. In contrast, client/server applications have static partitioning of code, data, and business logic between the server and the client, which is done in development phase. Actually, client/server applications can be seen as a special type of offloaded applications.

We developed a framework that integrates with the established Android application model for development of "offloadable" applications, a lightweight application partitioning, and a mechanism for seamless adaptive computation offloading [38]. We propose Mobile Augmentation Cloud Services (MACS), a services-based mobile cloud computing middleware. Android applications that use the MACS middleware benefit from seamless offloading of computation-intensive parts of the application into nearby or remote clouds. First, from a developer's perspective, the application model stays the same as on the Android platform. The only requirement is that computation-intensive parts are developed as Android services, each of which encapsulates specific functionality. Second, according to different conditions/parameters, the modules of program are divided into two groups: one group runs locally and the other group runs on the cloud side.

The decision for partitioning is done as an optimization problem according to the input on the conditions of the cloud side and devices, such as CPU load, available memory, remaining battery power on devices, and bandwidth between the cloud and the devices. Third, based on the solution of the optimization problem, our middleware offloads parts to the remote clouds and returns the corresponding results back. Two Android applications on the top of MACS demonstrate the potential of our approach.

The goal of our MACS middleware is to enable the execution of elastic mobile applications. Zhang et al. [20] consider elastic applications to have two distinct properties. First, an elastic application execution is divided partially on the device and partially on the cloud. Second, this division is not static but is dynamically adjusted during the run time of the application. The benefits of having such application model are that the mobile applications can still run independently on mobile platforms, but can also reach cloud resources on demand and availability. Thus, mobile applications are not limited by the constraints of the existing device capacities. The technical details of this approach are covered in a previous work [38].

The results show that the local execution time can be reduced a lot through offloading, which is sometimes not acceptable for the users to wait for, and by pushing the computation to the remote cloud can lower the CPU load on mobile devices significantly, thanks to the remote cloud, since most of the computations are offloaded to the remote cloud. Meanwhile, lots of energy can be saved, which indicates that the users can have more battery time compared to the local execution. The results also prove that the overhead of our framework is small.

3.5.5 Summary

The aforementioned prototype systems are developed as service-oriented architectures, which enables interoperability between them on the multimedia and metadata level. It means that the output from some services can be used as input for other services. For example, the multimedia content captured, tagged, and annotated with AnViAnno can easily be fed in the MVCS for further processing and semantics enrichment.

Table 3.4 summarizes the cloud multimedia services covered by the aforementioned prototypes. The symbol 'X' means that the respective prototype implements/uses this service. AnViAnno and SeViAnno applications provide support for basic mobile media operations. Typically, these applications are deployed on a cloud infrastructure, whereas the mobile and Web applications serve as front ends for the multimedia services.

TABLE 3.4   Summary of the Use-Case Prototypes, the Cloud Services They Implement, and the Applicable Mobile Cloud Model

	Cloud Multimedia Service											Applicable Mobile Cloud Model		
	Metadata	Ubiquitous Acquisition	Real-Time Communication	Transcoding	Indexing and Processing	Collaboration	Sharing and Tagging	Retargeting	Adaptive Streaming	MAR	Adaptive Streaming	Cloud-Based	Cloud-Aware	Fog Cloud
AnViAnno and SeViAnno	X	X					X					X		
MVCS	X	X		X	X			X	X		X	X	X	X
XMMC	X	X	X			X	X			X		X		X
MACS				X	X			X					X	X

The primary focus of MVCS is on the retargeting of video content for improving the UX during mobile video consumption. It uses different semantic and computer vision strategies to perform content-based video stream adaptation. Such operations can be performed on any type of mobile cloud model, whereas the local execution consumes most resources but is always available, and the fog cloud model provides benefits such as offloading and low latency. XMMC prototype focuses on field collaboration between coworkers. The collaboration is performed mostly on the multimedia metadata. In addition, XMMC provides immersive consumption of the produced multimedia via a MAR browser. MACS, on the contrary, has a goal to augment the processing capabilities of mobile devices, which based on the need can be applied in resource-intensive media operations such as transcoding, processing, indexing, and retargeting. All these example prototypes provide means for sharing knowledge represented in multimedia formats. The combined usage of these applications cover the media-theoretic operations described earlier. The use of cloud computing solves the issues with dynamic IS workloads and changing requirements. Moreover, higher level services such as semantic retargeting amend the UX of mobile multimedia applications.

3.6 CONCLUSIONS AND OUTLOOK

The aim of this chapter is to shed light on how mobile cloud computing can be applied to support the practices of professional communities, specifically their needs, from the point of view of multimedia and knowledge sharing ISs. Mobile professional communities exhibit complex structure and dynamic processes that reflect on the IS support. The utility-like resource provisioning of cloud resources suits perfectly to the dynamic membership nature of professional communities. Software-as-a-service (SaaS) concept of cloud computing provides means for unlimited configurations and mash-ups of community services to match any emerging IS requirement. In this chapter, we identified three models of mobile cloud computing, each with benefits and drawbacks, which affect UX, availability, responsiveness, and costs of the multimedia services. Furthermore, the multimedia services can be contemplated from a media-theoretic perspective and an organizational knowledge management theoretic perspective. This combination of these perspectives helps us to understand the knowledge creation and learning processes within professional communities. Finally, this chapter provided a brief overview of some prototype systems developed in our group, which cover most of the multimedia service needs identified for an IS support of professional communities. The experimental results provide a support for the mobile cloud approach applied to professional communities.

REFERENCES

1. Klamma, R. and Jarke, M. 2008. Mobile social software for professional communities. *UPGRADE,* IX(3): 37–43.
2. Mell, P. and Grance, T. 2009. *The NIST Definition of Cloud Computing.* http://csrc.nist.gov/groups/SNS/cloud-computing/cloud-def-v15.doc.
3. Armbrust, M., Fox, A., Griffith, R., Joseph, A. D., Katz, R. H., Konwinski, A., Lee, G. et al. 2009. *Above the Clouds: A Berkeley View of Cloud Computing.* Berkeley, CA: EECS Department, University of California.
4. Wenger, E., McDermott, R., and Snyder, W. M. 2002. *Cultivating Communities of Practice.* Boston, MA: Harvard University Press.
5. Brown, J. S. and Duguid, P. 2000. *The Social Life of Information.* Boston, MA: Harvard Business Press.
6. Etienne, W. 1998. *Community of Practice: Learning, Meaning, and Identity.* Cambridge: Cambridge University Press.
7. de Michelis, G., Dubois, E., Jarke, M., Matthes, F., Mylopoulos, J., Papazouglou, M., Schmidt, J. W., Woo, C., and Yu, E. 1998. A three-faceted view of information systems: The challenge of change. *Communications of the ACM,* 41(12): 64–70.

8. Hassenzahl, M. and Tractinsky, N. 2006. User experience—A research agenda. *Behaviour and Information Technology,* 25(2): 91–97.

9. Law, E. L.-C., Roto, V., Hassenzahl, M., Vermeeren, A. P., and Kort, J. 2009. Understanding, scoping and defining user experience: A survey approach. In *Proceedings of the 27th International Conference on Human Factors in Computing Systems,* April 4–9, Boston, MA, pp. 719–728.

10. *ISO FDIS 9241-210:2010.* 2010. Ergonomics of human system interaction—Part 210: Human-centered design for interactive systems (formerly known as 13407). Switzerland: International Organization for Standardization.

11. Cui, Y., Chipchase, J., and Jung, Y. 2007. Personal TV: A qualitative study of mobile TV users. In *Proceedings of the 5th European Conference on Interactive TV: A Shared Experience,* Amsterdam, The Netherlands. Berlin: Springer-Verlag, pp. 195–204.

12. Arthur, S. and Steven M. S. 2003. *Economics: Principles in Action.* Needham, MA: Pearson Prentice Hall.

13. Satyanarayanan, M., Bahl, P., Cáceres, R., and Davies, N. 2009. The case for VM-based cloudlets in mobile computing. *IEEE Pervasive Computing,* 8(4): 14–23.

14. Lagesse, B. J. 2011. Challenges in securing the interface between the cloud and mobile systems. In *Proceedings of the 1st IEEE PerCom Workshop on Pervasive Communities and Service Clouds,* Seattle, WA: IEEE.

15. Pearson, S. 2009. Taking account of privacy when designing cloud computing services. In *Proceedings of the 2009 ICSE Workshop on Software Engineering Challenges of Cloud Computing,* Washington, DC: IEEE Computer Society, pp. 44–52.

16. Barroso, L. A. and Hölzle, U. 2009. *The Datacenter as a Computer: An Introduction to the Design of Warehouse-Scale Machines.* San Rafael, CA: Morgan & Claypool.

17. Satyanarayanan, M. 1996. Fundamental challenges in mobile computing. In *Proceedings of the Fifteenth Annual ACM Symposium on Principles of Distributed Computing,* Philadelphia, PA: ACM, pp. 1–7.

18. Roelof, K., Nicholas, P., Thilo, K., and Henri, B. 2010. Cuckoo: A computation offloading framework for smartphones. In *Proceedings of the 2nd International ICST Conference on Mobile Computing, Applications, and Services,* Santa Clara, CA.

19. Zhang, X., Kunjithapatham, A., Jeong, S., and Gibbs, S. 2011. Towards an elastic application model for augmenting the computing capabilities of mobile devices with cloud computing. *Mobile Networks and Applications,* 16(3): 270–284.

20. Zhang, X., Jeong, S., Kunjithapatham, A., and Simon Gibbs. 2010. Towards an elastic application model for augmenting computing capabilities of mobile platforms. In *The Third International ICST Conference on MOBILe Wireless MiddleWARE, Operating Systems, and Applications,* Chicago, IL.

21. Giurgiu, I., Riva, O., Juric, D., Krivulev, I., and Alonso, G. 2009. Calling the cloud: Enabling mobile phones as interfaces to cloud applications.

In *Proceedings of the 10th ACM/IFIP/USENIX International Conference on Middleware,* Urbana Champaign, IL. Berlin: Springer, pp. 1–20.

22. Chun, B.-G. and Maniatis, P. 2010. Dynamically partitioning applications between weak devices and clouds. In *Proceedings of the 1st ACM Workshop on Mobile Cloud Computing & Services Social Networks and Beyond,* San Francisco, CA. New York: ACM Press, pp. 1–5.

23. Bonomi, F., Milito, R., Zhu, J., and Addepalli, S. 2012. Fog computing and its role in the internet of things. In *Proceedings of the ACM SIGCOMM 2012 Workshop on Mobile Cloud Computing.* ACM, pp. 13–16.

24. Nonaka, I. and Takeuchi, H. 1995. *The Knowledge-Creating Company.* Oxford: Oxford University Press.

25. Spaniol, M., Klamma, R., and Cao, Y. 2009. Media centric knowledge sharing on the Web 2.0. In *Knowledge Networks: The Social Software Perspective,* pp. 46–60.

26. Fohrmann, J. and Schüttpelz, E. 2004. *Die Kommunikation der Medien* [in German]. Tübingen: Niemeyer.

27. Kosch, H., Boszormenyi, L., Doller, M., Libsie, M., Schojer, P., and Kofler, A. 2005. The life cycle of multimedia metadata. *Multimedia, IEEE,* 12(1): 80–86.

28. Ellis, C. A., Gibbs, S. J., and Rein, G. 1991. Groupware: Some issues and experiences. *Communications of the ACM,* 34(1): 39–58.

29. Stockhammer, T. 2011. Dynamic adaptive streaming over HTTP: Standards and design principles. In *Proceedings of the Second Annual ACM Conference on Multimedia Systems.* ACM, pp. 133–144.

30. Cao, Y., Klamma, R., and Jarke, M. 2010. Mobile Multimedia Management for Virtual Campfire—The German Excellence Research Cluster UMIC. *International Journal on Computer Systems Science & Engineering,* 25(3): 251–265.

31. Cao, Y., Renzel, D., Jarke, M., Klamma, R., Lottko, M., Toubekis, G., and Jansen, M. 2010. Well-balanced usability and annotation complexity in interactive video semantization. In *2010 4th International Conference on Multimedia and Ubiquitous Engineering,* Cebu, Philippines, pp. 1–8.

32. Kovachev, D., Yiwei, C., and Klamma, R. 2012. Building mobile multimedia services: A hybrid cloud computing approach. *Multimedia Tools and Applications.*

33. Kosch, H. 2003. *Distributed Multimedia Database Technologies Supported by MPEG-7 and MPEG-21.* Boca Raton, FL: CRC Press.

34. Cao, Y., Klamma, R., and Khodaei, M. 2009. A multimedia service with MPEG-7 metadata and context semantics. In *Proceedings of the 9th Workshop on Multimedia Metadata.*

35. Kovachev, D., Cao, Y., and Klamma, R. 2013. Cloud services for improved user experience in sharing mobile videos. In *Proceeeedings of the 2013 IEEE International Symposium on Mobile Cloud, Computing and Service Engineering.* IEEE.

36. Dean, J. and Ghemawat, S. 2008. MapReduce: Simplified data processing on large clusters. *Communications of the ACM,* 51(1): 107–113.

37. Kovachev, D., Aksakali, G., and Klamma, R. 2012. A real-time collaboration-enabled mobile augmented reality system with semantic multimedia. In *Proceedings of the 8th International Conference on Collaborative Computing: Networking, Applications and Worksharing.* IEEE, pp. 345–354.
38. Kovachev, D., Yu, T., and Klamma, R. 2012. Adaptive computation offloading from mobile devices into the cloud. In *2012 IEEE 10th International Symposium on Parallel and Distributed Processing with Applications,* pp. 784–791.

CHAPTER 4

GPU and Cloud Computing for Two Paradigms of Music Information Retrieval

Chung-Che Wang, Tzu-Chun Yeh,
Wei-Tsa Kao, Jyh-Shing Roger Jang,
Wen-Shan Liou, and Yao-Min Huang

National Tsing Hua University
Hsinchu, Taiwan

CONTENTS

4.1 BASICS OF QUERY BY SINGING/HUMMING AND QUERY BY EXACT EXAMPLE

Query by singing/humming (QBSH) is an intuitive and successful paradigm for music information retrieval (MIR), allowing a user to retrieve a desired song by singing or humming a portion of the song. The user's humming or singing is recorded by a smartphone app or Web interface. The recording (or its pitch) is then sent to a server that returns a result within seconds. The typical workflow of such a QBSH system is shown in Figure 4.1.

As described in the figure, the workflow includes several key components:

1. *Pitch tracking.* In this step, the user's singing or humming is converted into a series of numbers representing the pitch (or melody) of the recording. The pitch is recorded as semitones, identical to those used

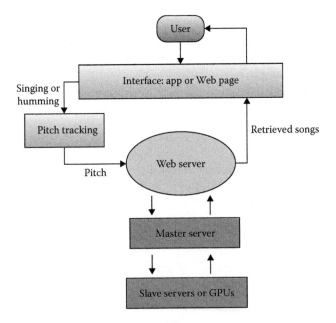

FIGURE 4.1   Typical QBSH system workflow.

in musical instrument digital interface (MIDI) files. This computation could be done on the client device, but performing it on the server allows for continuous improvement to the pitch tracking algorithm.

2. *Master server for job dispatch.* A cloud computing system usually requires a number of behind-the-scenes computing units, such as high-performance servers equipped with graphic processing units (GPUs). To reduce the required response time of QBSH system, comparison tasks can be divided and distributed evenly among the computing units. Thus, a master server is used to receive the client's request dispatch and assign the comparison tasks to other computing units (slave servers).

3. *Comparison to each song.* All computing units execute the same program to compare the input query (pitch of the singing or humming) to a disjoint part of the overall song database. Several different algorithms can be used to perform the comparison, assigning a score to each song and determining its similarity to the input query. In this chapter, we introduce a simple yet effective linear scaling (LS) comparison algorithm.

4. *Ranking and returning the result.* Once the scores of all songs in the database have been computed on the slave servers, the master sever collects the scores and ranks them, and then returns to the user a ranked list of songs most likely to match the input according to their scores.

Another successful MIR paradigm is query by exact example (QBEE), also known as audio fingerprinting (AFP). The goal of QBEE is to identify a noisy but exact audio clip of the original music. For example, suppose you hear an unfamiliar song on the radio. With QBEE, you can record a 10-second clip (in a noisy environment) using your mobile device and upload the clip to the server to identify the song. The workflow of the QBEE system is similar to that of the QBSH system shown in Figure 4.1, except that the pitch tracking part is replaced by landmark extraction. The basic two components of QBEE are described as follows:

1. *Feature extraction for database preparation.* The music clips in the database are first transformed to spectrograms by fast Fourier transform (FFT). A set of landmarks, defined as salient pairs of local

maxima over spectrograms, are then extracted from the spectrogram. These landmarks are then stored as a hash table for comparison during the retrieval stage.

2. *Retrieval stage.* During retrieval, the query clip obtains landmarks for comparison using the above-mentioned procedure. The system then performs an efficient table lookup to identify relevant hash entries in the hash table. Based on the difference in offset time of both the query and the database landmarks, we should be able to derive the number of matched landmarks for each song. The song with the most matching landmarks is more likely to be the desired song. The system then returns a ranked list according to the matched landmark counts.

Currently, several companies provide QBSH and/or QBEE as MIR services, including SoundHound [1] (snapshot of app is shown in Figure 4.2) and Shazam [2]. Companies such as IntoNow [3] also use QBEE to identify television programs so that extra information about the program can be used to promote related products on a second screen (i.e., mobile device).

FIGURE 4.2   Snapshot of the SoundHound app.

The following sections of this chapter describe our systems for QBSH and QBEE, and explain their deployment over a GPU-enabled cloud computing system.

4.2 BASICS OF GPU

The use of a GPU architecture can significantly accelerate both QBSH and QBEE for commercial or real-time applications. In this section, we briefly introduce the basics of GPU, using NVIDIA's GPU as an example.

As shown in Figure 4.3, GPU consists of several streaming multiprocessors (SMs), each composed of dozens of cores [streaming processors (SPs)], on-chip shared memories, and registers. The shared memory on each SM is usually 48 KB with 65,536 registers, with exact numbers depending on the type of GPU. It also features constant memory, texture memory, and global memory, which are shared by all of the SMs. Constant and texture memories can be accessed rapidly, but they are read-only by the GPU, and data can only be written to these memories by the central processing

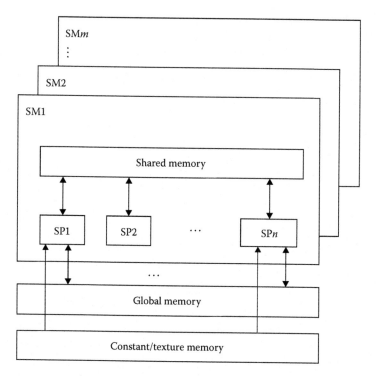

FIGURE 4.3 Basic GPU architecture.

unit (CPU). Global memory is much larger, and it can be written by the GPU, but the access time is usually several hundred times longer than that required for constant and texture memories. With unified virtual addressing (UVA), we can access the host memory directly from the GPU. Although such access might be slow, UVA still helps since putting data into global memory is even more inefficient. Our system uses the NVIDIA GeForce GTX 560 Ti, which contains 384 cores (48 cores per SM) sharing a global memory of 1 GB with 256-bit interface width, providing a throughput of 128 GB/second. Figure 4.3 shows the basic architecture of the GPU with arrows indicating the direction of data transfer.

Compute unified device architecture (CUDA) is a parallel computing framework developed by NVIDIA for their recent GPUs. It can be viewed as an extension of the C programming language, allowing programmers to define C functions (called kernels) for parallel execution by different CUDA threads. Several threads are grouped in a block. A block is executed on an SM; thus, data in shared memory are shared by all threads within the block. The number of threads within one block is limited to 1024 MB for GTX 560 Ti.

4.3 METHODS OF QBSH

This section describes the details of QBSH [4] and how to use GPU for acceleration.

4.3.1 Pitch Tracking

The autocorrelation function (ACF) is a commonly used method for pitch tracking of monophonic singing or humming input:

$$\text{acf}(\tau) = \sum_{i=0}^{n-1-\tau} s(i)s(i+\tau) \tag{4.1}$$

where:
 $s(i)$ is the element i of a frame s
 τ is the time lag in terms of sample points

The value of τ that maximizes $\text{acf}(\tau)$ over a reasonable range (determined by the range of human's pitch) is selected as the pitch period in the sample points. Figure 4.4 demonstrates the operation of ACF, where the sample rate is 16,000 Hz and the pitch is equal to 16,000/30 = 533 Hz or 72.3304 semitones.

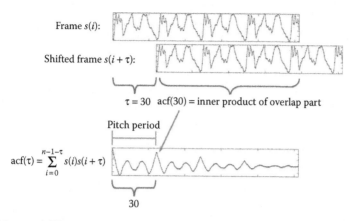

Frame $s(i)$:

Shifted frame $s(i + \tau)$:

$\tau = 30$ acf(30) = inner product of overlap part

Pitch period

$$\mathrm{acf}(\tau) = \sum_{i=0}^{n-1-\tau} s(i)s(i+\tau)$$

30

FIGURE 4.4   ACF operation.

4.3.2 Linear Scaling

The tempo of a user's singing or humming is usually different from that of the intended song in the database. Thus, the QBSH system needs to compress or stretch the pitch vector to match the songs in the database. Assuming that the query input is d-second long, we can compress or stretch the original vector to obtain r different versions, with their lengths equally spaced between $\mathrm{sf}_{\min} \times d$ and $\mathrm{sf}_{\max} \times d$, where $\mathrm{sf}_{\min}$ and $\mathrm{sf}_{\max}$ are the minimum and maximum of the scaling factor, respectively, with $0 < \mathrm{sf}_{\min} < 1 < \mathrm{sf}_{\max}$. The distance between the input pitch vector and a particular song is then the minimum of the r distances between the r vectors and that song, as shown in Figure 4.5, where we compress/stretch the d-second vector to obtain five vectors with lengths equally spaced between $0.5 \times d$ and $1.5 \times d$. The best result is obtained when the scaling factor is 1.25.

The user's singing/humming also differs from the target song in terms of key. Key transposition can be simply handled by shifting the mean (if L_2-norm is used) or the median (if L_1-norm is used) of the query input and a database song (of the same length) to the same value. The distance is usually based on L_p-norm, defined as follows:

$$\|x - y\|_p = \left(\sum_i |x_i - y_i|^p \right)^{1/p} \tag{4.2}$$

where:

x and y are two vectors of the same length

Figure 4.6 shows a typical example of LS using the L_1- and L_2-norms.

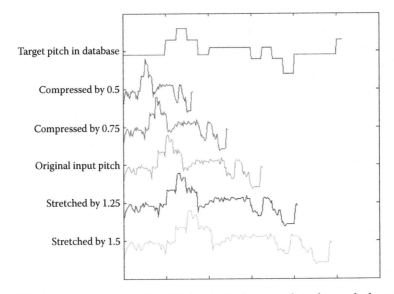

FIGURE 4.5   A typical example of LS, where the best match is obtained when the scaling factor is 1.25.

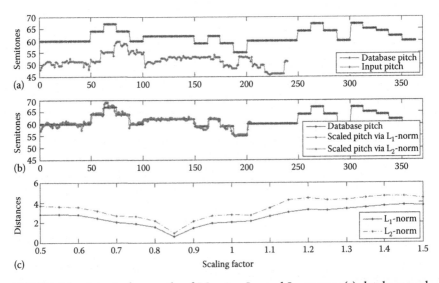

FIGURE 4.6   A typical example of LS using L_1- and L_2-norms: (a) database and input pitch vectors; (b) database and scaled pitch vectors; (c) normalized distances via L_1- and L_2-norms. (To save computation, the distance in L_2-norm is only the squared sum of the differences.)

4.3.3 Database Comparison by GPU

We now explain how to employ GPU for efficient comparison in QBSH. At system start-up, the whole song database is loaded into the global memory. Note that the database is not excessively large since only the pitch and duration are stored for each note of a song. To compress or stretch the input pitch vector, we simply launch r threads for each scaling factor, where r is 31 in our system. We then move the scaled vectors back to the main memory and then to the GPUs' constant memory to speed up the access. We have investigated three different schemes for database comparison [5]:

1. In scheme 1, we launch N threads to compare N different songs in the database. Recall that there are many notes in a song for comparison, and we have r versions of the input pitch vector, so the computational load of a thread is very heavy (heavy load threads may not be suitable to run on GPUs).

2. In scheme 2, we launch r threads for the comparison of one song. These r threads are grouped into a block, for a total of N blocks. Although the degree of parallelization is higher, the computation time is actually longer because each block only contains a few threads, leading to underutilized SPs.

3. In scheme 3, we still have N blocks for N songs, but now each block has k threads. Computation tasks starting at different notes in a song are equally distributed to the k threads. Since k could be much larger than r (e.g., 256 vs. 31), the SPs are more fully utilized than in scheme 2. Moreover, since there are multiple threads for one song, we could obtain the minimum distance between the input pitch vector and the song in parallel by using these threads directly.

After obtaining the distance between the query input and each of the songs in the database, we then sort all the distances on the CPU to obtain the top-n list. Sorting using the GPU provided unsatisfactory performance results due to excessive access time requirements over the global memory.

4.3.4 Performance Analysis

We used the public corpus MIR-QBSH [6] for our experiments. In this corpus, the beginnings of the corresponding songs serve as the anchor positions for all queries. To test the accuracy of "anchor anywhere," we duplicated the last fourth of each song in the database and prepended it to the beginning of the song. The corpus contains 6197 clips that correspond to 48 children's songs. To increase the complexity of the comparison, we added another set of 12,887 noise songs corresponding to Chinese, English, and Japanese pop songs to the database, resulting in a total of 12,935 songs in the database.

Figure 4.7 shows the distribution of song lengths in our database in terms of the number of music notes (Figure 4.7a) and the number of pitch points (Figure 4.7b). This plot indicates the complexity of our QBSH task. The number of music notes equals the number of positions needed to start the comparison, whereas the number of pitch points represents the length of the sequence required for comparison. In our QBSH system, the scaling factor varies from 0.6 to 1.5 to obtain 31 compressed or stretched versions of the original query input vector. The frame size is 256 points with no overlap; the sample rate is 8 kHz, leading to a pitch rate of 31.25 per second. The top-n recognition rate is shown in Figure 4.8, with a top-10 recognition rate of about 74%.

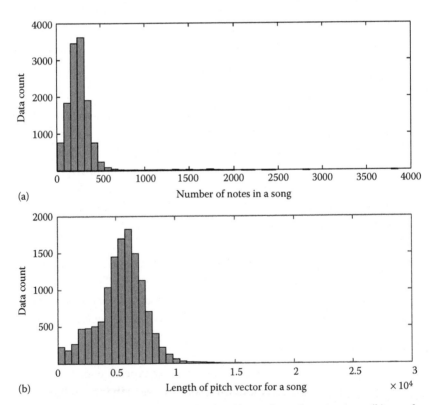

FIGURE 4.7   Distribution of song lengths: (a) number of music notes; (b) number of pitch points.

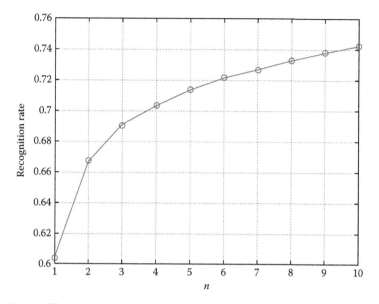

FIGURE 4.8 Top-n recognition rates.

Figure 4.9 shows the computation time per query with respect to database size for different parallelization schemes. In Figure 4.9a, schemes 1 and 3 have 1024 threads per block, whereas scheme 2 has 31. In Figure 4.9b, schemes 1 and 3 have 128 threads per block, whereas scheme 2 has 31. As shown in this figure, scheme 3 is the fastest, followed by schemes 1 and 2. More specifically, scheme 3 is about 10 times faster than the original CPU version (a PC with i7-2600 processor and 16 GB DDR3 1600 memory) demonstrating the effectiveness of the proposed method. Figure 4.10 shows the effect of the number of threads in a block for scheme 3. The best performance is achieved with 128 threads per block. Reducing the number of threads per block reduces GPU core utilization, and including more than 128 threads overutilizes SM resources. Moreover, the last 1000 songs include many for which the pitch vectors were obtained from human vocal singing alone. Pitch vectors of this kind tend to have a large number of small notes (due to the instability of the human singing voice), leading to many possible positions for the start of comparison and much longer computation time, which suddenly increases when the number of songs in the database exceeds 12,000.

Figure 4.11 shows the results screen of our QBSH system, named Music Information Retrieval Acoustically with Cluster and Parallel Engines (MIRACLE) [7,8]. The system is publicly accessible at http://mirlab.org/demo/miracle

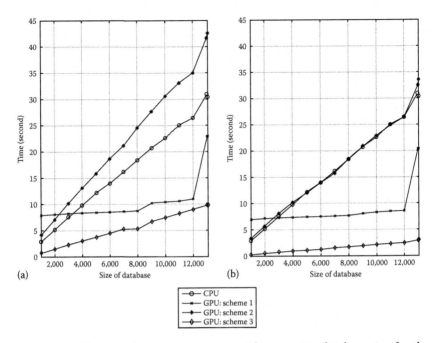

FIGURE 4.9   Computation time per query with respect to database size for the three parallelization schemes. See text for details.

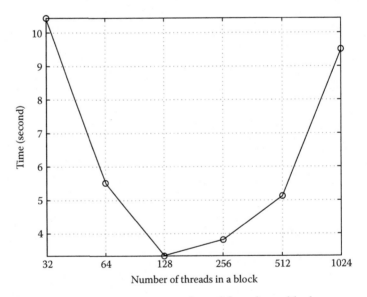

FIGURE 4.10   Computation time vs. number of threads per block.

Music Search Add a song to the database Song list Todo list How to use miracle web service
About Report

Record for 8 seconds

Sing or hum your tune for music search.
Click 'Record' button for recording.

Click 'Record' to record your tune.
Click 'Replay' to replay your recorded voice.
Click 'Options' for Advanced settings.

Select one to test:

Record **Replay** **Options**   A Time For Us, Any, Linear_scaling, Engli ▼

Click to hear what Miracle has heard

The settings are => Method:linear_scaling Compare:anywhere List_number:15 Language:english
Database:MIR

♪ Result ♪

Rank	Songs	Singer	Listen	Youtube NO.1	Youtube NO.2	Youtube NO.3	Scores
1	A Time For Us	Unknown	▶				97.62
2	A time for us	Unknown	▶				96.84
3	心心相印	不詳	▶				96.06
4	膏振情歌	動力火車	▶				96.06
5	The Inner Light	Beatles	▶				96.06
6	桑邛鐘琪亞	不詳	▶				96.06
7	SPIDER	Unknown	▶				96.06
8	Three Coins In a Fountain	Frank Sinatra	▶				96.06
9	給我溫情吧愛人	余天	▶				96.06
10	小詩	余天	▶				96.06
11	我是好男兒	若歌	▶				96.06
12	愛的源珠	不詳	▶				96.06
13	河流	柯以敏	▶				96.06
14	Lady In Red	Chris De Burgh	▶				96.06
15	Lady In Red	Chris De Burgh	▶				96.06

♪ Time ♪

	Host	Computing Time	No. of Compared Songs
1	CUDA (140.114.88.80)	2.418秒	13000 songs (0-12999)

FIGURE 4.11   Results interface of QBSH system—MIRACLE.

4.4 METHODS OF QBEE

QBEE (or AFP) is a fast, convenient, and noise robust method for MIR based on an exact but noisy example of the original music clip. Figure 4.12 shows the block diagram of a QBEE system based on Wang's seminal work [9]. In the offline stage, each music clip in the database is converted into a spectrogram, where pairs of salient peaks are selected to form landmarks. There are several approaches to the selection of salient peaks. One approach, as proposed by Ellis [10], is to create an energy threshold based on the maximum of the Gaussian functions centered at local maxima of the power spectrum for a specific frame. Thus, a local peak is considered to be salient if it is equal to the energy threshold, as shown in Figure 4.13, where thin line is the spectrum, the dots are the local peaks, and the thick line is the energy threshold. To remove the local peaks that are not salient along time, the energy threshold is linearly decayed and recomputed along the time in both directions. A typical 3D view of the energy threshold along the positive time axis is shown in Figure 4.14.

After obtaining the salient peaks, we pair the peaks to form landmarks. A peak is paired with other peaks following it in a specific range.

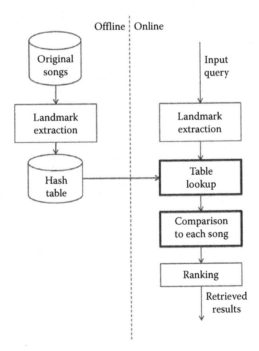

FIGURE 4.12   Block diagram of a QBEE system.

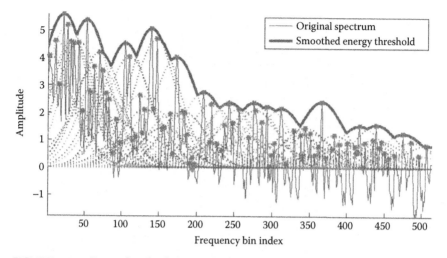

FIGURE 4.13 Example of salient peak extraction, where the energy threshold (thick line) is computed as the maximum of all Gaussians centered at local maxima (dots) of the power spectrum.

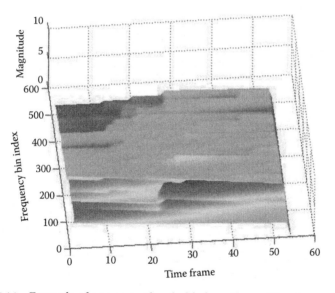

FIGURE 4.14 Example of an energy threshold along the positive time axis.

In Figure 4.15, the circles are salient peaks; the solid lines connect the peaks to form landmarks; and the dotted rectangle is an example of a specific range. We can convert each landmark to a 24-bit integer hash key, including the starting frequency, the ending frequency, and the difference in time. For each hash key, the hash value contains information regarding the corresponding

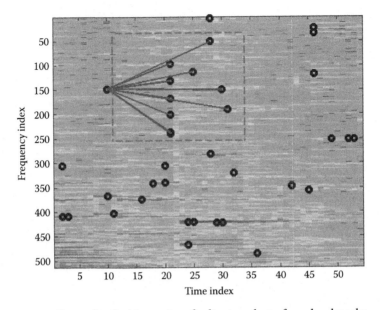

FIGURE 4.15   Example of taking pairs of salient peaks to form landmarks.

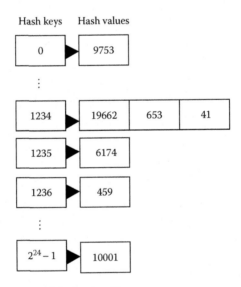

FIGURE 4.16   Structure of the hash table.

song index and the landmark's starting time. Figure 4.16 shows the structure of the hash table for mapping the hash keys to the hash values.

In the online stage, the same process of feature extraction is applied to the input query to obtain landmarks and their hash keys. We can then retrieve the hash values from the hash table and then obtain information

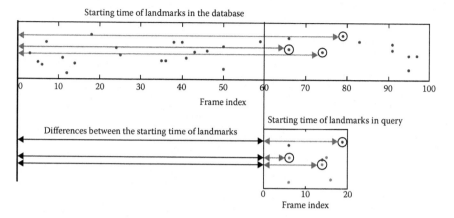

FIGURE 4.17 An example of finding matched landmarks, which demonstrates that if a database song and the input query are matched, the differences between the starting times of their landmarks should be similar.

for the corresponding songs and the starting time of the landmarks in the database songs. As shown in Figure 4.17, if a database song and the input query match, the differences between the starting times of their landmarks should be similar. If the time difference is ≤1 frame, it is called a matched landmark. A song with more matched landmarks will have a higher score since it is more likely to be the desired song. Once the scores of each song are obtained, we could rank the top-k likely songs and return a ranked song list as the final output of the QBEE system.

Based on Ellis' implementation [10], we developed a parallelized QBEE system over GPUs. Our initial results show that, using GPUs, the system has a response time of <1 second when comparing a 10-second query clip against a database of about 140,000 songs. Since this is still an ongoing study with all parameters that are yet to be optimized, we shall report the details elsewhere. The GPU-based QBEE system is publicly available at http://mirlab.org/demo/audioFingerprinting

4.5 CLOUD COMPUTING FOR QBSH

Cloud computing is an existing technology that allows a user to access remote devices or execute software on demand over an array of computing servers. Based on a server–slave paradigm, it is now widely used for business-oriented computation of massive amounts of data. Due to different types of demands, cloud providers offer several different fundamental service models, as shown in Figure 4.18. This section focuses on the

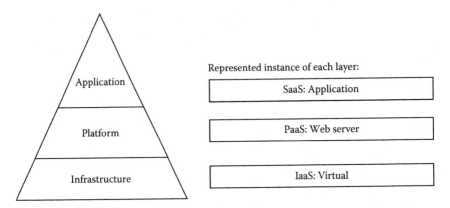

FIGURE 4.18   Basic cloud service models. IaaS, infrastructure as a service; PaaS, platform as a service; SaaS, software as a service.

application layer—the highest layer in the cloud service model. A QBSH system deployed with virtual machines is used to demonstrate how the application can be ported to the cloud.

For the application of QBSH and QBEE, cloud computing can be used to deploy computing units to balance computation loading for the comparison process, thus reducing response time. The computing unit in this study is referred to as the slave server, which runs in each virtual machine and is assigned tasks by the master server. The OpenStack [11] cloud computing platform is used to demonstrate the proposed cloud-enabled QBSH system.

There are several steps to deploy and manage the QBSH system on the cloud:

1. Package the image containing the autostart application and the operating system.

2. Upload the image to the cloud platform, allowing us to select the image containing the application later. OpenStack supports the command line console, but the user can also upload the image via Web browser.

3. Boot each virtual machine in the OpenStack management system to connect the computing units from the interface. A master server is used to monitor and maintain connections to the available computing units. Figure 4.19 shows the control table of the OpenStack Web system.

A key issue in traditional QBSH or QBEE system is that the system usually takes too long to return all the results when accepting many

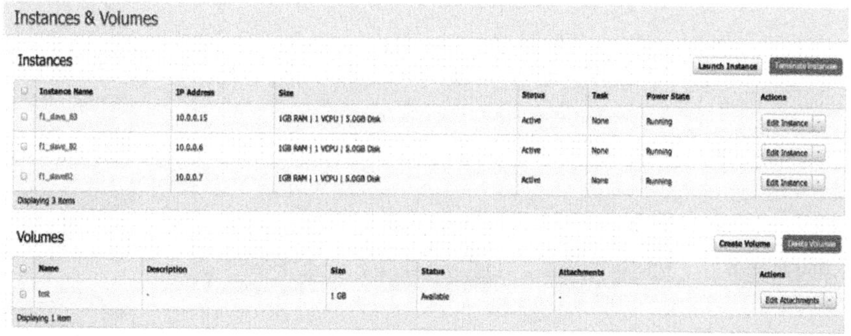

Instances & Volumes

Instances

Launch Instance | Terminate Instance

Instance Name	IP Address	Size	Status	Task	Power State	Actions
fi_slave_63	10.0.0.15	1GB RAM	1 VCPU	5.0GB Disk	Active	None	Running	Edit Instance
fi_slave_82	10.0.0.6	1GB RAM	1 VCPU	5.0GB Disk	Active	None	Running	Edit Instance
fi_slave82	10.0.0.7	1GB RAM	1 VCPU	5.0GB Disk	Active	None	Running	Edit Instance

Displaying 3 items

Volumes

Create Volume | Delete Volume

Name	Description	Size	Status	Attachments	Actions
test	-	1 GB	Available	-	Edit Attachments

Displaying 1 item

FIGURE 4.19 Status of the OpenStack Web system after scaling up three virtual machines.

simultaneous queries. This is caused by the traditional queue system and first-in, first-out (FIFO) mechanism, so the users must wait for all previous queries to be processed before receiving their results. We propose two methods to address this problem: autoscaling and parallel dispatching.

4.5.1 Autoscaling

Since the cloud platform provides resources on demand, users can control the virtual machines on their own. Autoscaling is developed based on the concept that applications can automatically allocate resources to maintain performance, and the cloud platform scales the virtual machine seamlessly at peak demand times. Animoto [12] used this mechanism to instantly scale from 40 to 4000 servers on Amazon Elastic Compute Cloud (Amazon EC2) for the launch of a Facebook plugin. This mechanism allows us to quickly and reliably scale up our application to deal with many simultaneous queries. A brief example is shown in Figure 4.20.

The fewer the computing units deployed on the cloud, the longer the response time for users. Therefore, the basic idea for autoscaling is to monitor the response time for each request. Server loading can be monitored if the cloud service provider supplies the relevant information (such as CPU or memory utilization). As shown in Figure 4.21, computation time is rapidly reduced when more virtual machines are deployed on the cloud to share the computational load.

4.5.2 Parallel Dispatching

Another method for dealing with many simultaneous requests is to dispatch queries to computing units over multiple cloud service providers. As long as the number of slave servers is increased, the response time will be

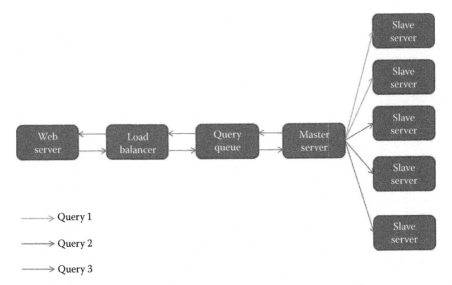

FIGURE 4.20 Flow of dynamically dispatched queries according to server loading.

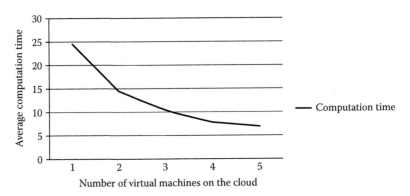

FIGURE 4.21 Average computation time vs. the number of virtual machines deployed on the cloud.

reduced. Figure 4.22 illustrates an example of scaling up slave servers to Dcloud, SScloud, and Amazon EC2 [13]—Dcloud and SScloud are experimental cloud platforms in our project.

4.6 SUMMARY

This chapter has described the use of GPU and cloud computing for QBSH and QBEE, two of the most successful paradigms of MIR. The basics of these two MIR paradigms are explained, along with how GPU and cloud computing can be used to accelerate retrieval when dealing with a large

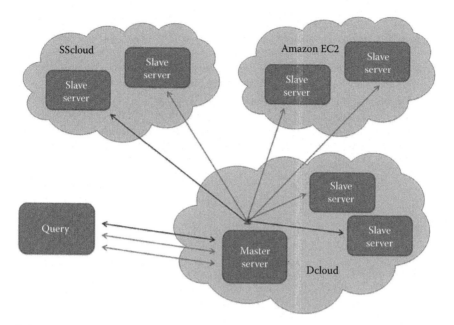

FIGURE 4.22 Structure of slave servers deployed over multiple cloud service providers.

database. For our current QBSH and QBEE systems, the database sizes are ~20,000 and ~200,000, respectively. Online access to our systems is provided to demonstrate the feasibility of the proposed methodologies.

REFERENCES

1. SoundHound Inc., available: http://www.soundhound.com/.
2. Shazam Entertainment, available: http://www.shazam.com/.
3. IntoNow, available: http://www.intonow.com/ci
4. Jyh-Shing Roger Jang, "Audio signal processing and recognition," available: http://neural.cs.nthu.edu.tw/jang/books/audioSignalProcessing/.
5. Chung-Che Wang, Chieh-Hsing Chen, Chin-Yang Kuo, Li-Ting Chiu, and Jyh-Shing Roger Jang, "Accelerating query by singing/humming on GPU: Optimization for web deployment," *The 36th International Conference on Acoustics, Speech, and Signal Processing*, Kyoto, Japan, March 2012.
6. Jyh-Shing Roger Jang, "MIR-QBSH Corpus," MIR Lab, CS Department, Tsing Hua University, Taiwan, available: http://mirlab.org/jang
7. Jyh-Shing Roger Jang, Jiang-Chun Chen, and Ming-Yang Kao, "MIRACLE: A music information retrieval system with clustered computing engines," in *Proceedings of the 2nd International Conference on Music Information Retrieval*, Indiana University, Bloomington, IN, 2001.
8. Miracle, available: http://mirlab.org/demo/miracle

9. Avery Li-Chun Wang, "An industrial-strength audio search algorithm," in *Proceedings of the 4th International Conference on Music Information Retrieval*, Maryland, 2003.

10. Dan Ellis, "Robust landmark-based audio fingerprinting," available: http://labrosa.ee.columbia.edu/matlab/fingerprint/, 2009.

11. OpenStack, "An open source cloud project," available: http://www.openstack.org

12. The case of Animoto using cloud computing, available: http://support .rightscale.com/06-FAQs/FAQ_0043_-_What_is_autoscaling%3F

13. Amazon EC2, "A scalable cloud computing solution," available: http://aws .amazon.com/ec2

CHAPTER 5

Video Transcode Scheduling for MPEG-DASH in Cloud Environments

Roger Zimmermann

National University of Singapore
Singapore

CONTENTS

YOUTUBE [1] HAS INDICATED that over 4 billion hours of videos are watched each month and 72 hours of video are uploaded every minute. Another study from Cisco [2] has indicated that the overall mobile data traffic reached 885 petabytes per month at the end of 2012, 51% of which are mobile video. Forecasts predict that mobile video will grow at a compound annual growth rate (CAGR) of 75% between 2012 and 2017 and reach at 1 exabyte per month by 2017. During video playback, mobile devices may encounter different wireless network conditions. The Dynamic Adaptive Streaming over Hypertext Transfer Protocol (HTTP; DASH or MPEG-DASH) [3] standard is designed to provide high-quality streaming of media content over the Internet delivered from conventional HTTP Web servers. The content, divided into a sequence of segments, is made available at a number of different bit rates so that an MPEG-DASH client can automatically select the next segment to download and play back based on the current network conditions. The task of transcoding media content to different qualities and bit rates is computationally expensive, especially in the context of large-scale video hosting systems. Therefore, it is preferably executed in a powerful cloud environment, rather than on the source computer (which may be a mobile device with limited memory, CPU speed, and battery life). In order to support the live distribution of media events and to provide a satisfactory user experience, the overall processing delay of videos should be kept to a minimum. In this chapter, we describe and explore various scheduling techniques for DASH-compatible systems in the context of large-scale media distribution clouds.

5.1 INTRODUCTION TO MPEG-DASH

With the development of the content delivery networks (CDNs) and multimedia technologies, HTTP streaming has emerged as a *de facto* streaming standard, replacing the conventional Real-time Transport Protocol (RTP) streaming or Real-Time Streaming Protocol (RTSP). Existing, commercial streaming platforms, such as Microsoft's Smooth Streaming (MSS) [4],

Apple's HTTP Live Streaming (HLS) [5], and Adobe's HTTP Dynamic Streaming (HDS) [6], all use HTTP streaming as their underlying delivery method. The commonalities [7] of these techniques are as follows:

1. Splitting an original encoded video into small pieces of self-contained media segments

2. Separating the media description into a single playlist file

3. Delivering segments over HTTP

In contrast, these techniques differ from each other as follows:

1. MSS is a compact and efficient method for the real-time delivery of MP4 files from Microsoft's Internet Information Services (IIS) Web server, using a fragmented, MP4-inspired ISO Base Media File Format (ISO BMFF).

2. HLS uses an MPEG-2 Transport Stream (TS) as its delivery container format and utilizes a higher segment duration than MSS.

3. HDS is based on Adobe's MP4 fragment format (F4F) and its corresponding Extensible Markup Language (XML)-based proprietary manifest file (F4M).

Once an HTTP client sends a request and establishes a connection between the server and itself, the progressive download is activated until the streaming is terminated [8]. Disadvantages of progressive download include the following:

1. Unstable conditions of a network, especially the wireless connection for mobile clients, may cause bandwidth waste due to reconnection or rebuffering events.

2. It does not support live streaming (e.g., a concert or a football match).

3. It does not support adaptive bit rate streaming.

MPEG-DASH or DASH addresses the above weaknesses. Published in April 2012, DASH [3] is known as video delivery standard that enables high-quality streaming of media content over HTTP. The video file is broken into a sequence of small playable HTTP-based segments and these segments are uploaded to the standard HTTP server sequentially. The visual content is then encoded at a variety of different bit rates and the HTTP client can automatically select the next segment from the alternatives to download and play back based on the current network conditions. The client selects the segment with the highest possible bit rate that can be downloaded in time for smooth and seamless playback, without causing rebuffering events. The main features of DASH are as follows:

- It splits a large video file into small chunks.

- It provides client-initiated flexible bandwidth adaptation by enabling stream switching among differently encoded segments.

- It supports on-demand, near-live, and time-shift application.

- It has segments with variable durations. With live streaming, the duration of the next segment can also be signaled with the delivery of the current segment.

In the following text, we will briefly introduce DASH standard and the two most important components of DASH: media presentation description (MPD) and the segment formats. Note that the vendor-specific adaptive HTTP streaming solutions are expected to converge toward the DASH standard in the future.

5.1.1 Scope of the DASH Standard

Figure 5.1 illustrates a simple streaming scenario between an HTTP server and a DASH client. The MPEG-DASH specification defines only the MPD and the segment formats. The delivery of the MPD and the media-transcoding formats containing the segments, as well as the client behavior for fetching, adaptation heuristics, and playing content, are outside of MPEG-DASH standard's scope. In this figure, the multimedia content is captured and uploaded to an HTTP server and is then delivered to end users using HTTP. The content exists on the server in two parts: MPD and the encoded video segments with various bit rates.

To play the content, the DASH client first obtains the MPD. The MPD can be delivered using HTTP (most likely), e-mail, thumb drive, broadcast, or other transports. By parsing the MPD, the DASH client learns about the program timing, media content availability, media types, resolutions, minimum and maximum bandwidths, and the existence of various

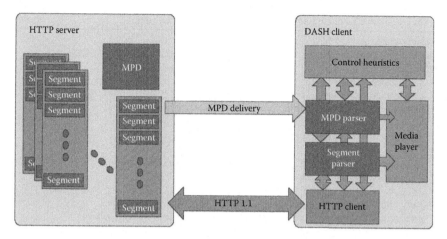

FIGURE 5.1 Scope of the MPEG-DASH standard (dark blocks). (From I. Sodagar, *IEEE Multimedia*, 18, 62–67, 2011. © 2011 IEEE. With permission.)

encoded alternatives of multimedia components, accessibility features and required digital rights management (DRM), media component locations on the network, and other content characteristics. Using this information, the DASH client selects the appropriate encoded alternative and starts streaming the content by fetching the segments using HTTP GET requests.

After appropriate buffering to allow for network throughput variations, the client continues fetching the subsequent segments and also monitors the network bandwidth fluctuations. Depending on its measurements, the client decides how to adapt to the available bandwidth by fetching segments of different alternatives (with lower or higher bit rates) to maintain an adequate buffer.

5.1.2 Multimedia Presentation Description

MPD is a document that contains metadata [e.g., a manifest of the available content, its various alternatives, their uniform resource locator (URL) addresses, and other characteristics] required by a DASH client to construct appropriate HTTP URLs to access segments and to provide the streaming service to the user. The MPD is an XML document that is formatted in the form of schema organized in a hierarchical data model (shown in Table 5.1) [3].

The hierarchical data model (highlighted with the rectangle in Table 5.1) is detailed in Table 5.2.

TABLE 5.1   XML Schema of the MPD

```
<?xml version = "1.0" transcoding = "UTF-8"?>
<xs:schema targetNamespace = "urn:mpeg:mpegB:schema:DASH:MPD:
DIS2011"
    attributeFormDefault = "unqualified"
    elementFormDefault = "qualified"
    xmlns:xs = "http://www.w3.org/2001/XMLSchema"
    xmlns:xlink = "http://www.w3.org/1999/xlink"
    xmlns = "urn:mpeg:mpegB:schema:DASH:MPD:DIS2011">

    <xs:import namespace = "http://www.w3.org/1999/xlink"
    schemaLocation = "xlink.xsd"/>

    <xs:annotation>
    <xs:appinfo>Media Presentation Description</xs:appinfo>
    <xs:documentation xml:lang = "en">
        This Schema defines Media Presentation Description for MPEG
        DASH.
    </xs:documentation>
    </xs:annotation>

    <!— MPD: main element—  >
    <xs:element name = "MPD" type = "MPDtype"/>

    <!— the remaining types, elements attributes are defined in the
    below—  >
    ...

</xs:schema>
```

5.1.3 Segment Formats

The segment formats specify the formats of the entity body of the request response when issuing a HTTP GET request or a partial HTTP GET with the indicated byte range through HTTP/1.1. In order to support the use of DASH, a delivery format should have the property that decoding and playback of any portion of the media can be achieved using a subset of the media, which is only a constant amount larger than the portion of the media to be played. To implement this functionality, each media segment is assigned a unique URL, an index, and an explicit or implicit start time and a duration. Each media segment contains at least one stream access point, which is a random access or switch to point in the media stream where decoding can start using only data from that point forward. Both ISO BMFF and MPEG-2 TS are supported in DASH.

TABLE 5.2   The XML Syntax of the Hierarchical Data Model

```
<!-  MPD Type-  >
<xs:complexType name = "MPDtype">
  <xs:sequence>
    <xs:element name = "ProgramInformation"
        type = "ProgramInformationType" minOccurs = "0"/>
    <xs:element name = "Period" type = "PeriodType"
        maxOccurs = "unbounded"/>
    <xs:element name = "BaseURL" type = "BaseURLType"
        minOccurs = "0" maxOccurs = "unbounded"/>
    <xs:any namespace = "##other" processContents = "lax"
        minOccurs = "0" maxOccurs = "unbounded"/>
  </xs:sequence>
  <xs:attribute name = "profiles" type = "URIVectorType"/>
  <xs:attribute name = "type" type = "PresentationType"
      default = "OnDemand"/>
  <xs:attribute name = "availabilityStartTime" type = "xs:dateTime"/>
  <xs:attribute name = "availabilityEndTime" type = "xs:dateTime"/>
  <xs:attribute name = "mediaPresentationDuration"
      type = "xs:duration"/>
  <xs:attribute name = "minimumUpdatePeriodMPD"
      type = "xs:duration"/>
  <xs:attribute name = "minBufferTime" type = "xs:duration"/>
  <xs:attribute name = "timeShiftBufferDepth" type = "xs:duration"/>
  <xs:anyAttribute namespace = "##other" processContents = "lax"/>
</xs:complexType>

<!-  Type of presentation - live or on-demand-  >
<xs:simpleType name = "PresentationType">
  <xs:restriction base = "xs:string">
    <xs:enumeration value = "OnDemand"/>
    <xs:enumeration value = "Live"/>
  </xs:restriction>
</xs:simpleType>

<!-  Supplementary URL to the one given as attribute-  >
<xs:complexType name = "BaseURLType">
  <xs:simpleContent>
    <xs:extension base = "xs:anyURI">
      <xs:anyAttribute namespace = "##other"
      processContents = "lax"/>
    </xs:extension>
  </xs:simpleContent>
</xs:complexType>


<!-  Type for space delimited list of URIs-  >
<xs:simpleType name = "URIVectorType ">
  <xs:list itemType = "xs:anyURI"/>
</xs:simpleType>
```

5.2 VIDEO TRANSCODING IN A CLOUD ENVIRONMENT

In the recent years, cloud computing has become an increasingly broad topic in computer science. It is an emerging technology aimed at sharing resources and providing various computing and storage services over the Internet. For multimedia applications and services over the Internet and mobile wireless networks, there are strong demands for cloud computing because of the significant amount of computation required for serving millions of Internet or mobile users simultaneously [10]. As stated in Section 5.1, to enable DASH streaming, the HTTP server has to prepare the MPD and multiple video segments with alternative bit rates as soon as the video is uploaded to the server. The complex nature of video transcoding (CPU intensive) and the bursty nature of streaming requirements have made cloud computing uniquely suitable for video transcoding. Cloud-based video transcoding exhibits the following key characteristics:

- Reliability: In a cloud environment, data backup and recovery are relatively easier. Currently, most cloud service providers usually provide service-level agreements to handle recovery of information. For example, the data (i.e., video segments in our case) are usually stored with multiple copies in different places and this improves the reliability of the whole system in case of disk failure, network disconnection, data center damage, and other unexpected events.

- Scalability: Cloud computing allows for immediate scaling, either up or down, at any time without long-term commitment as the computing requirements change. With cloud computing, the scheduler can easily run the video transcoding tasks in parallel according to the requirements. For example, if more video segments are uploaded to the server simultaneously, the video transcoding tasks can be deployed onto more processing nodes without installing any software or hardware by the end user for a new application to the new nodes and vice versa.

- Hardware: With cloud computing, it is easy to achieve large volume storage space and powerful computing units. As in the application of DASH, the HTTP server always needs to keep multiple formats and multiple bit rates of video copies, and processes multiple jobs in parallel, which pose extremely high demands on hardware resources. A cloud environment, instead of a single machine, can meet the requirements of the computing-intensive and time-consuming video transcoding jobs.

- Cost: The biggest advantage of cloud computing is the elimination of the investment in stand-alone software or servers by the user. With cloud computing, there are no separate overhead charges such as cost of data storage, software updates, management, and most importantly cost of quality control. Currently, anyone can use the services of cloud computing at affordable rates. In our case, video segments can be encoded within a fairly short time after being uploaded to the server by spending a small amount of money.

5.2.1 Existing Cloud Video Transcoding Service

There exist several commercial cloud services that support video transcoding.

- Amazon released their Elastic Transcoder [11] in January 2013. Amazon Elastic Transcoder manages all aspects of the transcoding process transparently and automatically. It also enables customers to process multiple files in parallel and organize their transcoding workflow using a feature called transcoding pipelines. With Amazon Elastic Transcoder's pipeline feature, customers set up pipelines for these various scenarios and ensure that their files are encoded when and how they want, thus allowing them to seamlessly scale for spiky workloads efficiently. It runs the transcoding jobs using Amazon Elastic Compute Cloud (Amazon EC2) [12] and stores the video content in the Amazon Simple Storage Service (Amazon S3) [13]. Developers can simply use the Web-based console or application programming interfaces (APIs) to create a transcoding job that specifies an input file, the transcoding settings, and the output file. They can also assign the priority of each pipeline to allow important videos to be encoded first. Amazon charges a video transcoding fee according to the output video duration and definition.

- Zencoder [14] provides video transcoding services as well. In addition to providing typical video transcoding services in the cloud, it also supports live cloud video transcoding. Its Web service accepts Real-Time Messaging Protocol (RTMP) input streams and encodes to RTMP and HLS at multiple bit rates for adaptive bit rate playback. Zencoder provides different price packages so that users can select the most appropriate one based on their own video duration and definition.

- EncoderCloud [15] also provides the same Web-based "pay-as-you-go" service. It helps to build application on top of other service providers: Amazon EC2 and Rackspace Cloud [16]. However, it offers a different pricing policy—charging by the volume of total amount of source video transfer in and encoded video transfer out.

5.2.2 Typical Scheduling Strategies

Cloud computing provides tremendous computing resources for applications. There is no universal "best" scheduling algorithm, and many operating systems use extended or combinations of the scheduling algorithms above. For example, Windows NT/XP/Vista uses a multilevel feedback queue, a combination of fixed-priority (FP) preemptive scheduling; round-robin (RR); and first in, first out (FIFO). Processes can dynamically increase or decrease in priority depending on whether they have been serviced already or whether they have been waiting extensively. Every priority level is represented by its own queue, with RR scheduling among the high-priority processes and FIFO among the lower ones. In this sense, response time is short for most processes, and short but critical system processes get completed very quickly. Since processes can only use one time unit of the RR scheme in the highest priority queue, starvation can be a problem for longer high-priority processes. To fully utilize computing resources in the cloud, an appropriate scheduling algorithm for the specific application is essential. In the following, a few widely used scheduling algorithms are described:

- FIFO: It is the simplest scheduling algorithm, which simply queues all the waiting jobs and processes them in the order that they arrived in the queue.

- Shortest job first (SJF): It enables the scheduler to assign the job, whose estimated processing time is the shortest among all the waiting jobs, to be processed in the first place.

- RR: It assigns a fixed processing time for each job in the queue. If the processor cannot finish the job in the current cycle, it preempts and allocates the resources to another job and reprocesses the current job when the next cycle comes.

- FP: The priority of each job is predefined, and the scheduler will choose to process the job with highest priority first. When a new job's priority is higher than the current processing job, the current

TABLE 5.3 Overview of the Existing Scheduling Algorithms

Scheduling Algorithm	CPU Overhead	Throughput	Turnaround Time	Response Time
FIFO[a]	Low	Low	High	Low
SJF[b]	Medium	High	Medium	Medium
RR[c]	High	Medium	Medium	High
FP[d]	Medium	Low	High	High
MQ[e]	High	High	Medium	Medium

[a] FIFO, first in, first out.
[b] SJF, shortest job first.
[c] RR, round-robin.
[d] FP, fixed priority.
[e] MQ, multilevel queue.

processing job gets interrupted. Processing will resume when the job with higher priority is finished.

- Multilevel queue (MQ): It is a hybrid scheduling algorithm by dividing jobs into groups and each group having different scheduling strategies.

Overall, these different scheduling algorithms have advantages in some aspects and disadvantages in other aspects when measuring performance. Table 5.3 shows an overview of these scheduling algorithms considering throughput, overhead, and time.

As presented earlier, the existing cloud video transcoding services have their own scheduling algorithms that are transparent to end users. The users do not know how the cloud schedulers are implemented. These services are appropriate for users who only want to encode all the videos as soon as possible without any specific requirement. However, in some cases (e.g., if one video is being watched but the required bit rate is not available on the server), the existing video transcoding services cannot handle these situations. Consequently, for specific purposes, we may want to develop our own application-orientated scheduling algorithm.

5.3 SCHEDULING ALGORITHM FOR DASH IN THE CLOUD ENVIRONMENT

We now focus on the scheduling methodology for video transcoding for DASH. To satisfy the requirements of DASH streaming, especially while a video is being requested, we use a hybrid of existing methods: If the

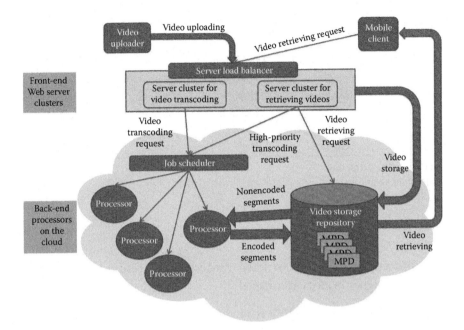

FIGURE 5.2 The architecture of scheduling on video transcoding for DASH in the cloud environment.

video is uploaded to the HTTP server, a typical scheduling algorithm (i.e., SJF) is used since the lengths of video segments for DASH streaming are always short and we can avoid the drawback that a long job must wait for a long time to be processed. However, if a video segment is requested by users, it is assigned with a higher priority and will be encoded as soon as possible.

Figure 5.2 shows the overall architecture of using DASH in a cloud environment. When the video is uploaded to the front-end Web server clusters, the server load balancer will accept the job and assign it to an HTTP server that can process this job with the consideration of balancing the workload. The HTTP server will then forward the job with a video transcoding request to the job scheduler and store the uploaded video in the video storage repository. In the back-end cloud environment, the job scheduler maintains two queues: One is for normally uploaded videos (referred as Nqueue), whereas the other one is for videos with high priority (referred as Pqueue). The job scheduler does not dispatch jobs in Nqueue until the Pqueue is empty.

When the job scheduler receives a job, it estimates the video transcoding time and inserts the job into Nqueue according to the estimated transcoding time. Once one processor starts to run a video transcoding job, it fetches the next nonencoded video segment from the storage repository. When the whole procedure finishes, the processor sends both the encoded video segment and the MPD, back to the storage repository for streaming purpose. When the mobile client sends a video retrieving request to the Web server, this job will be forwarded to the specific servers for video retrieving. If the requested video segment is available, an HTTP connection is set up between the mobile client and the storage repository for streaming. Otherwise, the HTTP server sends a high-priority transcoding request to the job scheduler to indicate that the targeted video has higher priority and needs to be processed immediately. Thus, the job scheduler moves the requested job from Nqueue to Pqueue and assigns it to a processor that is available. Note that the jobs in Pqueue are processed with the FIFO algorithm. The detailed scheduling algorithm is presented in Algorithm 5.1.

Algorithm 5.1: Scheduler()

Initial: The HTTP server receives video transcoding jobs $<J_0, J_1, ..., J_n>$ and has M processors $<P_0, P_2, ..., P_m>$

/*video uploading to server*/

1   for $s = 0 : n$

2   if (the segment linked to J_s is requested)

3   push_back(J_s, Pqueue);

4   else

5            $t_{est} = cal\_est\_time$ (length of (J_s));

6   insert (J_s, Nqueue, t_{est}); //insert job to Nqueue for SJF scheduling

7   end

8   end

/*jobs already in the queues*/

9   while there exists available processor P_t, $0 \leq t \leq m$

10   if (Pqueue is not empty)

11   assign(Pqueue.head, P_t); //assign the first job in Pqueue to Processor P_t

12   else

13   assign(Nqueue.head, P_t); //assign the first job in Nqueue to Processor P_t

14   end

15   end

5.4 EXPERIMENTAL EVALUATION

We first introduce the video transcoding time estimation method and metric based on a large number of videos and then evaluate different video transcoding algorithms with multiple video streams.

5.4.1 Experimental Settings

For all the experiments, we use the Gearman framework [17] to simulate the scheduling algorithms for DASH in a cloud environment. Gearman allows us to run jobs in parallel and to balance the workload between processors. All the experiments are conducted on a server with 24 Intel® Xeon® X5650 2.67 GHz CPUs and 48 GB memory running under Linux Red Hat 2.6.32. The storage repository is a QNAP TS-879U-RP network-attached storage system with 6 GB/s storage throughput and a 10 GB/s connection to the server. Considering that the video segments for DASH streaming are always short in duration within a few seconds and the fast connection between the storage repository and the processors, the time for fetching and storing a video segment is short compared to that of video transcoding. Therefore, we ignore the time spent on I/O at this point. To simulate multiple uploading streams and multiple processors in a cloud environment, we utilize 20 processors and 40 video uploaders on Gearman. Each video uploader will upload 25 video segments. The length of video segments varies from 0.3 s to 5 s and each segment is encoded with a low bit rate at 256 kbps and a medium bit rate at 768 kbps.

5.4.2 Video Transcoding Time Estimation

Next we introduce an estimation method for video transcoding time based on statistics. As presented in Section 5.3, the scheduling algorithm used for Nqueue is SJF. Therefore, we need to estimate the time to finish a job before it is inserted into Nqueue. We run the video transcoding jobs with 11,194 video segments recorded with Android devices at different resolutions. The length of these segments varies from 0.2 s to 6.5 s.

In order to show the independence among multiple CPUs of running video transcoding jobs in parallel, we measure the transcoding time for each individual segment using only one CPU and choose them as a baseline. We then run four streams simultaneously on the same video datasets and compare the video transcoding time to the baseline. Figure 5.3 shows the normalized transcoding deviation of the video

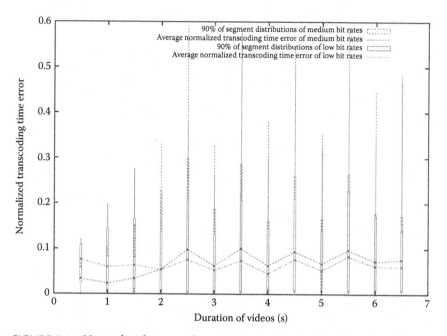

FIGURE 5.3 Normalized transcoding time error. The transcoding time for one stream on single CPU is considered as a baseline and that of four streams simultaneously is used to show the independence among CPUs.

transcoding time and we calculate the videos with 0.5 s duration differences together. For transcoding to both low and medium bit rates, the average error falls below 0.1% and 90% of the error distribution is below 0.3%. Considering the measurement errors and the time bias for each run, we establish that multiple video transcoding jobs can be considered CPU independent.

As explained above, we then choose the statistics on the video transcoding time on a single CPU to estimate the finishing time of a transcoding job based on its video duration. Based on the estimation of the transcoding time with respect to the video duration, we can predict the time to finish transcoding the uploaded video and place the job in the right place in Nqueue. In the following, the transcoding time estimation method is introduced.

5.4.2.1 Video Transcoding Time Estimation to Low Bit Rate

Figure 5.4 shows the video transcoding time to low bit rate with respect to the video duration. We draw a linear polynomial fitting curve using

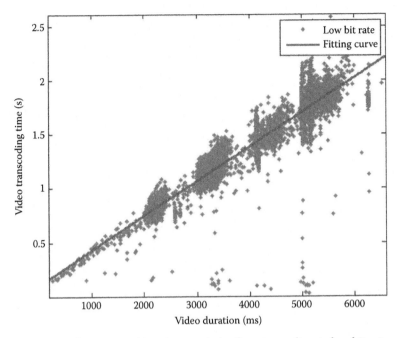

FIGURE 5.4   Video transcoding time statistics for transcoding to low bit rate and the fitting curve.

MATLAB® for estimating the video transcoding time. The coefficients of the fitting curve are as follows:

$$[0.00031878717297691053, 0.10858758089453366]$$

The estimated video transcoding time can be represented as follows:

$$t_{cal} = 0.0003 \times t_{dur} + 0.1086$$

where:
   t_{dur} is the video duration
   t_{cal} is the calculated time

Observed from Figure 5.4, given the duration of a segment, the actual transcoding time is biased to the estimated value calculated from the fitting curve, which is referred to as t_{err}. The actual video transcoding time (denoted as t_{est}) is then

$$t_{est} = t_{cal} + t_{err}$$

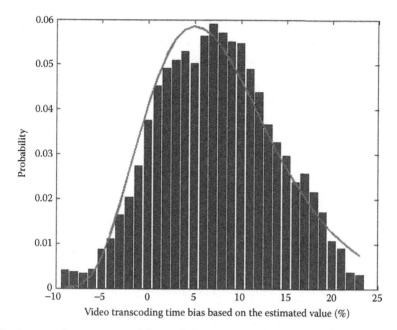

FIGURE 5.5   Fitting curve of the probability distribution of x_t for transcoding to low bit rate.

In order to study the probability distribution of t_{err}, we calculate the differences between the measured transcoding time and the value from the fitting curve for each segment. We then normalize the time error compared to the calculated value to show the percentage of time error. As shown in Figure 5.5, the probability distribution of x_t based on the calculated value follows a gamma distribution:

$$\text{prob}(x_t) \sim \frac{1}{\theta^k} \times \frac{1}{\Gamma(k)} \times x_t^{k-1} \times e^{-(x_t/\theta)}$$

where:
  $k = 6$
  $\theta = 3$

and

$$x_t = \text{round}\left(\frac{t_{err}}{t_{cal}} \times 100\right) + 10$$

where:
  $x_t \in [1, 33]$

Given this analysis, we can select the value of x_t according to its probability and predict the video transcoding time. In case of transcoding videos to low bit rates, the estimated transcoding time can be stated as follows:

$$t_{est} = t_{cal} + t_{err} = t_{cal} \times \left[1 + \frac{(x_t - 10)}{100}\right]$$

where:

t_{cal} is the calculated time

t_{est} is the estimated time

t_{err} is the error time

5.4.2.2 Video Transcoding Time Estimation to Medium Bit Rates

The estimation of transcoding time to medium bit rate is calculated in the same way. Figures 5.6 and 5.7 show the fitting curve on the transcoding time and the error distribution of x_t, respectively. The coefficient of the fitting curve for the medium bit rate is

[0.0006732551306103706, 0.070476737957686414]

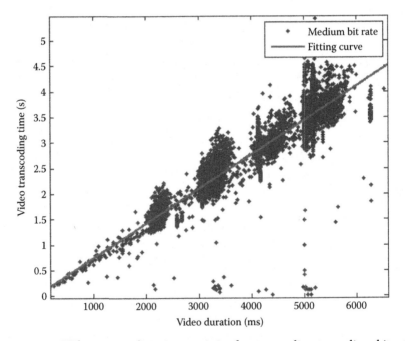

FIGURE 5.6 Video transcoding time statistics for transcoding to medium bit rate and the fitting curve.

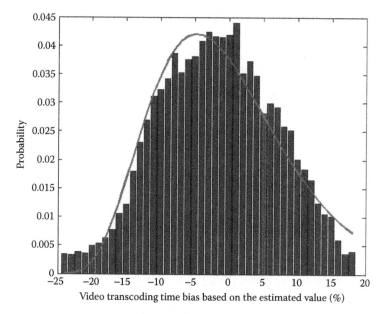

FIGURE 5.7 Fitting curve of the probability distribution for transcoding to medium bit rate.

And the estimated video transcoding time to the medium bit rate is

$$t_{\text{cal}} = 0.0006 \times t_{\text{dur}} + 0.0704$$

Compared with Figures 5.5 and 5.7, the probability distribution of x_t is wider for transcoding videos to a medium bit rate than that to a low bit rate, which means that the video transcoding time to a medium bit rate always varies in a larger range than that to a low bit rate. Note that the estimation of the video transcoding time is hardware dependent and we only intend to show the practicality of this method. The probability distribution of x_t based on the calculated value follows a gamma distribution:

$$\text{prob}(x_t) \sim \frac{1}{\theta^k} \times \frac{1}{\Gamma(k)} \times x_t^{k-1} \times e^{-(x_t/\theta)}$$

where:
$k = 5.7$
$\theta = 4.3$

and

$$x_t = \text{round}\left(\frac{t_{\text{err}}}{t_{\text{cal}}} \times 100\right) + 25$$

where:

$$x_t \in [1,43]$$

In case of transcoding videos to medium bit rates, the estimated transcoding time can be stated as follows:

$$t_{\text{est}} = t_{\text{cal}} + t_{\text{err}} = t_{\text{cal}} \times \left[1 + \frac{(x_t - 25)}{100} \right]$$

5.4.3 Experimental Results

We study the performance of different scheduling algorithms for video transcoding with DASH for multiple streams. The testing workloads consist of four uploading streams, including 5, 10, 20, and 30 videos, respectively, received by the HTTP server per second. Since it is not practical to interrupt the video transcoding procedure as it would result in large resource waste, we investigate the following existing scheduling algorithms: FIFO, SJF, and FP except RR.

5.4.3.1 First In, First Out

We study the performance of the FIFO algorithm on video transcoding. Figure 5.8 demonstrates the transcoding start-up latency with different workloads. In Figure 5.8, the start-up latency increases as the workload

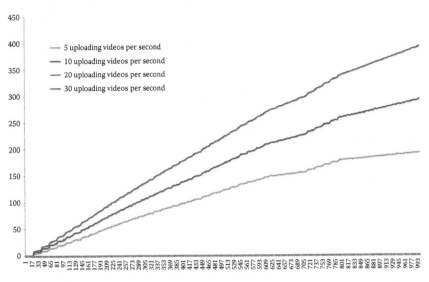

FIGURE 5.8   Start-up latency of FIFO with different workloads.

grows. When the workload reaches the threshold and all the processors are used for transcoding, the start-up latency reaches its peak and remains constant. In our case, since we have 20 processors, when the HTTP server receives more than 20 videos per second, all the incoming jobs will be inserted into the Nqueue and wait to be processed. According to this observation, we treat 20 uploading videos per second as the heaviest work-load in the experiments shown in Figure 5.8.

5.4.3.2 Shortest Job First

We present the start-up latency using SJF with different workloads. As shown in Figure 5.9, compared with the three streams, the latency trends are similar but small differences exist due to the interval gap of video uploading. The short start-up latency, except for the beginning of the curve, indicates that the uploaded videos are shorter than others. In our uploading streams, some large video segments are uploaded to the server at the beginning, so they have to wait until other, smaller ones are encoded first. Thus, the start-up latency for these large segments is long. Although the start-up latency for an individual segment might be large, the overall latency is contrarily small. Overall, the start-up latency of SJF is smaller compared to that of FIFO, but the gap decreases as the workload grows. The average start-up latencies of FIFO and SJF with different workloads are listed in Table 5.4.

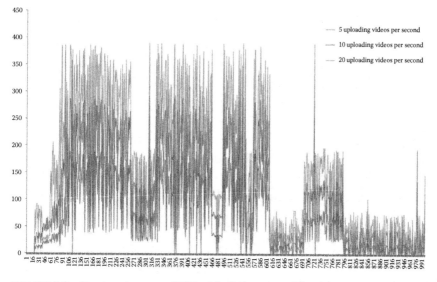

FIGURE 5.9   Start-up latency of SJF with different workloads.

TABLE 5.4   Average Start-Up Latency of Different Scheduling Algorithms

Scheduling Algorithm	5 Videos per Second (ms)	10 Videos per Second (ms)	20 Videos per Second (ms)
FIFO[a]	112.1439	161.9639	211.7924
SJF[b]	72.59118	120.9629	170.4255

[a] FIFO, first in, first out.
[b] SJF, shortest job first.

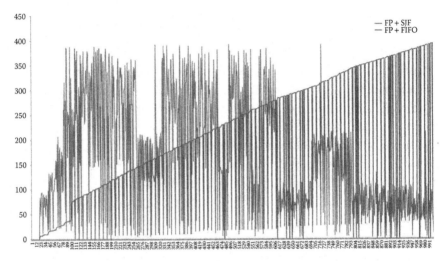

FIGURE 5.10   Start-up latency of hybrid methods with 20 uploading videos per second. FIFO, first in, first out; FP, fixed priority; SJF, shortest job first.

5.4.3.3 Hybrid Method

In order to test the performance of the scheduling methodology when videos are actively requested, we choose 100 video segments to be requested and have higher priority. In this experiment, we measure the start-up latency with the workload of 20 uploading videos per second and compare two hybrid methods: FP + FIFO and FP + SJF. Figure 5.10 displays the differences of start-up latency with these two methods. For both methods, the start-up latency of videos with higher priority is almost equal and quite small compared to the average. The overall difference still depends on the scheduling algorithms used in Nqueue.

5.5 CONCLUSIONS

In this chapter, we first presented the existing video streaming standard over HTTP and introduced the foundation of the MPEG-DASH standard. We then described the exiting platforms for video transcoding in a cloud environment and different scheduling algorithms. We further designed a framework for video transcoding with DASH in a cloud environment. Finally, utilizing the time estimation method of video transcoding, we explored different scheduling algorithms for DASH streaming. Experimental results show that the FP + SJF method achieves the lowest start-up latency for DASH in our test cloud environment.

REFERENCES

1. YouTube Press Statistics, 2012. http://www.youtube.com/t/press_statistics
2. Cisco Systems. Cisco visual networking index: Global mobile data traffic forecast update, 2012–2017. White paper, 2013. http://www.cisco.com/en/US/solutions/collateral/ns341/ns525/ns537/ns705/ns827/white_paper_c11-520862.pdf
3. ISO/IEC 23009-1: 2012. Information technology—Dynamic adaptive streaming over HTTP (DASH)—Part 1: Media presentation description and segment formats. http://www.iso.org/iso/iso_catalogue/catalogue_tc/catalogue_detail.htm?csnumber=57623
4. Microsoft Corporation. Smooth Streaming Protocol Specification, 2012. http://download.microsoft.com/download/9/5/E/95EF66AF-9026-4BB0-A41D-A4F81802D92C/[MS-SSTR].pdf
5. R. Pantos and W. May, Apple Inc. HTTP Live Streaming, 2012. http://tools.ietf.org/pdf/draft-pantos-http-live-streaming-10.pdf
6. Adobe Systems Inc. HTTP Dynamic Streaming. http://www.adobe.com/products/hds-dynamic-streaming.html
7. B. Seo, W. Cui, and R. Zimmermann. An experimental study of video uploading from mobile devices with HTTP streaming. In *3rd ACM Conference on Multimedia Systems*, Chapel Hill, NC, February 22–24, pp. 215–225, 2012.
8. T. Stockhammer. Dynamic adaptive streaming over HTTP—Standards and design principles. In *2nd ACM Conference on Multimedia Systems*, San Jose, CA, February 23–25, pp. 133–144, 2011.
9. I. Sodagar. The MPEG-DASH standard for multimedia streaming over the Internet. *IEEE Multimedia*, 18(4): 62–67, 2011.
10. W. Zhu, C. Luo, J. Wang, and S. Li. Multimedia cloud computing. *IEEE Signal Processing Magazine*, 28(3): 59–69, 2011.
11. Amazon Web services. Amazon Elastic Transcoder. http://aws.amazon.com/elastictranscoder/.
12. Amazon Web services. Amazon Elastic Compute Cloud (Amazon EC2). http://aws.amazon.com/ec2/.
13. Amazon Web services. Amazon Simple Storage Service (Amazon S3). http://aws.amazon.com/s3/.
14. Zencoder. http://zencoder.com/en/cloud
15. EncoderCloud. http://www.encodercloud.com/.
16. Rackspace. The rackspace cloud. http://www.rackspace.com/cloud/.
17. Gearman framework. http://gearman.org/.

CHAPTER 6

Cloud-Based Intelligent Tutoring Mechanism for Pervasive Learning

Martin M. Weng

Tamkang University
New Taipei, Taiwan

Yung-Hui Chen

Lunghwa University of Science and Technology
Taoyuan, Taiwan

Neil Y. Yen

University of Aizu
Aizuwakamatsu, Japan

CONTENTS

6.1 INTRODUCTION

"E-learning," which is based on the Internet technology, provides a new choice different from the past—the limitation of time and space—and learners require staying together in a regulation time, for the comportment of learning. "E-learning" is a process of studying through the digital media resources for learners, and these media include the Internet, computers, satellite broadcasts, tapes, videos, interactive TVs, and CDs . Recently, the technology of the network service in software and hardware becomes more mature, and the "e-learning" industry becomes a main promoter of the movement, which says: "lives old, learns old." In response to this concept, the distance learning pattern combined with the computer technology is getting more and more attention now; there are too many instances about the combine computer and multimedia technology to enumerate. But, with the development of wireless network technology, mobile networks such as WiFi, 3G, and WiMAX are getting universal; thus, the volume of mobile device is smaller than before. Therefore, the use of mobile devices such as smartphone is becoming the trend now (Laisheng and Zhengxia 2011). But there are many platforms for mobile devices; if we design different platforms for each application, it will face many constraints and limits, and waste much time on maintenance and series of tests for different devices. With the development of Web service applications, many programs transform the execution environment from the desktop into the Web world, for example, e-mail service that has been used many years through the Web service; users can arrange and back up their contact list and mails, do not need to remember the settings about POP3/SMTP (Pocatilu and Boja 2009), and even reinstall your computer system, which causes loss of the mails saved in the computer.

Moreover, no matter where you are, when you just need to launch the Web browser, all you can do is working through Internet connection; this is the charm of Web applications (Vaquero et al. 2009). This application mode is based on the cloud computing system; all the complex computing and data storage requirements are processed into the cloud server, and the client application is responsible to display the result, accept the simple operation, and increase the flexibility of use. This cloud computing architecture is the main trend of the future application-oriented service (Armbrust et al. 2010). Using this architecture, no matter what type of smart devices user uses, all can easily access to these services. Cloud computing architecture can also be used in the distance learning system, not just extend the use of the system from personal computer (PC) to mobile devices, meanwhile, learning on multiple devices neither the limitation environmental conditions and device capacity. The development of any information system includes the user interface, processing logic, and data storage. The technical report of Advanced Distance Learning (ADL) initiative, according to the functions of distance learning, divided into authoring tool, learning management systems (LMS), and repositories of three mutually overlapping parts.

With regard to the authoring tools of distance learning environment, it provides the editors, instructional designers, or teachers easily to design/revise the learning content, enable user to integrate an array of media to create professional, engaging and interactive learning content and also deliver the learning content to the learner with easily way in the distance learning environment. Emphasizing on sharing and exchanging of the information era, the learning content or materials follow some international standards or specifications to develop, and these usually focus on defining the level of architecture rather than the content itself (Tu and Chen 2011). Suppose the teacher must realize the process of using an authoring tool to construct a course, the difficulty will become the number one killer of the digitization of traditional teaching. Thus, another purpose of the authoring tools is to assist the author or instructional designer packaging the learning content corresponding to the international standards (Wu et al. 2011), which include Sharable Content Object Reference Model (SCORM), Aviation Industry Computer-Based Training (CBT) Committee (AICC), instructional management system (IMS) Question and test interoperability (QTI), or IMS Common Cartridge (CC).

LMS not only provides available courses, manages the learner and user, and maintains learning record, but also acts the role of the display platform of the learning component produced by the authoring tool mentioned earlier. Therefore, collecting the user information is quite important for further analysis, provide the author share the course, learning portfolio, record, and through the distributed learning systems to break the information exchange problem in the individual digital learning platform, allow the user to use one learning system and to get the different courses or information from the other systems. The storage is the platform to collect and store the learning component, and the system that stores the course component and provides the function such as research, publish, import, and export is called repository system. This research is based on the rich computing resource of cloud environment, aims to assist pervasive learning, and develops the core mechanism and related service. We use the HTML5 standard and Android operations systems to develop; escaping the dilemma of previous product development is limited by the terminal device.

The purpose of this research and its results are as follows:

1. Combination of the concept of cloud computing "anytime, anywhere, and any device" and smart device system develops a distance learning system that can remote use at any time. This system applies the SCORM and CC standards; compatible with other distance learning system, it applies the same standards and extends the applications from PCs to the mobile devices such as smartphone or tablet PC. The system of this research developed is fully using the mobility, execution, and touch screen interaction interface of the smartphone; when learner uses this system, it is based on the learner's demands and learning progress to access the learning resources on the cloud server and start the distance learning.

2. Learners are able to download the learning content to the handheld device and then learn offline without the Internet connection. The learner should connect to the cloud server when he or she needs to update or synchronize the learning progress. The system will determine the most need for course synchronization automatically and decrease the size of file transmission. Through the cloud service, the learner can upload and synchronize the learning progress to the cloud server and continue the learning activity on the other devices. For example, the learner can download the course from the cloud server to the smartphone and learn in the commuting time; after the learner enters into the classroom, the device will automatically update the learning progress to the cloud server, and then switch the learning platform to the PC and restart the learning. Since the course data are stored on the cloud, the learners only need to download the required part. The system developed will update and synchronize automatically; when the course has the new or updated version, the system will notify the learner by "push" and also notify the learner who has the permission to access the course or who has downloaded the course before it is updated.

6.2 CLOUD-BASED EDUCATIONAL SERVICE

In the past few years, cloud computing became a popular topic in many researches. Cloud computing is the idea of using a network to bring together the resources and information available in many places for problem solving and communication. In IBM's technical white paper, cloud computing (Boss et al. 2007) is defined as follows:

> Cloud computing term is used to describe both a system platform and a type of application. A cloud computing platform is to be deployed dynamically on-demand, configuration, reconfigure and deprovision, and so on. In the cloud computing platform, the server can be physical servers or virtual servers. A high-level calculation cloud usually contains a number of other computing resources, such as storage area networks, network devices, firewalls and other security equipment. In describing the application of cloud computing, it describes a scalable application through the Internet. Any user can access a cloud computing application through appropriate equipment and a standard browser.

Some researchers believe that cloud computing service has the potential to affect almost all the fields in the IT industry and make software and hardware even more attractive as a service (Bai et al. 2011; Dagger et al. 2007). Developers of cloud computing for novel interactive services do not require the large capital outlays in hardware to deploy their service, and also do not need the large human expense to operate it. Furthermore, the companies with large tasks can get their results as quickly as their programs can develop (Al-Zoube 2009). This extension of resources is unprecedented in the history of IT development.

Cloud computing comprises three layers, which are as follows (Figure 6.1):

- Infrastructure as a service (IaaS) delivers a full computer infrastructure via the Internet.

- Platform as a service (PaaS) offers a full or partial application development environment that users can access and utilize online, even in collaboration with others.

- Software as a service (SaaS) provides a complete, turnkey application including complex programs such as those for customer relationship management (CRM) or enterprise resource management via the Internet.

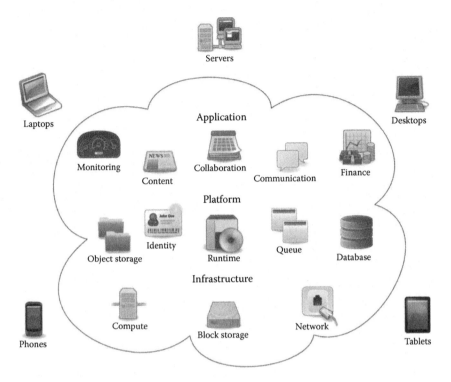

FIGURE 6.1 Cloud computing.

E-learning is an Internet-based learning process, using Internet technology to design, implement, select, manage, support, and extend learning, which will not replace the traditional education methods, but will greatly improve the efficiency of education (Buyya et al. 2009). E-learning does not simply use the Internet to learn but it provides the solutions containing the technology of system standardization and management means. E-learning systems usually require many hardware and software resources, education institutions cannot provide such investments, and the development of e-learning solutions based on cloud computing is the balance solution between the cost and the result for them (Hu and Zhang 2010). There are several computing services that offer support for educational systems.

Cloud learning is a new concept that inspired by cloud computing, also the cloud learning emphasizes learner-centered learning, resource sharing, collaboration among learners and to jointly build personalized learning environment. To some extent, within the cloud, learners take active and initiative roles in learning process, and they are not only enjoyers of learning resources, but also developers, organizers, and managers. Moreover, they collaborate with each other in the personalized learning environment. Cloud learning platform uses cloud computing, so all the required resources will be adjusted as needed.

Virtual learning environments (VLEs) are electronic platforms that can be used to provide and track e-learning courses and enhance face-to-face instruction with online components (Lainhart 2000). Primarily, they automate the administration of learning by facilitating and then recording the learner's activity. VLEs are the dominant learning environments in higher education institutions. Also known as LMSs and course management systems (CMSs), their main function is to simplify course management aimed for numerous learners. Traditional e-learning platforms, or LMSs, provide holistic environments for delivering and managing educational experiences (Brock and Goscinski 2008).

E-learning platforms can be divided into two different generations. The first generation of e-learning platforms provides, in essence, black box solutions. For the most part, these systems use proprietary formats to manage the courses directly. These platforms focus on the delivery and interoperability of the content designed for a specific purpose, such as a particular course. The second, or current, generation of e-learning platforms expands their predecessors' successes and begin to address their failures. Examples of these second-generation platforms include WebCT/Blackboard, Moodle, and Sakai. In terms of e-learning evolution, these platforms provide a shift toward modular architectural designs and recognize the need for semantic exchange.

6.3 SYSTEM ARCHITECTURE

In this chapter, we focus on the services of pervasive learning based on cloud environment and design a cloud-based server that provides distance learning management. On the user side, we implement a platform on Android-based smartphone. In this section, the system architecture and system design are discussed in detail.

The system architecture is shown in Figure 6.2. The cloud-based server has the following four roles:

- Identification of user
- Storage of learning progress
- Course data management
- Course database

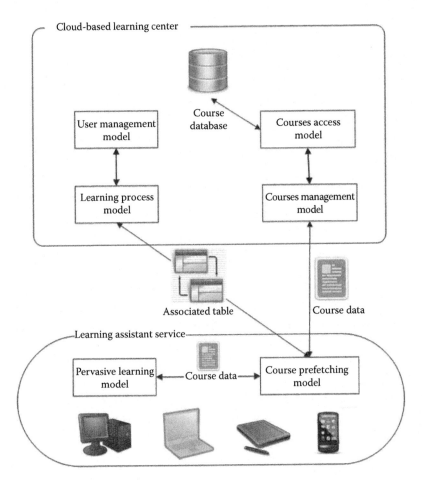

FIGURE 6.2   System architecture.

The applications on user devices [such as PC, notebook (NB), smartphone, and tablet PC] play the following three roles:

- Course downloading
- Learning devices
- Learning progress management

Summarizing the above roles in the user side and cloud server, we call the cloud server as "learning center" and the application on user devices as "learning assistant." In our system, we used the model-view-controller (MVC) model as the core architecture that has been widely used; in the

MVC model, the system has three layers of architecture: (1) data model layer, (2) view layer, and (3) controller layer. The data model layer administers business logic, the view layer takes charge of user interface, and the controller layer takes charge of the interaction between the users and the services. The controller layer also receives the requirement from the user, and then connects it to the specific business logic by event dispatcher and finds out the feedback to the user. In the traditional Web application of the MVC model, the three layers work on the Web server, but in our system, we use the cloud framework to deliver the computing ability of "anytime, anywhere, and any device" and the usability in offline status. Hence, this system integrates the data model layer and the controller layer into learning center and the view layer is integrated into learning assistant. The advantages of the design we proposed are as follows:

- With the cloud concept of "anytime, anywhere, any devices," the main data computing and the storage task will be assigned to the cloud server, which is called learning center. The performance and usability will be enhanced via the less computing and less restriction on devices.

- The learning assistant application on user devices only takes charge of interface and simple input operation. The advantage is that the developers get the balance between restricted operating performance and battery consumption, and have more flexible design on different user devices.

The relationship between learning center and learning assistant is shown in Figure 6.2. In the following, we will focus on the development of learning center and learning assistant.

6.3.1 Learning Center in Cloud

Learning center is a system that is constructed on the cloud computing technology; since building this system is not an easy task, we must have enough hardware to purchase and reliable network environment, which cannot be afforded by everyone. Thus, this research uses Google App Engine (GAE) to develop the related application. GAE is a PaaS platform and the designer only needs to focus on the programming. They do not need to worry about the other restriction. The hardware, network bandwidth, and computing power from GAE are provided from Google.

The system in this chapter implements the learning center based on the GAE architecture and uses the resources in GAE to provide related educational service that we design.

The main services of the learning center are as follows:

- Identification of user

- Storage of learning progress

- Course data management

- Course database

To satisfy the requirements, this research divides the learning center into the following four modules:

- User management module

- Learning progress recording module

- Course data access module

- Course data management module

In the next section, we will discuss about these modules and the problems that are met and how the research solves those problems.

6.3.1.1 User Management Module

This module is responsible for managing the accounts of the user such as adding, deleting, and modifying the permissions control of the user accounts. Learning center uses this module to decide the permissions in the system and manage the identity of the user by permissions control. The identification of the user in the learning center can simply be divided into two categories: teachers and students. Teachers in the learning center have a greater authority to manage the accounts, can access the functions in the system, and use the course data management module to set the access permissions of each course. Students in the learning center have less permission in the system. Both of these two identities use this module to maintain the personal data.

In addition, the devices that each user uses must be different; some users may use PC or NB at home but they use smart phone or tablet PC when they are outside of the home. Hence, this module will record the device type and number to recognize all learning devices. The learning progress recording module will use this information to decide how to synchronize the learning progress.

6.3.1.2 Learning Progress Recording Module

This module is responsible for recording the learning progress of every user, saving the details of the update of learning progress every time, such as account, materials, learning status, and transaction time, and synchronizing the course file to the learning assistant. Learning center uses this module to maintain the learning record; after the user uploads the learning record and then uses another device access the learning center, the system will check the status of synchronization and start to synchronize it. The data of synchronization that including the data of ongoing course data and learning progress, and the data also shares with course data management module at the same time. Inquiry-related course can allow the user predownload the files. This module records the user's use of each course status. Teacher can check the learning progress of each student. When the learning of the student is behind the progress, the teacher can ask the student to learn the specific course first. This module will send the request to the device on the user's hand when the device is requested to synchronize.

As the research needs to synchronize the course file on different devices, the instability of network is the first problem that is met. The learning assistant operates on the mobile devices and encounters the disconnection of the network. If the file is broken when disconnected, the incomplete file cannot be used normally. To solve this problem, we try to divide the file into several parts. The course file sends these parts to the learning center and learning assistant to store, and generates an association table. This table saves the name of the course file, the status, the access time, and the file name of each part in the XML format, and uses universal unique identifier (UUID) to ensure the uniqueness of the course. The system controls the size of each table in a certain range. According to this table, the learning assistant will download the course file to the learning center. When the user is in learning status, learning assistant will change the content of the table to monitor the status, and synchronize with the learning center in each connection.

6.3.1.3 Course Data Access Module

This module is responsible for storing the course files, deciding the format, and saving the path of the course data. When the learning assistant needs to access the course, it will get the association table from the learning progress recording module (Mikroyannidis et al. 2010). The table

TABLE 6.1   Description of Association Table

Table Name	Description
Course	Identify the course information
Unid	Unique identifier
Name	Course name
File	Course allocation in computer hard disk
Part	Identify how many parts of course are allocated, and also indicate the file address in the computer

records the list of all courses, the required files, and the access path. For example, when the learning center needs to assign a course named "Distance Learning," this module will generate the related data into the associated table (Table 6.1).

```
<course>
<unid>00010AC0100</unid>
<name>Distance Learning 001</name>
<file part="3">
<part seq="1" sum="2eb722f340d4e57aa79bb5422b94d556888cbf5f">
http://s3.tp.dl-center.tw/c/dl-001/p1&uuid=00010AC0100
</part>
<part seq="2" sum="45b522f540d4e57aa79bb5422b94d556888cbf34">
http://s3.tp.dl-center.tw/c/dl-001/p2&uuid=00010AC0100
</part>
<part seq="3" sum="75b722f344d4e57aa79bb5422b94d556888cbfac">
http://s3.tp.dl-center.tw/c/dl-001/p3&uuid=00010AC0100
</part>
</file>
</course>
```

In this table, the label <part></part> indicates the file address that is produced only for the synchronization at that time. The label uuid indicates the uniqueness of each course in the system. When the user learning progress recording module produces the associated table, it will send the synchronous attribute to the course data access module. For example, there is a portable device with learning assistant application, and the location of this device is in city A. The course data access module assign the cloud server near A that synchronizes the related information and downloads the related courses from this server, and the downloaded items are separated into several files. The label seq indicates the order of separated files. To reduce the file size, every separated file will be compressed by the zip file, and the learning assistant will restore the course when it receives all the zip files. After that, every file will be identified by sum attribute. The learning assistant will check the Checksum attribute to confirm that the course files just downloaded are correct.

6.3.1.4 Course Data Management Module

This module is responsible for managing the course data and recording the relations between the course data. The relations include the priority of the course and the connections. For example, course A is the leading course of course B. This module shows the dependencies of these two courses. According to this relation, the teacher can control the learning progress by adding, modifying, and deleting the course file, and decide which account can access the course by using the user management module. This module combined with the course prefetching module in the learning assistant is mainly responsible for the entire course prefetching strategy of this research.

6.3.2 Learning Assistant in Client

Learning assistant is an important tool that the user uses in his or her devices. With the connection between learning assistant and learning center, the learner can download the related courses on their devices and synchronize their learning progress to the learning center. The course prefetching module is the most important module in learning assistant.

6.3.2.1 Course Prefetching Module (Asynchronous Cache Mechanism)

Since large amounts of multimedia material will be applied to the course files in this research, the execution performance of the smart device and the status of the network connection will affect the learning fluency. When the learner is learning in an outdoor environment, he or she often encounters this situation. For example, if the learning activity needs a large size file, the learner will have to spend much time waiting for the transmission to complete; and the learner does not necessarily always stay in a permanent network environment. This situation will cause the interruption of learning. Thus, this module will construct an automatic learning content prefetching strategy, coordinate with the "course data management module" in the learning center, allow the learning assistant to predownload the course file that the learners need, reduce the interruption and unnecessary time of waiting, and help the learners to learn more smoothly.

Learning content prefetching strategy is an automatic mechanism in the mobile network environment. The purpose of this strategy is to conquer the limitation of the network environment and support the high mobility in the pervasive learning activity so that users can have smooth learning progress. With the learning content prefetching strategy, it doesn't need to consider the speed of network and performance, because the learning activity can still be completed.

The course information, user information, and device information are saved in the learning center and learning assistant. The learning content prefetching strategy is divided into course data access module in the learning center and course prefetching module in the learning assistant because it needs to collect enough information for generating the right decision of content prefetching. Thus, this strategy is based on this information, with virtual memory management and the disk cache technical concept, and offers the best learning environment to learners.

It neither uses virtual memory management to solve the network problem nor uses the disk cache technical concept to reduce the waiting time. The basic purpose of mobile learning content prefetching strategy is to predict the learning path and, according to the learning device, predownload the course resource. Thus, the mechanism of predicting the learning path will affect the performance of the strategy.

The learning prefetching strategy analyzes the following:

1. Specific relation of each learning content

2. Users' position information

3. Characteristics of the learning content

4. Efficacy of the mobile device

Learning center and learning assistant synchronize the learning progress by using the same course prefetching strategy. This strategy is divided into two parts: course data management module and replacement management module.

6.3.2.1.1 Course Data Management Module   This module is in the learning center, and according to the information from the learning progress recording module, it arranges the most likely part for the learner using the association table. Then it requests the learning assistant to download the file. It makes the system to predownload these parts using idle time, which reduces the access time of network connection.

6.3.2.1.2 Replacement Management Module   Learning assistant determines the learning time by replacement management module. The parts saved in the client will not be read again and release it. Then, the modified association table is sent to the learning progress recording module in the learning center to download the new parts assigned from the course management module. Make sure that the content saved in the device has high possibility of clicked and reading rate.

6.3.2.2 Course Cache Strategy

Figure 6.3 shows the operating process of course cache strategy. The course cache strategy considers the related information in the replacement management module, produces the drop order for the courses, and sends this message to the learner so that they can drop these unavailable courses. Then the devices start to synchronize the changed items in the associated table to the learning center. If the learner is a new user, the value of drop order is null. When the client side has enough storage space, the learning assistant on the user devices will send requests of course download. Meanwhile, the learning progress module in the server side will analyze the learning history and produce the recommended download order so that the learner may access in recent future. Then the learning center will send this download order to the learning assistant by the associated table, and the learning assistant can follow the associated table to download the related courses.

The accuracy of learning order prediction for learner is the main reason (Hu and Chen 2010) that may impact on the course prefetching strategy.

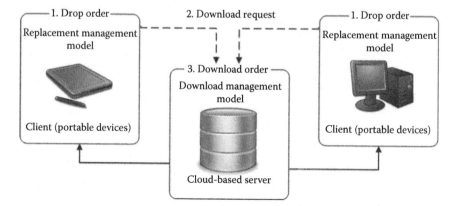

FIGURE 6.3   The process of course cache strategy.

In order to have accurate information to make precise prediction, there are many reasons that should be considered. The following items are the most important reasons that may impact on the course prefetching strategy.

- Relation of course order: The course order designed by the instructor has specific reasons why the instructor designs in this order? This is one of the reasons that should be considered.

- Location of learner: The associated table has related location of learner when they access courses. Hence, the system uses geography information as one of the reasons that may impact on the strategy because different location for learner may cause different learning motivation.

- Course content: The size of course, the path length of course, and the cluster number of course.

- Devices from users: Different devices have different restriction, such as storage size and network availability.

- Learning history of learner: Course download time and last access time of each learner.

6.3.2.3 Course Management Module

The course management module is designed in the cloud server side. This module works with the storage of learning progress module to predict the most possible course content that the learner may read in the recent future. After the prediction, the learning assistant will start to download the courses from prediction. This mechanism helps the system to pre-download the most possible courses that the learner may read before starting the next course. It will decrease the time when the learner downloads the courses from the server, and also decrease the system access time to the Internet. With the prefetching mechanism, the system performance has an obvious improvement.

The operation in the course management module is very simple as shown in Figure 6.4. The main task of the course management module is to produce the optimal download orders by all the predicted items when it receives the request for download. After the optimization, the system will send the optimized order to the client side and provide the cluster that includes the optimized order.

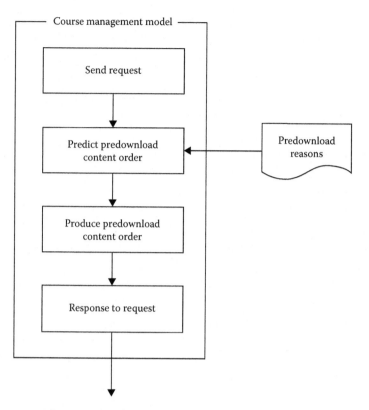

FIGURE 6.4   The progress of course management module.

6.3.2.4 Replacement Management Module

The learning assistant services in the client side can find out the possible unavailable course content in each learning time spot from the user and delete those unavailable course content from the user. After that, the system will download the new course content that is assigned by course management module. With this mechanism, we can confirm that all the course content in the user devices have high reading possibility. In the process of replacement management module, when the content prefetching strategy starts, this module will predict the possible content access order by a predicted factor, and then produce a drop order for learning content (Wang et al. 2011a). Meanwhile, the course management module sends the download requirement, and then gets the optimized course download order from the course management module. The system compares the drop order with the optimized course download order and removes the redundant parts, which will reduce the burden of system. For example, if the drop order is 3 and 4, the optimized course download order is 2, 5, and 3.

The predicted strategy will remove the item 3 from the drop list; hence, the drop order becomes 4 and the optimized course download order becomes 2 and 5. With this mechanism, it avoids the abandoned parts that will be downloaded in the recent future. After this step, the learning assistant starts to remove the learning content from the drop order, and the learning center starts to download the appropriate courses from the optimized course download order. In case of all the conditions that are mentioned earlier, the system will follow the traditional rules—first in, first out (FIFO)—to remove the oldest learning content in the client side.

6.4 CONCLUSION AND FUTURE WORK

This chapter proposed a cloud-based learning center and a learning assistant on the user devices that enhance the usability of pervasive learning via cloud computing. Except to improve the restriction to the user, it also improves the convenience of pervasive learning and provides a complete service and mechanism. The improvement satisfies the users that they can learn at anytime, anywhere, and any devices and accomplish the goal of pervasive learning. During the implementation, the synchronized mechanism based on the cloud server will develop in different devices that can satisfy the requirement from the user in distance learning scope.

ACKNOWLEDGMENTS

We thank the National Science Council. This research was supported in part by a grant from NSC 101-2221-E-240-004, Taiwan, Republic of China.

REFERENCES

Mohammed Al-Zoube, "E-learning on the cloud," *International Arab Journal of E-Technology*, 1: 58–64, 2009.

Michael Armbrust, Armando Fox, Rean Griffith, Anthony D. Joseph, Randy Katz, Andy Konwinski, Gunho Lee et al., "A view of cloud computing," *Communication of ACM*, 53(4): 50–58, 2010.

Yunjuan Bai, Shusheng Shen, Liya Chen, and Yongsheng Zhuo, "Cloud learning: A new learning style," *2011 International Conference on Multimedia Technology*, pp. 3460–3463, July 26–28, 2011.

Greg Boss, Padma Malladi, Dennis Quan, Linda Legregni, and Harold Hall, Cloud computing. IBM White Paper, 2007. http://download.boulder.ibm.com/ibmdl/pub/software/dw/wes/hipods/Cloud_computing_wp_final_8Oct.pdf

Michael Brock and Andrzej Goscinski, "State aware WSDL," *Sixth Australasian Symposium on Grid Computing and e-Research (AusGrid)*, Wollongong, NSW, pp. 35–44, January 24, 2008.

Rajkumar Buyya, Suraj Pandey, and Christian Vecchiola, "Cloudbus toolkit for market-oriented cloud computing," *Proceeding of the 1st International Conference on Cloud Computing (CloudCom)*, Springer, Germany, December 1–4, 2009.

Mariana Carroll, Paula Kotzé, and Alta van der Merwe, "Securing Virtual and Cloud Environments." In I. Ivanov et al. (eds.) *Cloud Computing and Services Science, Service Science: Research and Innovations in the Service Economy.* Springer Science+Business Media, 2012.

Declan Dagger, Alexander O'Connor, Séamus Lawless, Eddie Walsh, and Vincent P. Wade, "Service-oriented e-learning platforms: From monolithic systems to flexible services," *Internet Computing, IEEE*, 11(3): 28–35, 2007.

Shueh-Cheng Hu and I-Ching Chen, "A mechanism for accessing and mashing-up pedagogical Web services in cloud learning environments," *2010 3rd IEEE International Conference on Computer Science and Information Technology*, pp. 567–570, July 9–11, 2010.

Zhong Hu and Shouhong Zhang, "Blended/hybrid course design in active learning cloud at South Dakota state university," *2010 2nd International Conference on Education Technology and Computer*, pp. V1-63–V1-67, June 22–24, 2010.

John W. Lainhart, "COBIT: A methodology for managing and controlling information and information technology risks and vulnerabilities," *Journal of Information Systems*, 14: 21–25, 2000.

Xiao Laisheng and Wang Zhengxia, "Cloud computing: A new business paradigm for e-learning," *2011 3rd International Conference on Measuring Technology and Mechatronics Automation*, pp. 716–719, January 6–7, 2011.

Alexander Mikroyannidis, Paul Lefrere, and Peter Scott, "An architecture for layering and integration of learning ontologies, applied to personal learning environments and cloud learning environments," *2010 IEEE 10th International Conference on Advanced Learning Technologies*, pp. 92–93, July 5–7, Sousse, Tunisia, 2010.

Paul Pocatilu and Catalin Boja, "Quality characteristics and metrics related to m-learning process," *The Amfiteatru Economic Journal*, 11(26): 346–354, 2009.

Luis M. Vaquero, Luis Rodero-Merino1, Juan Caceres, and Maik Lindner, "A break in the clouds: Towards a cloud definition," *SIGCOMM Computer Communication Review*, 39(1): 50–55, 2009.

Hsing-Wen Wang, Jin-Sian Ji, Tse-Ping Dong, Chin-Mu Chen, and Jung-Hsin Chang, "Learning effectiveness of science experiments through cloud multimedia tutorials," *2011 2nd International Conference on Wireless Communication, Vehicular Technology, Information Theory and Aerospace & Electronic Systems Technology (Wireless VITAE)*, pp. 1–6, February 28–March 3, 2011a.

CHAPTER 7

Multiple-Query Processing and Optimization Techniques for Multitenant Databases

Li Jin, Hao Wang, and Ling Feng

Tsinghua University
Beijing, China

CONTENTS

7.1 INTRODUCTION

Cloud computing explains the phenomenon of the integration among multiple devices, which refers to both the applications delivered as services over the Internet and the hardware or systems software in the data centers that can provide those services. Here, cloud means a metaphor for the Internet where data are stored and costly operations are performed to generate the necessary response delivered to each client. Meanwhile, database applications are becoming increasingly complex, with demand for handling the distributed data that are often heterogeneous in nature. In order to provide efficient Web services, multitenancy comes up to achieve high scalability, which incurs the research issues of multiple-query processing and optimization.

In the data management field, multiple-query processing and optimization problem has long historically extensively been investigated, spanning from relational databases, deductive databases, semistructured Extensible Markup Language (XML) databases, to streaming databases. Given a set of database queries, instead of separately processing the queries one at a time, multiple-query optimization (MQO) aims to perform the queries together by taking advantage of common data and common query subexpressions that these queries may have. In this way, redundant data access can be avoided and the overall execution time can be reduced. In this chapter, the aim is to survey two main bodies of research activities in traditional databases and streaming databases. Based on the previous classic models, we discuss some possible extensions in the multitenant database domain.

The remainder of the chapter is organized as follows. Sections 7.2 and 7.3 review multiple-query processing techniques in traditional databases and streaming databases, respectively. Section 7.4 discusses their possible extensions to multitenant databases. Section 7.5 concludes the chapter and explores the future work.

7.2 MULTIPLE-QUERY PROCESSING AND OPTIMIZATION IN TRADITIONAL DATABASES

In the early literature, Grant and Minker [1] describe the optimization of sets of queries in the context of deductive databases and propose a two-stage optimization procedure. During the first preprocessor stage, the system obtains at compile time the information on the access structures that can be used in order to evaluate the queries. Then, at the second stage, the optimizer groups queries and executes them in groups instead of one at a time. During that stage, common tasks are identified and sharing the results of such tasks is used to reduce the processing time. References 2 and 3 show how to improve an incoming query evaluation by deriving the result based on the results of the earlier queries through the identical common subexpressions. Roussopoulos [4,5] provides a framework for inter-query analysis, aiming to find fast access paths for view processing. The objective of his analysis is to identify all possible ways to produce the result of a view, given other view definitions and base relations. Indexes are then built as data structures to support fast processing of views.

Overall, the major tasks of MQO in traditional databases are (1) identifying possibilities of shared computation through common operations or subexpressions and (2) constructing a global optimal plan, taking shared computation into account [6,7]. Through the first phase of common subexpressions' identification among a set of queries, the alternative plans for each query can be obtained. These alternative plans will be used in the second phase for the selection of exactly one plan for each query, which will then generate a globally optimal execution plan that will produce the answers for all the queries. A taxonomy of multiple-query processing and optimization in traditional databases is shown in Figure 7.1.

7.2.1 Identifying Common Subexpressions for Multiple Database Queries

A subexpression is a part of a query that defines an intermediate result used during the process of query evaluation [6,8]. For two queries, four possible relationships may hold [9]: (1) nothing in common, (2) identical, (3) subsumption, and (4) overlap. Taking the select condition on a relational attribute x, for example, $x > 5$ subsumes $x > 10$ because the result of $x > 5$ can be used for evaluating $x > 10$; meanwhile $x > 10$ overlaps with $x < 20$. Except for case (1), the two queries can be regarded to have common subexpressions. The problem of identifying common subexpressions is proved to be nondeterministic polynomial (NP)-hard [8,10]. Therefore, Jarke [8] indicates that

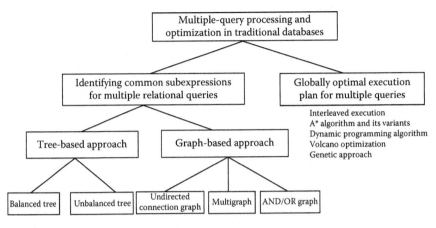

FIGURE 7.1 MQO in traditional databases.

multirelation subexpressions can only be addressed in a heuristic manner. He discusses the common subexpression isolation under various query language frameworks such as relational algebra, domain relational calculus, and tuple relational calculus, and shows how common subexpressions can be detected and used according to their types such as single-relation restrictions and joins. Park and Segev [11] process multiple queries, utilizing subset relationships between intermediate results of query executions, which are inferred employing both semantic knowledge on data integrity and information on predicate conditions of the access plans of queries.

7.2.1.1 Approach Based on Operator Tree
Hall presents a bottom-up heuristic method of using algebraic operator tree (expression tree) to detect the common subexpressions in a single query [12,13]. Specifically, he views the existence of common subexpressions in terms of lattices formed from the expression trees. Considering tree structure, it can be divided into balanced and unbalanced trees to deal with different kind of queries [12]. Identification of the common subexpressions is equivalent to converting from a simple query tree to a lattice in which no nodes are redundant. This conversion can be achieved by working from the bottom level of the tree, and the leaves go upward a level at a time, considering all convergences recursively with the pruning operations based on idempotency and null relations.

7.2.1.2 Approach Based on Query Graph
The query graph approach is designed to take advantage of common intermediate results among plenty of queries. Chakravarthy and Minker [14]

identify the equivalence and subsumption of two expressions at the logical level using heuristic rules. They first present a multiple-query connection graph for a set of queries by extensions of the query graphs introduced by Wong and Youssefi [15]. The connection graph contains nodes (representing relations) and edges (representing conditions between attributes of the nodes they are connected with). They then detect and exploit the common subexpressions to decompose the connection graph into a single plan for the set of queries through two transformations: instantiation and iteration. The transformation consists of selecting a node, writing an expression for evaluating the graph in terms of the selected node and the edges associated with the node, and rewriting the graph in terms of a simpler plan. Such a process is recursively applied till the graph consists only of isolated nodes. Here, a node chosen for instantiation is based on the following five heuristics: (1) it has several edges from different queries incident on it and the conditions associated with all the edges are identical; (2) it has several edges from different queries incident on it and the edges can be partitioned into two groups (each having identical conditions within the group) such that the condition of one group subsumes (intuitively more general) the condition of the other group; (3) it has several edges from different queries incident on it and these edges can be partitioned into sets, where each set consists of edges from the same subset of queries; (4) whenever it satisfies; and (5) the number of partitions is small (either 1 or 2), then perform the iteration immediately followed by instantiation [14].

Chakravarthy and Rosenthal [16] also use an AND/OR graph to represent queries and detect subsumption by comparing each pair of operator nodes from different queries.

Apart from the identical and subsumption operations, Chen and Dunham [9] propose a multigraph processing technique to cover the overlap case. First, they decompose a set of queries into unrelated sets. Once these sets are created, queries in each subset are executed separately. This decomposition step is performed before the determination of common subexpressions to reduce the size of query sets. The later can then be easily handled by selecting multiple edges between the same pair of nodes in the multigraph. With a set of heuristics for selecting common operations for processing (e.g., delay selection in one query in order to take advantage of the common join operations with another query, leave all the project operations to the final stages, and select edges with the identical/subsumed/overlapped select and join conditions), high scalability among queries can be achieved.

In the distributed databases, Yi et al. proposed a cost-based dynamic method to identify the correlation among the queries, where I/O and communication costs are on the same order of magnitude [17]. To speed up the performance, an index-based vector set reduction is performed at data node level in parallel with a start–distance-based load balancing scheme.

7.2.2 Constructing a Globally Optimal Execution Plan for Multiple Queries

As sharing of common subexpressions during execution is not always better than independent execution of multiple queries, blindly using common subexpressions may not always lead to a globally optimal execution plan [6,7]. Instead, the use of common subexpressions should be determined based on cost–benefit analysis. It has been shown that for small problem sizes up to 10 queries, near-optimal global plans can be generated very efficiently through exhaustive strategies. However, as the number of queries increases, the exhaustive MQO solutions become impractical, calling for heuristic methods [7].

Sellis [6] gives two alternative architectures for generating a global access plan (GAP). Architecture 1 (plan merger) makes the use of existing query optimizers. A conventional local optimizer generates one locally optimal access plan per query, and a plan merger examines all access plans and generates a larger GAP for the system to execute. Architecture 2 (global merger) is not restricted to solely using locally optimal plans. It relies on a global optimizer to process the set of queries and generates a GAP.

Interleaved algorithm. Sellis [6] gives an exhaustive interleaved execution (IE) algorithm. It decomposes queries into smaller subqueries and runs them in some order, depending on the various relationships among the queries. Then the results of subqueries are assembled to get the answers of the original queries. It proceeds as follows: First, the queries that possibly overlap on some selections or joins are identified by checking the base relations that are used. For any query that overlaps with some other query, its corresponding local access plan is considered. A directed labeled graph, GAP representing the union of all such local plans, is then built. Some transformation rules (i.e., proper implications, identical nodes, and recursive elimination) are enforced on the graph, taking the effects of common subexpressions into account. The transformed final direct graph is the GAP that is generated by IE algorithm. Note that IE algorithm keeps the partial order defined on the execution of tasks that a local access plan must be preserved in the GAP.

A* algorithm and its variants. While the IE algorithm considers only one locally optimal plan per query [6], Grant and Minker's branch-and-bound algorithm with a depth-first search method uses more locally optimal plans [18]. However, it is limited to the case of identical relationships. This algorithm is modified in Reference 19 by using a new lower bound function and a breadth-first search method to reduce the search space in a stochastic sense. It also extends [18] to the case of implied relationships. Chakravarthy and Rosenthal [16] addressed the MQO problem at various levels of detail, depending on the cost measure used. Sellis gave a state space search formulation and search algorithm A* with bounding functions and intelligent state expansion, based on query ordering, to eliminate the states of little promise rapidly [6,19,20].

Dynamic programming algorithm. Further improvement of Sellis's effort is discussed in References 11, 17, and 21. Park and Segev [11] present a dynamic programming algorithm based on the cost–benefit analysis for GAP selection, which has a lower computational complexity than that in References 19 and 22. Cosar et al. revise Sellis' A* algorithm by having an improved heuristics function that prunes search space more effectively while still guaranteeing an optimal solution. Simulated annealing technique has also been experimentally analyzed to handle larger MQO problems that cannot be solved using A* in a reasonable time with the currently available heuristics [17]. Furthermore, they adopt a set of dynamic query ordering algorithms so that the order in which plans are merged with the multiplan dynamically changes based on the current partial multiplan to be augmented by a new plan [23].

Volcano optimization. The volcano optimizer [24] is a cost-based query optimizer on account of equivalence rules on query algebras. It represents a query as an AND–OR directed acyclic graph (DAG), which can compactly represent the set of all evaluation plans. The nodes can be divided into AND nodes and OR nodes. AND nodes have only OR nodes as children and OR nodes have only AND nodes as children. An AND node in the AND–OR DAG corresponds to an algebraic operation, such as the join operation $\bowtie$ or a select operation σ, and an OR node in the AND–OR DAG represents a set of logical expressions that generate the same result set. Henceforth, the AND nodes can be referred to as operation nodes and the OR nodes as equivalence nodes. New operations and equivalences can be easily added into the graph. The key ideas of the volcano optimizer are listed as follows [7,24]: (1) Hashing scheme is used to efficiently detect duplicate expressions. Each equivalence node has an ID. The hash function of operation nodes is based on the IDs of child equivalence nodes. The strategy can avoid creating duplicate equivalence nodes due to cyclic derivations. (2) Physical algebra can be represented by DAG. (3) The best plan for each equivalence node is calculated by three heuristic rules: using the cheapest of child operation nodes, dynamically caching the best plans, and branch-and-bound pruning when searching.

To apply MQO to a batch of queries, the queries are represented together in a single DAG, sharing common subexpressions. Considering two queries (A ⋈ B) ⋈ C and A ⋈ (B ⋈ C) that are logically equivalent but syntactically different, the initial query DAG would contain two different equivalence nodes representing the two subexpressions. Through applying join associativity rules, the volcano DAG generation algorithm searches the logically equivalent nodes and replaces them by a single equivalence node.

Volcano-SH and volcano-RU. As the best plans produced by the volcano optimization algorithm may have common subexpressions, the consolidated best plan for the root of the DAG may contain nodes with more than one parent. Working on the consolidated best plan, References 7 and 25 introduce two heuristics algorithms: volcano-SH and volcano-RU. Volcano-SH decides which of the nodes to materialize and share in a cost-based manner. Volcano-RU is a volcano variant considering the case when optimizing a query, treating the subparts of plans for earlier queries as available. The algorithm is based on local decisions and the plan quality is sensitive to query sequence. A greedy strategy that iteratively picks the subexpression which gives the maximum benefit (reduction in cost) if it is materialized and reused is also given in Reference 7.

Genetic approach. Since MQO is an NP-hard problem, Bayir et al. [25] present an evolutionary genetic technique [25]. A chromosome corresponds to a solution instance for the set of queries of the MQO problem. In a chromosome, each gene of a chromosome represents a plan to the corresponding query. The value of the gene is the plan selected for the evaluation of the corresponding query. To select the chromosomes for the next generation, the quality of the solution represented by the chromosome is used. This quality is represented by the fitness function, which is simply the inverse of the total execution time of all the tasks in the selected plans for the queries. Under this modeling, MQO is also very suitable for genetic operations. Mutation and crossover operations can easily be defined to produce new valid solution instances. Since a gene in a chromosome represents the plan selected for the query corresponding to the gene position, the mutation operation is to replace the plan number with another randomly selected valid plan's number for that query. Therefore, a mutation operation always generates valid solutions.

Different crossover operations can also be applied to chromosomes, such as one-point, segmented, and multipoint crossovers. If two chromosomes represent two valid solutions for the same MQO problem, then any crossover operation on these two chromosomes produces new chromosomes representing valid solutions for the same MQO problem. Regardless of the crossover type and positions, since all chromosome segments that are going to be exchanged to produce a new chromosome represent valid plans for their corresponding queries, the new chromosome obtained by appending these segments represents a valid solution for the MQO problem.

7.3 MULTIPLE-QUERY PROCESSING AND OPTIMIZATION IN STREAMING DATABASES

Traditional databases have been used in applications that store sets of relatively static records with no predefined notion of time, unless timestamp attributes are explicitly added. A data stream is a real-time, continuous, ordered sequence of items; meanwhile, it is infeasible to locally store a stream in its entirety. Queries are executed continuously over streams during a selected time window and incrementally return new results as new data arrive. Therefore, a static batch-oriented approach is unsuitable in real-world environments where queries join and leave the system in an *ad hoc* fashion. Similar to multiple-query processing in the traditional databases, multiple streaming query processing proceeds in two steps: (1) to optimize each individual query and then find out sharing opportunities in the access plans, and (2) to globally optimize all the queries to produce a shared access plan. For single-site processing, many dynamic "on-the-fly" approaches have been put forward for different kinds of queries: (1) joins with individual predicates or varying windows, (2) aggregates with individual predicates or varying windows, and (3) joins and aggregates with individual predicates and varying windows. Similar strategies are also applied to distributed environment by taking the optimal allocation of each node resource into account. A taxonomy of multiple-query processing and optimization in streaming databases is shown in Figure 7.2.

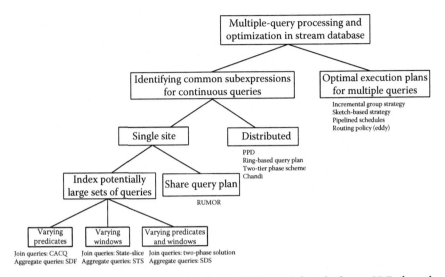

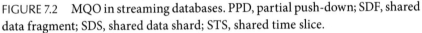

FIGURE 7.2   MQO in streaming databases. PPD, partial push-down; SDF, shared data fragment; SDS, shared data shard; STS, shared time slice.

7.3.1 Identifying Shared Computation for Continuous Queries

In contrast to traditional sharing strategy of just finding common subexpressions in nonstreaming systems, MQO in streaming databases has generally been accomplished through "indexing" potentially large sets of queries to efficiently process the incoming data. Different approaches are developed to cope with different types of streaming queries.

Join queries with no aggregate operators. These kinds of queries are identical except for individual predicates. Continuous adaptive continuous queries processing (CACQ) [27] proposes a solution of using the eddy to route tuples among all the continuous queries currently in the system. Eddy [28] is a typical query processing mechanism, which continuously reorders the application of pipelined operators in a query plan based on tuple-by-tuple granularity. Each tuple is extended to carry the "lineage" consisting of "steering" and "completion" vectors in the memory. Through encoding the work performed on a tuple's lineage, the operators from many queries can be applied in a single tuple. Meanwhile, an efficient predicate index called group filter is applied to reduce computation when selection predicates have commonalities. For instance, when a selection operator is inserted into a new query, its source field is checked whether it matches an already instantiated grouped filter. If the requirements are met, its predicate is merged into the filter; otherwise, a new filter will be created with a single predicate. To avoid redundancy, the scheme splits joins into unary operators called state modules (SteMs) [29] to achieve the sharing of state between joins in different queries. SteMs is a half-join that encapsulates a dictionary data structure over tuples from a table, and handles build and probe requests on that dictionary. CACQ is especially useful for access method adaptation by providing a shared data structure for materializing and probing the data accessed from a given table.

Join queries with varying windows. State-slice [30] is developed to solve the restriction on sharing of window-based join operators. Unlike CACQ [27] and PSoup [31], which design a special tuple storage structure based on the maximum window size for continuous queries, State-slice partitions multiple join queries into fine-grained window slices and form a chain to share the stateful operators. For example, a sliced one-way window join on streams A and B is denoted as $A[W^{\text{start}}, W^{\text{end}}] \overset{s}{\ltimes} B$, where stream A has a sliding window of range: $W^{\text{start}} - W^{\text{end}}$. The execution contains four steps: (1) insert, (2) cross-purge, (3) probe, and (4) propagate. When the new tuple b from stream B arrives, tuple a will be added into a sliding window $A[W^{\text{start}}, W^{\text{end}}]$. Then b will first purge the state of stream A with W^{end}, before probing is attempted. Selection can be pushed down into the middle of the join chain to avoid unnecessary probings. During the processing, the one-way window join between A and B is sliced as a sequence of pipelined N sliced joins, denoted as $A[W_0, W_1] \overset{s}{\ltimes} B$, $A[W_1, W_2] \overset{s}{\ltimes} B, \ldots, A[W_{N-1}, W_N] \overset{s}{\ltimes} B$. Based on a lemma to prove that the chain of sliced joins provides the complete join answer, the scheme shows that the join results of $A[W_i, W_j] \overset{s}{\ltimes} B$ and $A[W_j, W_k] \overset{s}{\ltimes} B$ are equivalent to the results of a regular sliding window join $A[W_i, W_k] \overset{s}{\ltimes} B$. The order of the join results is restored by the merge union operator. Meanwhile, binary window join denoted as $A[W_A^{\text{start}}, W_A^{\text{end}}] \bowtie B[W_B^{\text{start}}, W_B^{\text{end}}]$ can be viewed as a combination of two one-way sliced window joins. Each input tuple from stream A or B is captured as two reference copies, which can divide the join into two dependent processes: $A[W_A^{\text{start}}, W_A^{\text{end}}] \overset{s}{\ltimes} B$ and $A \overset{s}{\rtimes} B[W_B^{\text{start}}, W_B^{\text{end}}]$. Through adopting a copy-of-reference instead of a copy-of-object strategy, two copies of a tuple will not require double system resources. Because states of the sliced window joins in the chain are disjoint with each other, it will not waste extra state memory.

Aggregate queries with identical predicates and varying windows. Shared time slice (STS) [32] is a basic technique applied in TelegraphCQ [33]. With unequal periods, windows are stretched to the common period by repeating their slice vectors. Through chopping an input stream into nonoverlapping sets of contiguous tuples (called slices), the tuples can be locally combined to form partial aggregates, which can then be used to answer each individual aggregate query computed with the common sliced window.

Aggregate queries with identical windows and varying predicates. Shared data fragment (SDF) [32,34] is developed to divide a stream into disjoint groups of tuples (called fragments). Fragments are where all tuples behave identically with respect to predicates of the queries. During the processing, the tuples can then be aggregated to form partial fragment aggregates, which in turn can be processed to produce the results for the various queries. For cases where both windows and predicates can vary, shared data shard (SDS) [32,35] is proposed to combine STS and SDF innovatively. Shards are fragmented slices that partition an input dataset into chunks of tuples from a high level. During the processing, the approach uses a slice manager to be aware of the paired windows of each query and demarcate slice edges. Paired windows split a window into a "pair" of exactly two unequal slices, which are superior to paned windows as they can never lead to more slices. With a shared selection operator (e.g., GSFilter), the sliced augmented tuples are then sent to an SDF-style fragment manager to compute partial aggregates of shards and finally produce the appropriate per-query overlapping window aggregates.

Aggregate and join queries with varying windows and predicates. Krishnamurthy [35] integrates SDS with TULIP plans and proposes a two-phase solution. In the first phase, an arbitrary operator tree can be substituted in place of the GSFilter in the chain of operators of the SDS approach, as long as the tree also produces tuples augmented with lineage vectors. The second phase is to use the TULIP plan to search for the common parts through the operator tree using lineage vectors. However, such kind of queries still requires higher computational complexity. Therefore, another randomized sampling strategy has been put forward in the preprocessing stage to improve the efficiency through reducing the precision.

7.3.2 Constructing Optimal Execution Plans for Multiple Queries

Five types of techniques are used for optimal plan construction as follows:

Cost-based method. In terms of share query plans, incremental group strategy deals with query asynchrony by adding new query incrementally to the existing query plan with heuristic local optimization. Not just like a naive approach for grouping continuous queries that apply these methods directly by reoptimizing all queries once a new query is added, Chen et al. [36] consider the existing groups as potential optimization choices by using either cost-based heuristics or a slightly modified cost-based query optimizer. In the cost model for incremental group optimization, the cost of maintaining materialized views is included since the intermediate query results are materialized. Groups are created for the existing queries according to their expression signatures, which represent similar structures among the queries. For example, with regard to three grouping strategies: PushDown, PullUp, and filtered PullUp, a selection expression signature represents the same syntax structure of a selection predicate with potentially different constant values. In general, a group consists of three parts: a group signature, a group constant table, and a group plan. Each part is defined as follows: The group signature is the common expression signature of all queries in the group, the group constant table contains the distinct signature constants of all queries, and the group plan is the query plan shared by all queries. Through employing a query-split scheme, the system can add a significant part of the query and avoid redundancy, which is also very significant to analytical cost models.

Sketch-based method. Based on randomized sampling on data stream tuples, Cesario et al. [37] propose an approach to solve the MQO problem. It computes sketch sharing and space allocation so as to process each stream concurrently. With the definition of frequency moments, sketch can be regarded as special-purpose lossy data compression using four-wise independent generating schemes. Given a query workload $\ell = \{Q_1, Q_2, ..., Q_q\}$ comprising multijoin COUNT aggregate queries, there can be large well-formed join graphs for ℓ. The output of count query Q_{count} is to calculate the number of tuples that satisfy the constraints from $R_1, ..., R_n$'s vector product. Therefore, the solution of multiple-query processing is to build disjoint join graphs $\zeta(Q_i)$ for each query $Q_i \in \ell$ and construct independent atomic sketches for the vertices of each $\zeta(Q_i)$. Based on Karush–Kuhn–Tucker (KKT) optimality conditions, Cesario et al. [37] gives an approach to partition the join graph into equivalence classes, which can solve the convex optimization problem in the graph and minimize the query error. Through sharing atomic-sketch computations among the vertices for stream R_i $(i = 1, ..., n)$, it can reduce the space requirement and avoid the drawback of ignoring a relation r_i that appears in multiple queries.

Pipelining method. For sliced or fragmented stream tuples, pipelining schedules are useful to integrate them and improve the efficiency. Pipelining techniques, which require that all operators need to be operated in an unblocked fashion, provide real-time response for continuous queries. Therefore, Dalvi et al. [38] present a general model for pipelining schedules and determine validity with a necessary and sufficient condition. Assume that the input is a multiple-query graph. A pipelining schedule is a Plan-DAG with each edge labeled either pipelined or materialized. Through a heuristic greedy algorithm, the least cost pipeline schedule can be obtained to realize shared read optimization.

Routing policy method. In terms of predicate index and physical operators, eddy [28] is an efficient routing policy to continuously reorder the application of pipelined operators in a query plan. Given a set of input streams, the approach routes tuples of each stream to operators and the operators run as independent threads to return tuples to the eddy. Once the tuples have been handled by all the operators, the eddy will send the result to the output. During query processing, three properties of run-time fluctuations, respectively, are the costs of operators, their selectivities, and the rates at which the tuples arrive from the inputs. The implementation used two ideas for routing: the first approach, called Backpressure, limits the size of the input queues of operators, capping the rate at which the eddy can route tuples to slow operators. This causes more tuples to be routed to fast operators early in query execution. The second approach augments backpressure with a ticket scheme, whereby the eddy gives a ticket to an operator whenever it consumes a tuple and takes a ticket away whenever it sends a tuple back to the eddy. In this way, higher selectivity operators accumulate more tickets. The priority scheme learning the varied selectivity is implemented via lottery scheduling, which is a novel randomized resource allocation mechanism and probabilistically fair. Each time the eddy gives a tuple to an operator, it credits the operator one "ticket." A single physical ticket may represent any number of logical tickets. This is similar to monetary notes, which may be issued in different denominations. Lottery tickets encapsulate resource rights that are abstract, relative, and uniform. When an eddy plans to send a tuple to be processed, it "holds a lottery" among the operators eligible for receiving the tuple. The number of lotteries won by a client has a binomial distribution. The chance of winning a lottery and receiving the tuple for an operator corresponds to the owned count of tickets. Meanwhile, by characterizing the moments of symmetry and the synchronization barriers, the eddy tracks an ordering of the operators that improves the overall efficiency using the lottery scheme.

Channel-based method. In order to adapt to varied workloads and build different implementation models, a rule-based MQO framework (RUMOR) [39] extends the rule-based query optimization and query plan-based processing model. Inspired by the classical query graph model (QGM) [1], the optimization process involves the application of transformation rules, which maps one query plan to another semantically equivalent plan. To model a set of operators with shared computation, RUMOR proposes an abstraction called physical multioperator (m-op). The state of m-op is a vector conceptually, which mainly executes all its operators from input stream and writes the output produced by the relevant operators to the corresponding output streams. Meanwhile, transformation rules on m-ops can be extended as multiple query transformation rules or m-rules for short, which consist of a pair of condition and action functions. The condition of an m-rule is a Boolean side effect-free function from the power set of all possible m-ops to true and false set. The action of an m-rule maps a set of m-ops to a single m-op, which is referred to as the target m-op. Therefore, a query plan is composed of m-ops, and the representative m-rules can be used to express the existing and new MQO techniques. Considering the sharing in the case where "similar" streams are processed by identical operators, RUMOR also defines *channels* to form them, which have compatible schemas. A channel is defined as the union of its streams, and each stream tuple has an additional attribute called membership component to label the source with a bit vector. When identical tuples from different streams are encoded as a single channel tuple, their space can be shared. When multiple streams are encoded into the same channel, the computation of operators can be shared. For example, considering a case that an m-op $\pi_{\{1,...,n\}}$ implements n projections on different input streams $\{S_1, ..., S_n\}$, n input streams are encoded by channel C, and n output streams are encoded by channel D. Then $\pi_{\{1,...,n\}}$ can perform projection only once for each input channel t from C and produce only one output channel tuple in D, keeping the membership component of t intact in the output D tuple. To decide the similar query plans encoded by channels, RUMOR builds the cost model and proposes a heuristic algorithm based on a set of criteria.

7.3.3 Optimization of Multiple Stream Queries in Distributed DSMSs

Partial push-down strategy. In terms of aggregate queries with varying windows in a scenario where the data sources are widely distributed and managed in a hierarchical system such as HiFi [40], partial push-down (PPD) [32] is developed to share the communication resources and promote efficiency. The technique first extracts the nonoverlapping parts of each query and then composes them to form a common subquery with query caching technique [41], which is used to speed up the access to remote data and also reduce the monetary costs of charge for access. A query submitted to the root node is optimized in a recursive fashion, where each node is only aware of its children. When a new query is sent to the cache, it checks to see if it can be answered using the cached results of earlier queries. Finally, the strategy dynamically adjusts the execution order that pulls up the overlapping parts of each query and pushes down the nonoverlapping common subquery using the partitioned parallelism of the data stream management systems (DSMS).

Ring-based strategy. In terms of join queries with varying windows, Wang et al. [30] provide a ring-based query plan that makes the state slicing and join ordering orthogonal. Ring structure is a virtual machine, which is formed by partitioning the states into disjoint slices in the time domain and distributes the fine-grained states in the cluster. The sliced join containing the window size that equals zero in the ring is called the head of the ring, and the one containing the largest end window is called the tail of the ring. The ring-based query plan is formed first by searching for a shortest path among all nodes and then uses the regular pipelined parallelism of the DSMS.

Two-tier strategy. In terms of queries in the sensor network, Shili et al. [42] proposes a two-tier phase scheme to minimize the average transmission time and the communication cost. Since sensor nodes are resource constrained, a lightweight but effective greedy algorithm is designed to support multiple queries running inside a wireless sensor network.

View tree strategy. In terms of multidimensional queries, Chandi [43] analyzes the interrelations between the cluster sets identified by queries with different parameter settings, including both pattern-specific and window-specific parameters. The share solution is to build a predicted view tree (PVT), which can integrate multiple predicted view hierarchies as branches into a single tree structure.

7.4 EXTENSIONS OF MULTIPLE-QUERY PROCESSING AND OPTIMIZATION TECHNIQUES TO MULTITENANT DATABASES

With the advent of the ubiquitous Internet, a new trend has emerged: cloud computing, which explains the phenomenon of the integration among multiple devices. From an enterprise perspective, a very modern form of application hosting is software-as-a-service (SaaS) motivation [44]. As opposed to traditional on-premises solutions, the way SaaS customers just need to pay the hosting provider a monthly fee, where service charges are paid for those really consumed resources. Based on the service maturity model, multitenancy is a significant paradigm shift to make configuring applications simple and easy for the customers, without incurring extra operation costs.

In the following, we discuss the extensions of the above multiple-query processing and optimization techniques to the domain of multitenant databases.

7.4.1 Sharing in Multitenant Query Processing

Queries for a single tenant have to contend with data from all tenants. However, previous query methods have been inefficient for multitenant databases because it is very difficult for such methods to understand or account for the unique characteristics of each tenant's data. While one tenant's data includes numerous short records with just fewer indexable fields, another may include fewer longer records with numerous indexable fields [44]. Apart from the structural differences, each tenant's data distribution may also be different compared with the similar schemas. This brings a challenge for existing relational databases that just gather an aggregate or average statistics of all tenants periodically. Therefore, the approach for MQO can lead to incorrect assumptions and query plans for any given tenant.

A natural way to ameliorate the problem is to share tables among tenants [45,46]. Through mapping multiple single-tenant logical schemas to one multitenant physical schema using query transformation rules, the logical tables can be divided into fixed generic structures, such as universal and pivot tables, to avoid the interference with each tenant's ability. For each table, queries are generated to filter the correct columns and align the different chunk relations based on each TenantId. Then shared process offers bulk execution of administrative queries by allowing them to be parameterized over the domain of each table.

In addition, each tenant database may encounter various query expressions (QEs) over different data sources, such as relational and structured XML data. Therefore, a multigraph-based approach is proposed to introduce edges that navigate both the XML nodes and the relational dot notations. Through utilizing the intrasegment compression techniques and adding new edges, similar nodes can form a subgraph that consists of identical or subsumed conditions.

7.4.2 Multitenant Querying Plans

More efficient execution plans of multitenant databases are to adopt a two-phase solution with dynamic tuning of database indices [47]. A layer of meta-data associates the data items with tenants via tags and the meta-data are used to optimize searches by channeling processing resources during a query to only those pieces of data bearing relevant unique tag. In certain aspects, each tenant's virtual schema includes a variety of customizable fields, some or all of which may be designated as indexable. One goal of traditional multiple query optimizer is to minimize the amount of data that must be read from disk and choose selective tables or columns that will yield the fewest rows during the processing. If the optimizer knows that a certain column has a very high cardinality, it will choose to use an index on that column instead of a similar index on a lower cardinality column. However, consider in a multitenant system that a physical column has a large number of distinct values for most tenants, but a small number of distinct values for specific tenant. Then, the overall high-cardinality column strategy will not get a better performance because the optimizer is unaware that for this specific tenant, the column is not selective. Furthermore, by using system-wide aggregate statistics, the optimizer might choose a query plan that is incorrect or inefficient for a single tenant that does not conform to the "normal" average of the entire database as determined from the gathered statistics. Therefore, the first phase typically includes generating tenant-level and user-level statistics to find the suitable tables or columns for the common subexpressions. The statistics gathered includes the information in entity rows for tenants being tracked to make decisions about query access paths and a list of users to have access to privileged data. The second phase constructs an optimal plan based on query graph. The difference is that some edges are labeled directed and single node consists of multiple relations considering the private security model to keep data or application separate. The common subexpressions of the first phase are stored by building many-to-many (MTM) physical table, which can also specify whether a user has access to a particular entity row. When handling multiple queries for entity rows that the current user can see, the optimizer must choose between accessing MTM table from the user and the entity side of the relationship.

7.5 CONCLUSION

In this chapter, we overviewed multiple-query processing and optimization techniques in traditional databases and streaming databases. We also discussed their possible extensions to multitenant multiple-query processing and optimization.

As an interesting future work, we view three major issues. First, without data integration engine in cloud computing environment, how to build a cost-based heuristic model to selectively materialize the candidate common subexpressions over diverse data source needs some efficient algorithms. Second, the recent studies focus on accurate query evaluation for multitenant database. It is worthwhile to study the approximate query processing and obtain error-energy trade-offs, especially for stream data. We would like to adapt nowadays techniques to multipath aggregation or join methods that can provide more fault tolerance. Third, there are still research issues to better employ schema knowledge or integrity constraints to perform query optimization at compile time. And it is very significant to detect "unsafe" queries considering data privacy for multi-tenant database.

REFERENCES

1. J. Grant and J. Minker. Optimization in deductive and conventional relational database systems. In *Advances in Data Base Theory*, H. Gallaire, J. Minker, and J. M. Nicholas (eds.). New York: Springer, pp. 195–234, 1981.
2. S. Finkelstein. Common expression analysis in database applications. *Proceedings of the 1982 ACM SIGMOD International Conference on Management of Data*. June 2–4, ACM Press, Orlando, FL, pp. 235–245, 1982.
3. P. Larson and H. Yang. Computing queries from derived relations. *Proceedings of the 11th International Conference on Very Large Data Bases*. August 21–23, Morgan Kaufmann, Stockholm, Sweden, pp. 259–269, 1985.
4. N. Roussopoulos. View indexing in relational databases. *ACM Transactions on Database System*, 7(2): 258–290, 1982.
5. N. Roussopoulos. The logical access path schema of a database. *IEEE Transactions on Software Engineering*, 8(6): 563–573, 1982.
6. T.K. Sellis. Multiple-query optimization. *ACM Transactions on Database System*, 13(1): 23–52, 1988.
7. P. Roy, S. Seshadri, S. Sudarshan, and S. Bhobe. Efficient and extensible algorithms for multi query optimization. *Proceedings of the 2000 ACM SIGMOD International Conference on Management of Data*. May 16–18, ACM Press, Dallas, TX, pp. 249–260, 2000.
8. M. Jarke. Common subexpression isolation in multiple query optimization. In *Query Processing in Database Systems*, W. Kim, D. S. Reiner, and D. S. Batory. Berlin: Springer-Verlag, 1985.
9. F.C. Fred Chen and M.H. Dunham. Common subexpression processing in multiple-query processing. *IEEE Transactions on Knowledge and Data Engineering*, 10(3): 493–499, 1998.
10. D.J. Rosenkrantz and H.B. Hunt. Processing conjunctive predicates and queries. *Proceedings of IEEE International Conference on Data Engineering*. 1980.

11. J. Park and A. Segev. Using common subexpressions to optimize multiple queries. *Proceedings of the IEEE International Conference on Data Engineering.* February 1–5, Los Angeles, CA, IEEE Computer Society, pp. 311–319, 1988.

12. P.V. Hall. Optimization of a single relational expression in a relational data base system. *IBM Journal of Research and Development*, 20(3): 244–257, 1976.

13. P.V. Hall. Common subexpression isolation in general algebraic systems. *Technical Report UKSC 0060*, IBM United Kingdom Scientific Centre, 1974.

14. U.S. Chakravarthy and J. Minker. Multiple query processing in deductive databases using query graphs. *Proceedings of the 12th International Conference on Very Large Data Bases.* August 25–28, Kyoto, Japan. ACM Press, pp. 384–391, 1986.

15. E. Wong and K. Youssefi. Decomposition: A strategy for query processing. *ACM Transactions on Database System*, 223–241, 1976.

16. U.S. Chakravarthy and A. Rosenthal. Anatomy of a modular multiplier query optimizer. *Proceedings of International Conference on Very Large Data Bases.* August 29–September 1, Los Angeles, CA: Morgan Kaufmann, 1988.

17. T. Sellis. Global query optimization. *Proceedings of the 1986 ACM SIGMOD International Conference on Management of Data.* May 28–30, Washington, DC, 1986.

18. T. Sellis and S. Ghosh. On the multiple query optimization problem. *IEEE Transactions on Knowledge and Data Engineering*, 2(2): 262–266, 1990.

19. E.-P. Lim, J. Srivastava, and A. Cosar. An extensive search for optimal multiple query plans. *International Conference on Management of Data.* June 2–5, San Diego, CA, 1992.

20. A. Cosar, J. Srivastava, and S. Shekhar. On the multiple pattern multiple object match problem. *International Conference on Management of Data.* May 29–31, Denver, CO, 1991.

21. A. Cosar, E.-P. Lim, and J. Srivastava. Multiple query optimization with depth-first branch-and-bound and dynamic query ordering. *Proceedings of the 2nd International Conference on Information and Knowledge Management.* Washington, DC. November 1–5, ACM, New York, pp. 433–438, 1993.

22. G. Graefe and W.J. McKenna. The volcano optimizer generator: Extensibility and efficient search. *Proceedings of the 9th International Conference on Data Engineering.* April 19–23, Vienna, IEEE Computer Society, pp. 209–218, 1993.

23. H. Mistry, P. Roy, S. Sudarshan, and K. Ramamritham. Materialized view selection and maintenance using multi-query optimization. *Proceedings of the 2001 ACM SIGMOD International Conference on Management of Data.* Santa Barbara, CA. May 21–24, ACM, New York, pp. 307–318, 2001.

24. A.B. Murat, H.T. Ismail, and C. Ahmet. Genetic algorithm for the multiple-query optimization problem. *IEEE Transactions on Systems, Man, and Cybernetics, Part C: Applications and Reviews*, 37(1): 147–153, 2007.

25. Z. Yi, L. Qing, and C. Lei. Multi-query optimization for distributed similarity query processing. *28th International Conference on Distributed Computing Systems.* June 17–20, Beijing, IEEE Computer Society, pp. 639–646, 2008.

26. J. Grant and J. Minker. On optimizing the evaluation of a set of expressions. *Technical Report TR-916*, University of Maryland, College Park, MD, July 1980.

27. S. Madden, M. Shah, J.M. Hellerstein, and V. Raman. Continuously adaptive continuous queries over streams. *Proceedings of the 2002 ACM SIGMOD International Conference on Management of Data*. Madison, WI. June 3–6, ACM, New York, pp. 49–60, 2002.

28. R. Avnur and J.M. Hellerstein. Eddies: Continuously adaptive query processing. *Proceedings of the 2000 ACM SIGMOD International Conference on Management of data*. Dallas, TX. May 14–19, ACM, New York, pp. 261–272, 2000.

29. R. Vijayshankar, A. Deshpande, and J.M. Hellerstein. Using state modules for adaptive query processing. *Proceedings of the 19th International Conference on Data Engineering*. March 5–8, Bangalore, India, IEEE Computer Society, pp. 353–364, 2003.

30. S. Wang, E. Rundensteiner, S. Ganguly, and S. Bhatnagar. State-slice: New paradigm of multi-query optimization of window-based stream queries. *Proceedings of the 32nd International Conference on Very Large Data Bases*. September 12–15, ACM Press, Seoul, Republic of Korea, pp. 619–630, 2006.

31. S. Chandrasekaran and M.J. Franklin. PSoup: A system for streaming queries over streaming data. *The VLDB Journal*, 12(2): 140–156, 2003.

32. S. Krishnamurthy, C. Wu, and M.J. Franklin. On-the-fly sharing for streamed aggregation. *Proceedings of the 2006 ACM SIGMOD International Conference on Management of Data*. June 27–29, ACM Press, Chicago, IL, pp. 623–634, 2006.

33. S. Chandrasekaran, O. Cooper, and A. Deshpande. TelegraphCQ: Continuous dataflow processing. *Proceedings of the 2003 ACM SIGMOD International Conference on Management of Data*. San Diego, CA. June 9–12, ACM, New York, pp. 668–674, 2003.

34. R. Zhang, N. Koudas, B.C. Ooi, and D. Srivastava. Multiple aggregations over data streams. *Proceedings of the 2005 ACM SIGMOD International Conference on Management of Data*. Baltimore, MD. June 13–16, ACM, New York, pp. 299–310, 2005.

35. S. Krishnamurthy. Shared query processing in data streaming systems. University of California, Berkeley, CA, 2006.

36. J. Chen, D.J. DeWitt, and J.F. Naughton. Design and evaluation of alternative selection placement strategies in optimizing continuous queries. *Proceedings of the 18th International Conference on Data Engineering*. February 26–March 1, San Jose, CA, IEEE Computer Society, pp. 345–356, 2002.

37. E. Cesario, A. Grillo, C. Mastroianni, and D. Talia. A sketch-based architecture for mining frequent items and itemsets from distributed data streams. *2011 11th IEEE/ACM International Symposium on Cluster, Cloud and Grid Computing*. May 23–26, Newport Beach, CA, IEEE Computer Society, pp. 245–253, 2011.

38. N.N. Dalvi, S.K. Sanghai, P. Roy, and S. Sudarshan. Pipelining in multi-query optimization. *Proceedings of the 20th ACM SIGMOD-SIGACT-SIGART Symposium on Principles of Database Systems.* Santa Barbara, CA. May 21–24, ACM, New York, pp. 59–70, 2001.

39. M. Hong, M. Riedewald, C. Koch, J. Gehrke, and A. Demers. Rule-based multi-query optimization. *Proceedings of the 12th International Conference on Extending Database Technology: Advances in Database Technology.* Saint Petersburg, Russia. March 24–26, ACM, New York, pp. 120–131, 2009.

40. O. Cooper, A. Edakkunni, and M.J. Franklin. HiFi: A unified architecture for high fan-in systems. *Proceedings of the 30th International Conference on Very Large Data Bases.* August 31–September 3, Morgan Kaufmann, Toronto, ON, pp. 1357–1360, 2004.

41. D. Kossmann, M.J. Franklin, and G. Drasch. Cache investment: integrating query optimization and distributed data placement. *ACM Transactions on Database Systems,* 25(4): 517–558, 2000.

42. X. Shili, B.L. Hock, T. Kian-Lee, and Z. Yongluan. Two-tier multiple query optimization for sensor networks. *27th International Conference on Distributed Computing Systems.* June 25–29, Toronto, ON, IEEE Computer Society, pp. 39–49, 2007.

43. A.P. Boedihardjo, C.-T. Lu, and F. Chen. A framework for estimating complex probability density structures in data streams. *Proceedings of the 17th ACM Conference on Information and Knowledge Management,* Napa Valley, CA. October 26–30, ACM, New York, pp. 619–628, 2008.

44. H. Mei, J. Dawei, L. Guoliang, and Z. Yuan. Supporting database applications as a service. *IEEE 25th International Conference on Data Engineering.* March 29–April 2, Shanghai, China, IEEE Computer Society, pp. 832–843, 2009.

45. S. Aulbach, T. Grust, and D. Jacobs. Multi-tenant databases for software as a service: Schema-mapping techniques. *Proceedings of the 2008 ACM SIGMOD International Conference on Management of Data.* Vancouver, BC. June 10–12, ACM, New York, pp. 1195–1206, 2008.

46. F.S. Foping, I.M. Dokas, and J. Feehan. A new hybrid schema-sharing technique for multitenant applications. *4th International Conference on Digital Information Management.* November 1–4, Michigan, IEEE Computer Society, pp. 1–6, 2009.

47. C. Weissman and S. Wong. Query optimization in a multi-tenant database system. US Patent 7,529,728 B2, salesforce.com, 2009.

CHAPTER 8

Large-Scale Correlation-Based Semantic Classification Using MapReduce

Fausto C. Fleites, Hsin-Yu Ha,
Yimin Yang, and Shu-Ching Chen

Florida International University
Miami, Florida

CONTENTS

8.1 INTRODUCTION

The ubiquitous reach of social sites coupled with the proliferation of mobile devices has brought forth an explosion in multimedia data. This fact has motivated the research community to develop systems that allow the meaningful retrieval of multimedia data. One of the requirements of the meaningful retrieval is the understanding of the semantics embedded in the data. However, understanding of the semantics is a very challenging task. The reason is the well-known problem of the semantic gap between low-level features and high-level semantic concepts (Shyu et al. 2007). Multimedia data are usually modeled using low-level features such as color, shape, and texture information, but such information may not be discriminative enough at the concept level. Two images with similar low-level features may represent different semantic concepts. For example, querying for an image with a cloudy sky may return an unrelated image depicting a blue car parked in front of a white house. Moreover, a textual representation, for example, tags and file names of the semantics, is not a feasible solution. Tagging requires a significant amount of human involvement that is prone to errors and inconsistent labeling (Shyu et al. 2007).

Content-based multimedia information retrieval (CBMIR) focuses on the understanding of the semantics and the retrieval of multimedia data. One important task in the realm of CBMIR is semantic concept detection, the purpose of which is to classify multimedia data into semantic concepts. Its importance is highlighted by the TRECVID conference series sponsored by the National Institute of Standards and Technology (NIST 2012; Over et al. 2012; Smeaton et al. 2009). Usually, this task involves training a classifier using ground truth data, followed by the classification of unlabeled test data. There are several classifier options proposed in the relevant literature, but in this work, we utilize multiple correspondence analysis (MCA) as it has been utilized not only as an effective classification method (Lin et al. 2008) but also as a discretization (Zhu et al. 2012), feature selection (Zhu et al. 2010), data pruning (Lin et al. 2009), and ranking mechanism (Chen et al. 2012). Based on the correlation information, MCA is a data analysis technique that allows us to find correspondences between feature values and a target concept that are helpful in bridging the gap between low-level features and high-level semantic concepts.

Nevertheless, even though MCA has shown its effectiveness in CBMIR, its direct application to big data is not scalable. To train an MCA classification model, it is required to manipulate large matrices extracted from the training data, which impedes the useful utilization of MCA in today's pervasive big data environments. Existing works that utilize MCA for CBMIR tasks do not take into account scalability problems that arise when processing large amounts of data nor provide a framework that effectively utilizes multiple computers to speed up processing. The pertinent question is then how to improve the scalability of MCA and bring it onto the big data scale.

In the domain of big data, MapReduce (MR; Dean and Ghemawat 2008) is the framework of choice for data-intensive applications. It provides an easy-to-use programming model and supporting processing framework for large-scale distributed applications and is actively used by top technology companies to process big data. Recent works in the literature have shown that MR can be utilized to scale CBMIR tasks, such as semantic classification (Basilico et al. 2011; Palit and Reddy 2010; Panda et al. 2009; Wu et al. 2009; Yang et al. 2009; Zhao et al. 2012), retrieval (Raj and Mala 2012; Zhang et al. 2010), and feature extraction (Wang et al. 2012; White et al. 2010).

In this work, we propose an MR-based MCA classification framework to support CBMIR in big data environments. The goal of the framework is to bring the usefulness of MCA in discretization, feature selection, data pruning, and classification tasks to large-scale CBMIR. The proposed system leverages the MR framework to provide distributed processing in a cluster of computers and presents a novel way of building MCA models that eliminates the need to process large matrices in memory. To the best of our knowledge, this is the first attempt to implement MCA as a CBMIR mechanism to process large-scale multimedia data using MR. Moreover, we show the usefulness of the system in big data environments by providing experiments that demonstrate the scalability of the system in the task of semantic concept classification.

The work in this chapter is organized as follows: Section 8.2 describes the related work. Section 8.3 introduces MCA and describes how it is used as a classifier. Section 8.4 details the implementation of the proposed MR-based MCA classification framework. Section 8.5 presents the experiments and results. Finally, Section 8.6 concludes the work.

8.2 RELATED WORK

Related literature to the work presented in this chapter consists of (1) previous works that apply MCA to CBMIR tasks and (2) recent works that apply the MR framework in CBMIR tasks. In CBMIR tasks, MCA has been successfully utilized for discretization (Zhu et al. 2011, 2012), feature selection (Zhu et al. 2010), data pruning (Lin et al. 2009), classification (Lin et al. 2008; Yang et al. 2011, 2012; Zhu et al. 2013), and ranking (Chen et al. 2012; Lin and Shyu 2010). In this section, we review relevant examples. Such works demonstrate the usefulness of MCA in CBMIR and show the importance of having a framework that can support the same tasks in a big data environment. With regard to big data, recent works (Basilico et al. 2011; Panda et al. 2009; Yang et al. 2009) apply the MR framework to classification tasks. These works utilize classification/modeling methods different from MCA and their MR implementations are orthogonal to the work presented herein.

Zhu et al. (2012) propose a discretization mechanism that discretizes numerical data based on the criteria of maximizing the correlation between feature intervals and class labels, utilizing MCA to measure the correlation between the feature values and the class labels. Their discretization strategy follows a similar recursive pattern to that of decision tree building. They compare the MCA-based discretization method against four other discretization methods on six classifiers and show their method producing the best discretization in terms of the final classification results.

Lin et al. (2008, 2009) utilize MCA for data pruning, an important task when the training data are imbalanced, and classification. They propose a framework that prunes training data based on the instances' transaction weights. Computed using MCA, these weights consider the correlation information from the feature value pairs (feature-level discretization intervals) and the class labels. The pruning threshold is obtained based on an iterative process that selects the best threshold based on the F1 score. They validate the MCA-based data pruning framework by noting the improvement in accuracy on four classifiers using the pruned training data. As a classification method, Lin et al. (2008) utilize correlation information from MCA to generate classification rules. The classification model utilizes the correlation coefficients between the feature values and the concept labels to obtain classification rules that map a feature value to a concept label. For an unknown data instance, the final classification result is given by a majority vote on the class labels generated by the rules corresponding to the feature values of the instance. They compare the classification performance of the MCA-based classifier against decision trees, support vector machine (SVM), and naive Bayes—three of the most popular classifiers.

Chen et al. (2012) demonstrate MCA as a re-ranking mechanism. Their proposed method utilizes relationships between semantic concepts to re-rank classification results. Based on the correlation information obtained from MCA, such relationships between semantic concepts are categorized as inclusive and exclusive relationships. The former represents high co-occurrence between a target semantic concept and a reference semantic concept, that is, an inclusive relationship means that the two semantic concepts are likely to appear together. The exclusive relationship represents low or nonexisting co-occurrence, that is, the appearance of one concept indicates a low chance for the appearance of the other concept. Using the average precision measure, they compare the MCA-based re-ranking method against the ranking produced by the subspace model proposed by Lin and Shyu (2010).

Yang et al. (2009) present an MR-based, semantic modeling framework called robust subspace bagging (RB-SBag) algorithm. The RB-SBag is an ensemble learning method that combines random subspace bagging with forward model selection. The latter is utilized to create composite classifiers based on the most effective base models. They compare the RB-SBag method with SVM and show RB-SBag features a speedup by an order of magnitude on learning with competitive performance. The MR-based implementation consists of a two-state MR process. In the first stage, an MR job partitions the training data and builds a pool of base models, without using a reduce function. In the second-stage MR job, the map function computes the classification results of the base models on a validation set and conducts forward model selection. The reduce function then combines the selected base models with composite classifiers.

Panda et al. (2009) introduce an MR-based framework for learning tree models over large datasets. Their framework is called PLANET, which stands for parallel learner for assembling numerous ensemble trees. To construct one tree model over the entire training data, PLANET basically performs several iterations of MR jobs, where each job computes the node splits for the current tree model over the training data. The first MR job in the iteration receives an empty tree model, so it computes the best split for the top node of the tree model. A controller thread schedules the execution of MR jobs. Moreover, PLANET can use boosting or bagging to build ensemble of tree models by instructing the controller thread to schedule more than one tree model. They showcase the scalability of PLANET against in-memory implementation of tree models.

Basilico et al. (2011) propose an MR-based, mega-ensemble learning method that combines multiple random forests. The framework is called COMET, which stands for cloud of massive ensemble trees. Different from

PLANET, COMET is a single-pass MR framework. It is able to just use one MR pass because the tree models are built on partitions of the training data, not the entire training dataset. In the map phase of the MR job, COMET builds a random forest on each partition of the training dataset and combines them into mega-ensemble in the reduce phase. They compare COMET with serially built random forests on large datasets and show that COMET compares favorably in both accuracy and training time.

8.3 MULTIPLE CORRESPONDENCE ANALYSIS

Being a natural extension of the standard correspondence analysis to more than two variables, MCA is an exploratory data analytic technique designed to analyze multiway tables for some measures of correspondence between the rows and the columns (Greenacre and Blaslus 2006). In this section, we explain the inner workings of MCA and how it has been used for classification (Lin et al. 2008).

The observations used for MCA are a set of nominal values. For the task at hand, each feature in which the multimedia data instances are represented, which usually consists of numerical values, is discretized into several intervals that represent nominal values. We refer to these intervals as feature values. For example, the multimedia instances shown in Table 8.1 are numerically represented by two features F^1 and F^2, whose class labels are either C_1 or C_2, and Table 8.2 shows the discretization results. The discretized values are denoted by F_k^j, where F^j represents the jth feature, F_k^j the kth value of F^j, and $\{F_k\}^j$ denotes the set of values of feature F^j. Let $|F^j|$ and $|C|$ denote the number of discretized values in feature F^j and the class C, respectively. Having the instances discretized, MCA is applied for each feature and class combination. The discretized instances are utilized to construct an indicator matrix Y, whose dimensions are $N \times (|F^j| + |C|)$, where N is the number of instances. The indicator matrix maps the instances to a feature value-based binary representation. An example derived from Table 8.2 is shown in Table 8.3.

TABLE 8.1   Example of Multimedia Data Representation

Instance Id	Feature F^1	Feature F^2	Class C
1	0.212	0.190	C_1
2	0.256	0.798	C_1
3	0.173	0.125	C_2
4	0.141	0.972	C_2

TABLE 8.2   Example of Multimedia Data Representation after Discretization

Instance Id	Feature F^1	Feature F^2	Class C
1	F_2^1	F_1^2	C_1
2	F_2^1	F_2^2	C_1
3	F_1^1	F_1^2	C_2
4	F_1^1	F_2^2	C_2

TABLE 8.3   Example of Indicator Matrix

Instance Id	F_1^1	F_2^1	F_1^2	F_2^2	C_1	C_2
1	0	1	1	0	1	0
2	0	1	0	1	1	0
3	1	0	1	0	0	1
4	1	0	0	1	0	1

The ensuing MCA analysis is based on the square matrix $B = Y^T Y$, called the Burt matrix, whose dimensions are $(|F^j| + |C|) \times (|F^j| + |C|)$ and the elements are denoted as b_{ij}. Let $n = \Sigma b_{ij}$ be the grand total of the Burt matrix, $Z = B/n$ the probability matrix derived from B whose elements are z_{ij}, M the $1 \times (|F^j| + |C|)$ mass matrix where $m_{1j} = \Sigma_i z_{ij}$, and D the $(|F^j| + |C|) \times (|F^j| + |C|)$ diagonal mass matrix where $d_{ii} = m_{1i}$ and zero otherwise. Since B is symmetric, the solution for the rows and columns is identical, and the analysis stemming from B corresponds to the columns of the indicator matrix. MCA then computes the singular value decomposition (SVD) of the following normalized chi-square distance matrix:

$$A = D^{-\frac{1}{2}}(Z - M^T M)(D^T)^{-\frac{1}{2}} = P \Sigma Q^T \qquad (8.1)$$

where:
Σ is the diagonal matrix containing the singular values
Σ^2 is the matrix of the eigenvalues
The columns of P are the left singular vectors
The columns of Q are the right singular vectors

Since over 95% of the total variance encoded in A can be captured by the first two principal components, we can project A on its first two principal components and analyze the correlations between the feature values in this 2D space. The closer a feature value is to a class label, the more correlation there is between the feature value and the class label. We can measure such correlations between feature values and class labels by computing their inner product, that is, the cosine of the angle between each feature value and a class label. The larger the cosine value of the angle between a feature value and a class label, the stronger the correlation between them.

8.3.1 MCA-Based Classification

Having explained how MCA is used to find correspondences between the discretized feature values and the class labels of a multimedia dataset, we now explain how such information has been used for classification in previous works (Lin and Shyu 2010; Lin et al. 2008) and describe the reason why a direct application of MCA is unfeasible in big data environments. Such an explanation is necessary to understand our proposed MR-based MCA classification system. Since this work is oriented toward the task of semantic detection, we assume a training dataset with ground truth information is provided, and the goal is to determine the class labels for the instances of a test dataset. Moreover, we consider the multimedia data that have already been consistently discretized, that is, the same cut points used to discretize the training instances have been applied to the test instances.

First, the MCA-based semantic classification proceeds by obtaining the cosines of the angles between $\{F_k\}^j$ and C for each feature F^j in the training dataset. Let $w_{k,l}^j$ denote the cosine of the angle between the feature value F_k^j and the class label C_l. Each $w_{k,l}^j$ is termed an MCA-based classification rule. Second, the class label for a test instance i with feature values $\{\{F_k\}^j\}_{j=1}^F$, where F is the number of features, is estimated by $\arg \max_{C_l}(\{\sum_{j,k} w_{k,l}^j\}_l)$, that is, a majority vote is taken on the class, which yields the maximum sum of cosine values with respect to the feature values of i. We term the set of MCA-based classification rules as the MCA model, which constitutes the MCA-based classifier. In addition, for ease of description, we term the MCA-based classification process described so far as serial MCA (S-MCA) classification.

The described S-MCA classification process is not suitable for big data environments because the dimension indicator matrix Y is proportional to N, which is the number of data instances. In a big data environment, it is unfeasible to load Y into the memory and perform the MCA steps previously described. We term this obstacle the MCA scalability problem. To the best of our knowledge, this work is the first one that provides a solution to this problem using MR. The experimental results we provide support our claim.

8.4 MR MCA-BASED CLASSIFICATION

MR is a data-processing programming model and supporting framework introduced by Dean and Ghemawat (2008). Its open source implementation, Hadoop (Apache 2013), is utilized by a large number of companies for distributed data processing. Hadoop offers a distributed file system called Hadoop distributed file system (HDFS) on which applications store large data as well as an MR framework on which distributed applications are built. Furthermore, the MR framework abstracts the programmer from the details of parallelization, fault tolerance, data distribution, and load balancing, which greatly reduces the complexity of developing distributed applications.

As shown in Figure 8.1, the execution of an MR job consists of three phases: a map phase, a shuffle phase, and a reduce phase. The map and reduce phases execute multiple map and reduce tasks, respectively, where a task is an invocation of the map or reduce function. The MR programming paradigm is based on these two functions, which receive input and generate output in terms of key–value pairs. The map and reduce functions can be abstractly represented as follows:

$$map(K_1, V_1) \rightarrow list(K_2, V_2)$$

$$reduce(K_2, list(V_2)) \rightarrow list(K_3, V_3)$$

The processing of an input dataset begins with the map phase, in which the MR framework partitions the data into chunks called splits and invokes a map task to process each split. The size of a split is by default that of an HDFS block, although users can define their own criteria for splitting. The MR framework abstracts a split as a list of records, where a record is a

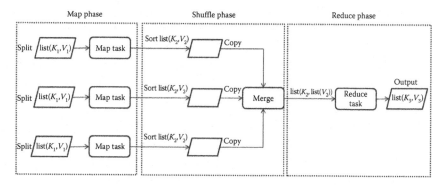

FIGURE 8.1   Execution of an MR job.

key–value pair $\langle K_1, V_1 \rangle$. After executing the map function, the map tasks generate intermediate key–value pairs $\langle K_2, V_2 \rangle$. The shuffle phase sorts the generated key–value pairs by key, copies them to the node where the reduce operation is performed, and merges the sorted pairs from all the map tasks. Subsequently, the reduce phase spawns one or more reduce tasks, which invoke the reduce function. If there is more than one reduce task, the shuffle phase invokes also a partition function to separate the keys between the reduce tasks. Each invocation of the reduce function receives as input $K_2, \text{list}(V_2)$, a map-generated key paired with all the map-generated values for the key. The output key–value pairs of the reduce tasks make up the output data of the MR job. Additionally, users can also specify a combine function of the form $\text{combine}(K_2, \text{list}(V_2)) \rightarrow \text{list}(K_2, V_2)$ that is used by the MR framework to reduce the data transfer between the map and reduce tasks. The combine function is run on the map output key–value pairs, and its output key–value pairs are used as input for the reduce function. It is worth noting that the combine function is an optimization feature of the MR framework, and there are no guarantees of how many times it will be called, if at all.

With the aforementioned description of the MR framework, we proceed to explain the MR-based MCA classification framework and our solution for the MCA scalability problem. We term our proposed framework as distributed MCA (D-MCA).

The S-MCA classification process previously described can be divided into model training and data classification procedures. The proposed D-MCA classification framework consists of two MR jobs, as shown in Figure 8.2. The first MR job carries the MCA model training to generate the MCA model, that is, the MCA-based classification rules that are used to classify the test data. Once the MCA model is built, the second MR job loads the MCA model as side data and classifies the test dataset, which consists of unknown data instances. We point out the steps of model training and data classification consist of a single-pass MR.

The scalability problem of S-MCA can be overcome by interpreting the b_{ij} values of the Burt matrix $B = Y^T Y$. The value of $b_{k_1 k_2}$, for some k_1, k_2, and feature F^j, is the number of data instances that have both $F_{k_1}^j$ and $F_{k_2}^j$. This interpretation can be proven since $b_{k_1 k_2} = \sum_{m=1}^{N} (y_{k_1 m}, y_{k_2 m})$, where $y_{k_1 m} = 1$ if and only if instance m has the feature value $F_{k_1}^j$ and $y_{k_2 m} = 1$ if and only if instance m has the feature value $F_{k_2}^j$; $y_{k_1 m}$ and $y_{k_2 m}$ are zero otherwise. Another insight is that for $F_{k_1}^j$ and $F_{k_2}^j$, $b_{k_1 k_2} = 0$ since the same instance can have $F_{k_1}^j$ or $F_{k_2}^j$ but not both simultaneously. With this result, our

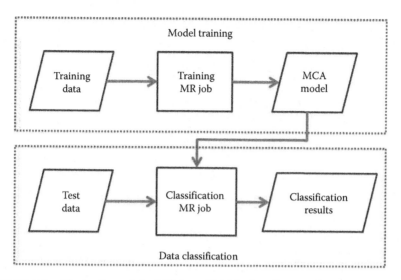

FIGURE 8.2   Illustration of the MR-based MCA classification framework.

proposal is to compute the values of the Burt matrix without having to rely on the indicator matrix Y. This can be achieved by iterating over the data instances and counting for each feature F^j the number of instances that have the combination F_k^j, C_l for all k's and j's. Our proposal can be efficiently implemented in MR because MR offers an efficient, distributed mechanism to process large datasets, and it is able to efficiently do the counts for the combinations F_k^j, C_l given suitable definitions for the key–value pairs.

8.4.1 Model Training

The map and reduce functions of the MR job that carries out the model training are depicted in Algorithms 8.1 and 8.3, respectively.

Algorithm 8.1: Model training map$(i, [\{\{F_k\}^j\}_{j=1}^F, C_l])$
1.   let $c = C_l$
2.   **for all** $F_k^j \in \{\{F_k\}^j\}_{j=1}^F$ **do** // iterate through the record's feature values
3.   output $\langle F^j, [F_k^j, c, 1] \rangle$
4.   **end for**

The map function's purpose is to transform the training data into a representation that makes easy the computation of the counts needed for the Burt matrix. The function receives as input pairs of the form

$\langle i, [\{\{F_k\}^j\}_{j=1}^F, C_l]\rangle$, where i is the instance id, $\{\{F_k\}^j\}_{j=1}^F$ is the set of feature values for the instance, and C_l is the class label of the instance. Subsequently, in lines 2–4, the map function outputs all the combinations of feature values and class labels for the instance to the MR framework. The output key–value pairs are of the form $\langle F^j, [F_k^j, c, 1]\rangle$, where the key represents the feature id of the jth feature. The value $[F_k^j, c, 1]$ is a tuple of three items that holds a feature value, the class label $c = C_l$, and the number 1, which signifies that the framework has "seen" one combination of the feature value and the class label.

With this key–value representation, the shuffle-phase MR framework will link all the values $[F_k^j, c, 1]$ across all the map tasks that correspond to feature F^j, efficiently allowing the reduce function to count all the data instances that have the combination F_k^j, c. The counts for the combination F_k^j, F_k^j can be obtained from the counts of $F_k^j c$ as the set of instances that have feature value F_k^j is the same as that for F_k^j, c, for all possible values of c. Furthermore, an invocation of the reduce function will process only the counts for a single feature as the keys generated by the map function and feature ids.

Algorithm 8.2: Model training combine(F^j, list($[F_k^j, C_l, n]$))

1. Create hash map H to aggregate counts
2. **for all** $[F_k^j, C_l, n] \in$ list($[F_k^j, C_l, n]$) **do** // iterate through the list of values
3. $\quad H[F_k^j, c] = H[F_k^j, c] + n$
4. **end for**
5. **for all** $[F_k^j, c] \in H$ **do** //iterate through the keys in H
6. $\quad$ Output $\langle F^j, [F_k^j, c, H[F_k^j, c]]\rangle$
7. **end**

Since the number of key–value pairs generated by the map tasks is significantly large for big datasets, we include a combine function that will aggregate the counts for the combinations of feature values and class labels. The combine function is depicted in Algorithm 8.2. It takes as input pairs of the form $\langle F^j, \text{list}([F_k^j, C_l, n])\rangle$, where list($[F_k^j, C_l, n]$) is a list of intermediate values that correspond to the jth feature. When the combine function is executed for key–value pairs directly generated by the map tasks, the value of n is 1. However, as previously mentioned, the MR framework can invoke the combine function with key–value pairs generated by previously combine invocations, and in this case, n holds previously aggregated values. In step 1, the function creates a hash map H used to efficiently aggregate the counts. Subsequently, in lines 2–4, the function iterates over the list of values and utilizes H to aggregate the counts for each combination of feature value and class label. Finally, in lines 5–7, the function iterates over the combinations of feature value and class label found in the list of values and outputs $\langle F^j, [F_k^j, c, H[F_k^j, c]] \rangle$, which represents the number of instances seen so far that have the combination F_k^j, c.

Algorithm 8.3: Model training reduce$(F^j, \text{list}([F_k^j, C_l, n]))$

1. Allocate Burt matrix B for F^j and C, with size $J \times J$, where $J = |F^j| + |C|$
2. **for all** $[F_k^j, C_l, n] \in \text{list}([F_k^j, C_l, n])$ **do** // iterate through the list of values
3.   let $c = C_l$
4.   set $B[F_k^j, F_k^j] = B[F_k^j, F_k^j] + n$
5.   set $B[F_k^j, c] = B[F_k^j, c] + n$
6.   set $B[c, c] = B[c, c] + n$
7.   set $B[c, F_k^j] = B[c, F_k^j] + n$
8. **end for**
9. Let n be the grand total of matrix B
10. Compute the probability matrix $Z = B / n$
11. Compute the $1 \times J$ mass matrix M
12. Compute the $J \times J$ diagonal mass matrix D
13. Compute the normalized chi-square distance matrix $A = D^{-1/2}(Z - M^T M)(D^T)^{-1/2}$
14. Perform SVD: $A = P \Sigma Q^T$
15. Project A onto its two principal components
16. Compute the similarity between every feature value and class label $W^j = \{w_{k,l}^j : 1 \le k \le |F^j|, 1 \le l \le |C|\}$, where $w_{k,l}^j$ is the inner product of F_k^j and C_l in the projected A matrix
17. Output $\langle F^j, W^j \rangle$

The reduce function, shown in Algorithm 8.3, obtains the MCA classification rules for a feature F^j. The function receives as input pairs of the form $\langle F^j, \text{list}([F_k^j, C_l, n]) \rangle$, which have the same meaning as in the combine function. First, in line 1, the function allocates the Burt matrix B for the current feature. This is possible as we can know the set of possible feature values for the feature and the class. As previously covered, the dimension of the Burt matrix is $(|F^j| + |C|) \times (|F^j| + |C|)$. Second, in lines 2–8, the function iterates through the list of input values and increments by n the Burt matrix cells corresponding to the combinations of F_k^j and c. Consequently, after step 8, the matrix B will be exactly as if we had computed it as $B = Y^T Y$. Notwithstanding, we computed B without manipulating $Y^T Y$ in the memory. In lines 9–16, the function performs the same computations of S-MCA after the computation of the Burt matrix. The reduce function finally outputs the MCA classification rules for the current feature. One benefit provided by the algorithm of the reduce function is that lines 9–16 follow the same steps as in S-MCA, except for the computation of the Burt matrix. Hence, it is possible to reuse most of the S-MCA code for these steps, which is beneficial as it allows using data mining libraries such as Weka (Witten et al. 2005).

We can then abstractly represent the map and reduce functions of the model training MR job as follows:

$$\text{map}\left(i, \left[\left\{\{F_k\}^j\right\}_{j=1}^F, C_l\right]\right) \rightarrow \text{list}(F^j, [F_k^j, C_l, 1])$$

$$\text{reduce}\left(F^j, \text{list}\left(\left[F_k^j, C_l, 1\right]\right)\right) \rightarrow \text{list}(F^j, W^j)$$

8.4.2 Data Classification

The MR job that carries out the classification of test data utilizes only the map function, which is depicted in Algorithm 8.4. Since this job does not utilize the reduce function, the output of the job is that of the map tasks. The function receives as input pairs of the form $\langle i, [\{\{F_k\}^j\}_{j=1}^F] \rangle$, where i is the test instance id and $\{\{F_k\}^j\}_{j=1}^F$ is the set of feature values for the instance. Before processing records from the input split, in line 1, the map tasks loads the MCA model generated by Algorithm 8.3 as side data and applies it to obtain the classification result for the test instance. In lines 2–6, the function iterates over the instance's feature values and over the class labels to compute the score s_l of class C_l. The final classification result, line 7, for the test instance is the class with the highest aggregated score.

Algorithm 8.4: Data classification map $(i, [\{\{F_k\}^j\}_{j=1}^F])$

1. Let $m = \{F^j, W^j\}_{j=1}^F$ be the pre-loaded classification model
2. **for all** $F_k^j \in \{\{F_k\}^j\}_{j=1}^F$ **do** // iterate through the record's feature values
3.   **for** $l = 1$ to $|C|$ **do** //iterate over the class labels
4. $s_l = s_l + \text{abs}(w_{k,l}^j)$

5.   **end for**
6.   **end for**
7.   Output $\langle i, \arg\max_{c_l}(\{s_l\}) \rangle$

It is worth mentioning that in case the classification task requires the ranking of the results according to the test instances' scores, we can efficiently accomplish this by changing the output key–value pairs of the map function to $\langle \arg\max_{c_l}(\{s_l\}), i \rangle$, which would cause the MR framework to sort the generated key–value pairs by classification score.

8.5 EXPERIMENTS AND RESULTS

To evaluate the proposed D-MCA classification framework, we conducted experiments that assess the scalability of system and compared it to the S-MCA. In this work, we did not evaluate the accuracy of the classification results as previous works (Chen et al. 2012; Lin and Shyu 2010; Lin et al. 2008, 2009; Zhu et al. 2013) have already proven the usefulness of MCA for CBMIR tasks; instead, our concern in this work is scaling MCA to big data environments. The scalability is measured separately on the model training and data classification tasks, comparing D-MCA with S-MCA. The S-MCA code utilized for the experiments is that used in the work of Lin et al. (2008), which is written in Java and utilizes the Weka (Witten et al. 2005) and JAMA libraries (NIST 2012). The D-MCA framework was implemented in Java using Hadoop 1.0.4. Except for the Burt matrix, steps 9–16 of Algorithm 8.3 were implemented using the S-MCA code.

The dataset we utilized in the experiments comes from the semantic indexing (SIN) task (Smeaton et al. 2009) of the TRECVID 2012 project (Over et al. 2012). We picked one concept that has 262,912 training instances and 137,327 test instances. We refer to these datasets as Train1× and Test1×, respectively. All instances were previously discretized using Weka's minimum description length (MDL) discretization method (Fayyad and Irani 1993). Each instance is represented by a sequential numerical id and 563 low-level visual features. In addition, the training instances have an extra feature that specifies the class labels. Since a few hundred thousand records do not make up a significantly large dataset, we created the datasets shown in Table 8.4. An increase factor of k means that we repeated each instance k times.

The cluster on which the D-MCA was executed consists of 10 nodes that allow the concurrent execution of 83 MR tasks simultaneously, considering the number of CPU cores in each node and available main memory.

TABLE 8.4 Datasets Created for Scalability Experiments

Increase Factor	Train Dataset	Number of Training Instances	Test Dataset	Number of Test Instances
1×	Train1×	262,912	Test1×	137,327
2×	Train2×	525,823	Test2×	274,654
4×	Train4×	1,051,645	Test4×	549,308
8×	Train8×	2,103,289	Test8×	1,098,616
16×	Train16×	4,206,577	Test16×	2,197,232
32×	Train32×	8,413,154	Test32×	4,394,464
64×	Train64×	16,826,308	Test64×	8,788,928
128×	Train128×	33,652,616	Test128×	17,577,856
256×	Train256×	67,305,232	Test256×	35,155,712
512×	Train512×	134,610,464	Test512×	70,311,424

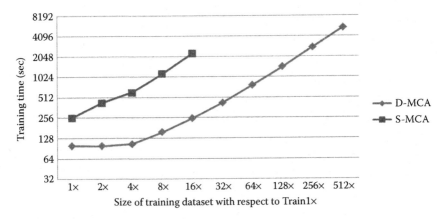

FIGURE 8.3 Training time vs. data size.

S-MCA was run in one of the nodes of the cluster. Moreover, in the following experiments, the number of reduce tasks was set to five.

The first experiment compares the training time in seconds between S-MCA and D-MCA as we increase the sizes of the training set. Figure 8.3 shows the results in a $\log_2$ scale. As expected, the training times increase linearly as we duplicate the size of the training datasets. Two observations can be drawn from Figure 8.3. First, the training times for S-MCA stop at Train16×. The reason is that S-MCA consumed the node's available main memory at Train16×. This fact supports our claim that S-MCA is not suitable for big data environments as it has to manipulate all the training data in the memory. D-MCA, however, utilizes a fixed amount of memory per map task. Second, the training times for D-MCA exhibit a linear increase,

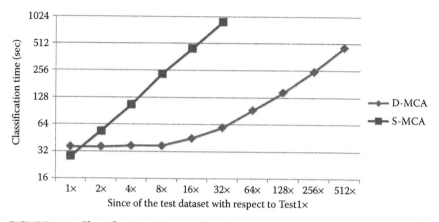

FIGURE 8.4   Classification time vs. data size.

except for Train1×, Train2×, and Train4×. The reason is that these three datasets were not large enough, and the number of map tasks required to process them was less than that of the task capacity of the cluster; hence, all the map tasks executed concurrently.

Similarly, the second experiment compares the classification time in seconds between S-MCA and D-MCA as we increase the sizes of the training set. Figure 8.4 shows the results in $\log_2$ scale. Echoing the results of the previous experiment, the classification times exhibit a linear relationship with the size of the test dataset. In this case, the numbers for S-MCA go up to Test32×, as this test dataset was the one that reached the memory limit. With regard to D-MCA, the classification times remain constant up to Test8× because Test1× has fewer instances than Train1×.

The third experiment evaluates the effect of the split size in the model training task as we increase the size of the training dataset. This experiment can yield valuable information with regard to optimizing the performance of model training. The results are shown in Figure 8.5 in $\log_2$ scale. The plot D-MCA-block-split uses the default split size, that is, the size of the block in HDFS, whereas the plot D-MCA-2.5K-split uses a split size that only contains 2500 input training instances. As such, a split of D-MCA-block-split is much larger than that of D-MCA-2.5K-split since the former holds ~162,000 instances. One interesting observation to be made from the results is that from Train1× through Train32×, the larger split size yields larger but constant training times (case 1), whereas from Train64× through Train512×, both split sizes yield training times

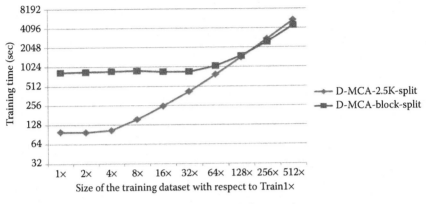

FIGURE 8.5   Training time vs. data size for different split sizes.

that increase linearly and quickly become virtually the same, and then D-MCA-block-split tends to produce lower training times (case 2). The reason behind case 1 is as follows: D-MCA-block-split uses less map tasks than the available 83 tasks; hence, the training time is relatively constant and is driven by how much time one map task takes to process its 162,000 training instances. However, the training times for D-MCA-2.5K-split increase linearly because D-MCA-2.5K-split uses much more map tasks than the available 83 tasks; therefore, all the available map slots in the MR cluster are constantly executing, and the total training time is proportional to the number of training instances. In case 2, both D-MCA-2.5K-split and D-MCA-block-split generate more map tasks than the available map slots, and the total training times are proportional to the number of training instances. However, the slope of D-MCA-2.5K-split becomes steeper than that of D-MCA-block-split because the former spawns much more map tasks than the latter, causing the MR framework to spend too much time on the setup and scheduling of map tasks.

Based on these results, it is recommended for smaller dataset to fully utilize the task capacity of the cluster by reducing the size of the splits. On the contrary, for very large datasets, the split size should not be small to avoid the aggregating overhead of task setup and scheduling. However, not only the size of the input dataset has to be considered but also the amount of key–value pairs generated by map tasks for deriving an optimal split size.

Figure 8.6 shows the effect of the split size in classification times as we increase the size of the test datasets. Cases 1 and 2 described for Figure 8.6 now appear at both sides of 64× of the initial dataset instead of at 32×.

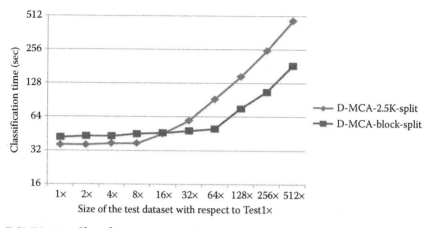

FIGURE 8.6   Classification time vs. data size for different split sizes.

The reason of this new "split" dataset is that test datasets are smaller for the same data increase level. The relevant observation to be made is that D-MCA-block-split yields lower times than D-MCA-2.5K-split after 16× in data classification, instead of 256× in model training. The reason behind this effect is the difference in the number of key–value pairs generated by map tasks in model training and data classification. For example, in model training, map tasks generated 563 key–value pairs for each input record, whereas map tasks in data classification generated 1 key–value pair for each input record. Consequently, the overhead caused by the setup and scheduling of map tasks for D-MCA-2.5K-split is much more significant with respect to the overhead of the shuffle phase in data classification than in model training.

8.6 CONCLUSIONS

This chapter has introduced an MR-based MCA framework for semantic classification of multimedia data. The previous applications of MCA in CBMIR tasks, such as discretization, feature selection, data pruning, classification, and ranking, cannot directly scale to big data environments because MCA requires the processing of large matrices in memory. The proposed D-MCA framework overcomes this scalability limitation by bypassing the processing of large matrices in the memory, which is achieved by counting combinations of attribute values in a distributed approach using MR. The CBMIR task chosen to showcase the proposed MCA framework is semantic classification. Additionally, the proposed distributed implementation of MCA lays the foundation for the application of MCA to the aforementioned CBMIR tasks. The experiment results show that the proposed MR-based MCA framework is suitable for big data environments.

ACKNOWLEDGMENTS

The research undertaken for this chapter was supported by the US Department of Homeland Security under grant award number 2010-ST-062-000039, the US Department of Homeland Security's VACCINE Center under award number 2009-ST-061-CI0001, and NSF HRD-0833093.

REFERENCES

Apache. *Hadoop.* 2013. http://hadoop.apache.org (accessed February 2013).

Basilico, Justin D., Arthur M. Munson, Tamara G. Kolda, Kevin R. Dixon, and Philip W. Kegelmeyer. "COMET: A recipe for learning and using large ensembles on massive data." *Proceedings of the 2011 IEEE International Conference on Data Mining.* December 11–14, IEEE, Vancouver, BC, pp. 41–50, 2011.

Chen, Chao, Lin Lin, and Mei-Ling Shyu. "Re-ranking algorithm for multimedia retrieval via utilization of inclusive and exclusive relationships between semantic concepts." *International Journal of Semantic Computing* 6(2): 135–154, 2012.

Dean, Jeffrey and Sanjay Ghemawat. "MapReduce: Simplified data processing on large clusters." *Communications of the ACM* 51(1): 107–113, 2008.

Fayyad, Usama M. and Keki B. Irani. "Multi-interval discretization of continuous valued attributes for classification learning." *13th International Joint Conference on Artificial Intelligence.* September 1, Morgan Kaufmann, Chambéry, France, pp. 1022–1027, 1993.

Greenacre, Michael and Jörg Blaslus. *Multiple Correspondence Analysis and Related Methods.* Boca Raton, FL: Chapman & Hall/CRC. 2006.

Lin, Lin, Guy Ravitz, Mei-Ling Shyu, and Shu-Ching Chen. "Correlation-based video semantic concept detection using multiple correspondence analysis." *IEEE International Symposium on Multimedia.* December 15–17, IEEE, Berkeley, CA, pp. 316–321, 2008.

Lin, Lin and Mei-Ling Shyu. "Correlation-based ranking for large-scale video concept retrieval." *International Journal of Multimedia Data Engineering and Management* 1(4): 60–74, 2010.

Lin, Lin, Mei-Ling Shyu, and Shu-Ching Chen. "Enhancing concept detection by pruning data with MCA-based transaction weights." *IEEE International Symposium on Multimedia.* December 14–16, IEEE, San Diego, CA, pp. 14–16, 2009.

NIST. *JAMA: A Java Matrix Package.* 2012. http://math.nist.gov/javanumerics/jama/ (accessed February 2013).

Over, Paul et al. "TRECVID 2012—An overview of the goals, tasks, data, evaluation mechanisms and metrics." *Proceedings of TRECVID 2012.* NIST, Gaithersburg, MA, 2012.

Palit, Indranil and Chandan K. Reddy. "Parallelized boosting with map-reduce." *Proceedings of the 2010 IEEE International Conference on Data Mining Workshops.* December 13, Sydney, NSW. IEEE Computer Society, Washington, DC, pp. 1346–1353, 2010.

Panda, Biswanath, Joshua S. Herbach, Sugato Basu, and Roberto J. Bayardo. "PLANET: Massively parallel learning of tree ensembles with MapReduce." *Proceedings of the VLDB Endowment* 2(2): 1426–1437, 2009.

Raj, Arockia Anand D. and T. Mala. "Cloud press: A next generation news retrieval system on the cloud." *2012 International Conference on Recent Advances in Computing and Software Systems.* April 25–27, IEEE, Chennai, India, pp. 299–304, 2012.

Shyu, Mei-Ling, Shu-Ching Chen, Qibin Sun, and Heather Yu. "Overview and future trends of multimedia research for content access and distribution." *International Journal of Semantic Computing* 1(1): 29–66, 2007.

Smeaton, Alan F., Paul Over, and Wessel Kraaij. "High-level feature detection from video in TRECVid: A 5-Year retrospective of achievements." In *Multimedia Content Analysis, Theory and Applications.* edited by A. Divakaran. Berlin: Springer-Verlag, pp. 151–174, 2009.

Wang, Hanli, Yun Shen, Lei Wang, Kuangtian Zhufeng, Wei Wang, and Cheng Cheng. "Large-scale multimedia data mining using MapReduce framework." *Proceedings of IEEE 4th International Conference on Cloud Computing Technology and Science.* December 3–6, IEEE, Taipei, Taiwan, pp. 287–292, 2012.

White, Brandyn, Tom Yeh, Jimmy Lin, and Larry Davis. "Web-scale computer vision using MapReduce for multimedia data mining." *Proceedings of the 10th International Workshop on Multimedia Data Mining.* July 25–28, Washington, DC, 2010.

Witten, Ian H., Eibe Frank, and Mark A. Hall. *Data Mining: Practical Machine Learning Tools and Techniques.* 2nd edition. Boston, MA: Morgan Kaufmann. 2005.

Wu, Gong-Qing, Hai-Guang Li, Xue-Gang Hu, Yuan-Jun Bi, Jing Zhang, and Xindong Wu. "MReC4.5: C4.5 ensemble classification with MapReduce." *Proceedings of the 4th ChinaGrid Annual Conference.* August 21–22, Yantai, China, IEEE Computer Society, Washington, DC, pp. 249–255, 2009.

Yang, Ron, Marc-Oliver Fleury, Michele Merler, Apostol Natsev, and John R. Smith. "Large-scale multimedia semantic concept modeling using robust subspace bagging and MapReduce." *Proceedings of the 1st ACM Workshop on Large-scale Multimedia Retrieval and Mining.* ACM, New York, pp. 35–42, 2009.

Yang, Yimin et al. "MADIS: A multimedia-aided disaster information integration system for emergency management." *8th IEEE International Conference on Collaborative Computing: Networking, Applications and Worksharing.* October 14–17, Pittsburgh, PA, 2012.

Yang, Yimin, Hsin-Yu Ha, Fausto Fleties, Shu-Ching Chen, and Steven Luis. "Hierarchical disaster image classification for situation report enhancement." *12th IEEE International Conference on Information Reuse and Integration.* August 3–5, IEEE, Las Vegas, NV, pp. 181–186, 2011.

Zhang, Jing, Xianglong Liu, Junwu Luo, and Bo Lang. "DIRS: Distributed image retrieval system based on MapReduce." *5th International Conference on Pervasive Computing and Applications.* December 1–3, IEEE, Maribor, Slovenia, pp. 93–98, 2010.

Zhao, Jun, Zhu Liang, and Yong Yang. "Parallelized incremental support vector machines based on MapReduce and bagging technique." *IEEE International Conference on Information Science and Technology.* March 23–25, Hubei, 2012.

Zhu, Qiusha, Lin Lin, and Mei-Ling Shyu. "Correlation maximization-based discretization for supervised classification." *International Journal of Business Intelligence and Data Mining* 7: 40–59, 2012.

Zhu, Qiusha, Lin Lin, Mei-Ling Shyu, and Shu-Ching Chen. "Feature selection using correlation and reliability based scoring metric for video semantic detection." *4th IEEE International Conference on Semantic Computing.* September 22–24, IEEE, Pittsburgh, PA, pp. 462–469, 2010.

Zhu, Qiusha, Lin Lin, Mei-Ling Shyu, and Shu-Ching Chen. "Effective supervised discretization for classification based on correlation maximization." *The 12th IEEE International Conference on Information Reuse and Integration.* August 3–5, IEEE, Las Vegas, NV, pp. 390–395, 2011.

Zhu, Qiusha, Mei-Lin Shyu, and Shu-Ching Chen. "Discriminative learning assisted video semantic concept classification." In *Multimedia Security: Watermarking, Steganography, and Forensics.* edited by Frank Snih. Boca Raton, FL: CRC Press, pp. 31–49, 2013.

CHAPTER 9

Efficient Join Query Processing on the Cloud

Xiaofei Zhang and Lei Chen

Hong Kong University of Science and Technology
Hong Kong, China

CONTENTS

JOIN QUERY IS ONE of the most expressive and expensive data analytic tools in traditional database systems. Along with the exponential growth of various data collections, NoSQL data storage has risen as the prevailing solution for big data. However, without the strong support of heavy index, the join operator becomes even more crucial and challenging for querying against or mining from massive data. As reported from Facebook [1] and Google [2], the underlying data volume is of hundreds of terabytes or even petabytes. In such scenarios, solutions from the traditional distributed or parallel databases are infeasible due to unsatisfactory scalability and poor fault tolerance. There have been intensive studies on different types of join operations over distributed data, for example, similarity join, set join, fuzzy join, all of which focus on efficient join query evaluation by exploring the massive parallelism of the MapReduce computing framework on the cloud platform. In this chapter, we explore the efficient processing of multiway generalized join queries, namely, the "complex join," which are widely employed in various practical data analytic scenarios, that is, querying resource description framework (RDF), feature selection from biochemical data, and so on. The substantial challenge of complex join lies in, given a number of processing units, mapping a complex join query to a number of parallel tasks and having them executed in a well-scheduled sequence such that the total processing time span is minimized. In this chapter, we focus on the evaluation of complex join queries on the cloud platform and elaborate with case studies on the efficient Simple Protocol and RDF Query Language (SPARQL) query processing and the multiway theta-join evaluation.

9.1 INTRODUCTION

Multiway join is an important and frequently used operation for numerous applications including knowledge discovery and data mining. Since the join operation is expensive, especially on large datasets and/or in multidimensions, multiway join becomes even more a costly operation. Although quite some researchers have devoted to the efficient evaluation of different types of pairwise joins, surprisingly few efforts have been made on the evaluation strategy of multiway join queries. Especially, with the fast increase in the scale of the input datasets, processing large data in a parallel and distributed fashion is becoming a popular practice. Even though a number of parallel algorithms for equi-joins in relational engines in MapReduce have been designed and implemented, there has been little work on parallel evaluation of multiway joins in large data, which is a challenging task and becoming increasingly essential as datasets continue to grow at an exponential rate.

When dealing with extreme-scale data, parallel and distributed computing using shared-nothing clusters, which typically consist of a number of commodity machines, is quickly becoming a dominating trend. MapReduce was introduced with the goal of providing a simple yet powerful parallel and distributed computing paradigm. The MapReduce architecture also provides good scalability and fault tolerance mechanisms. In the past few years, there has been an increasing support for MapReduce from both industry and academia, making it one of the most actively utilized frameworks for parallel and distributed processing of large data today.

Moreover, the (*key,value*)-based MapReduce programming model substantially guarantees great scalability and strong fault tolerance property. It has emerged as the most popular processing paradigm in a shared-nothing computing environment. Recently, devoting research efforts toward an efficient and effective analytic processing over immense data have been made within the MapReduce framework. Currently, the database community mainly focuses on two issues: (1) the transformation from certain relational algebra operator, such as similarity join, to its (*key,value*)-based parallel implementation and (2) the tuning or redesign of the transformation function such that the MapReduce job is executed more efficiently in terms of less time cost or computing resources consumption. Although various relational operators, such as pairwise theta-join, fuzzy join, and aggregation operators, are evaluated and implemented using MapReduce, there is little effort exploring the efficient processing of multiway join queries, especially more general computation, namely, theta-join, using MapReduce. The reason is that the problem involves more than just a *relational operator* → (*key,value*) pair transformation and the tuning. There are other critical issues to be addressed: (1) How many MapReduce jobs should we employ to evaluate the query? (2) What is each MapReduce job responsible for? (3) How should multiple MapReduce jobs be scheduled?

It is not trivial to efficiently process complex join queries using MapReduce. There are two challenging issues needed to be resolved. First, the number of available computing units is in fact limited, which is often neglected when mapping a task to a set of MapReduce jobs. Although the *pay-as-you-go* policy of cloud computing platform could promise as many computing resources as required, however, once a computing environment is established, the allowed maximum number of concurrent Map and Reduce tasks is fixed according to the system configuration. Even taking the *autoscaling* feature of Amazon Elastic Compute Cloud (EC2) platform [3] into consideration, the maximum number of involved computing units is predetermined by the user-defined profiles. Therefore, with the user-specified Reduce task number, a multiway theta-join query is processed with only a limited number of available computing units.

The second challenge is that the decomposition of a multiway theta-join query into a number of MapReduce tasks is nontrivial. The work of Wu et al. [4] targets at the multiway equi-join processing. It decomposes a query into several MapReduce jobs and schedules the execution based on a specific cost model. However, it only considers the pairwise join as the basic scheduling unit. In other words, it follows the traditional multiway join processing methodology, which evaluates the query with a sequence of pairwise joins. This methodology excludes the possible optimization opportunity to evaluate a multiway join in one MapReduce job. Our observation is that, under certain conditions, evaluating a multiway join with one MapReduce job is much more efficient than with a sequence of MapReduce jobs conducting pairwise joins. The work of Lee et al. [5] reports the same observation. One dominating reason is that the I/O costs of intermediate results generated by multiple MapReduce jobs may become unacceptable overheads. The work of Afrati and Ullman [6] presents the solution for evaluating a multiway join in one MapReduce job, which only works for the equi-join case. Since the theta-join cannot be answered by simply making the join attribute the partition key, the solution proposed in Reference 6 cannot be extended to solve the case of multiway theta-joins. The work of Okcan et al. [7] demonstrates an effective pairwise theta-join processing using MapReduce. However, its solution is designed for partitioning a two-dimensional result space formed by the cross-product of two relations. In the case of multiway join, the result space is a hypercube, the dimensionality of which is the number of the relations involved in the query. The solution in Reference 7 cannot be intuitively extended to handle the partition in high dimensions.

Moreover, the question about whether we should evaluate a complex query with a single MapReduce job or several MapReduce jobs is not clear yet. Therefore, there is no straightforward solution for combining the techniques in the existing literatures to evaluate a multiway theta-join query.

Meanwhile, assume a set of MapReduce jobs are generated for the query evaluation. Given a limited number of processing units, it remains a challenge to schedule the execution of MapReduce jobs such that the query can be answered with the minimum time span. These jobs may have dependency relationships and intercompetition for resource consumptions during the concurrent execution. Currently, the MapReduce framework requires the number of Reduce tasks as a user-specified input. Thus, after decomposing a multiway theta-join query into a number of MapReduce jobs, one challenging issue is how to specify each job a proper Reduce task number such that the overall scheduling achieves the minimum execution time span.

9.2 BACKGROUND

9.2.1 MapReduce Essentials

MapReduce is a programming framework introduced by Google to perform parallel computation on very large datasets, for example, crawled documents, or Web request logs. Large amount of data requires that it is distributed over a large number of machines (nodes). Each participating node contributes storage sources and all are connected under the same distributed file system. Additionally, each machine performs the same computations over the same locally stored data, which results in large-scale distributed and parallel processing. The MapReduce framework takes care of all the underlying details regarding parallelization, data distribution, and network partition tolerant while the user can concern only about the local computations executed on every machine. These computations are divided into two phases: the Map phase and the Reduce phase. The nodes assigned with the Map phase takes their local data as input and processes it producing intermediate results that are stored locally. The nodes performing Reduce computation receive the Map intermediate outputs, and combine and process them to produce the final result, which in turn is stored in the distributed file system.

As shown in Figure 9.1, the workflow of naive MapReduce is as follows: Generally, a *Master* node invokes Map tasks on computing nodes that possess input data, which guarantees the locality of computation. Map tasks

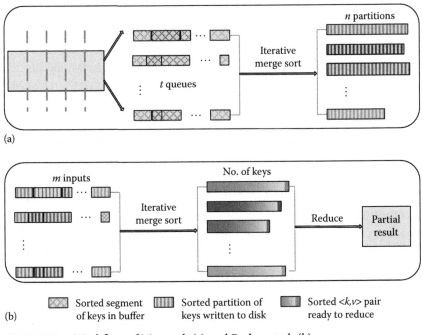

FIGURE 9.1 Workflow of Map task (a) and Reduce task (b).

transform the input (*key,value*) pair (k^1, v^1) to, for example, n new pairs: $(k_1^2, v_1^2), (k_2^2, v_2^2), \dots, (k_n^2, v_n^2)$. The output of Map tasks is then by default hash partitioned to different Reduce tasks according to k_i^2. Reduce tasks receive (*key,value*) pairs grouped by k_i^2, then perform user-specified computation on all the values of each *key*, and write results back to the storage. For the ease of presentation, we use MRJ to denote a MapReduce job in the rest of this chapter.

9.2.2 A Cost Model of Join Processing Using MapReduce

Considering the processing cost of join operations with MRJs, we elaborate a generalized analytic study of an MRJ's execution time given in Reference 8. Generally, most of the CPU time for join processing is spent on simple comparison and counting; thus, system I/O cost dominates the total execution time. For MapReduce jobs, heavy cost on large-scale sequential disk scan and frequent I/O of intermediate results dominate the execution time. Therefore, we shall build a model for an MRJ's execution time based on the analysis of I/O and network cost.

General MapReduce computing framework involves three phases of data processing: Map, Reduce, and the data copying from Map tasks to

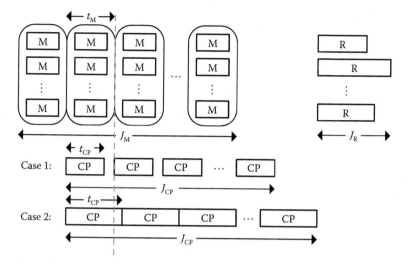

FIGURE 9.2 MapReduce workflow.

Reduce tasks, as shown in Figure 9.2. In the figure, each M stands for a Map task; each CP stands for one phase of Map output copying over network, and each R stands for a Reduce task. Since each Map task is based on a data block, we assume that the unit processing cost for each Map task is t_M. Moreover, since the entire input data may not be loaded into the system memory within one round [9,10], we assume that these Map tasks are performed round by round (we have the same observation in practice). However, the size of Reduce task is subjected to the (*key,value*) distribution. As shown in Figure 9.2, the makespan of an MRJ is dominated by the most time-consuming Reduce task. Therefore, we only consider the Reduce task with the largest volume of inputs in the following analysis: Assume the total input size of an MRJ is S_I. The total intermediate data copied from Map to Reduce are of size S_{CP}, and the number of Map tasks and Reduce tasks are m and n, respectively. In addition, as a general assumption, S_I is considered to be evenly partitioned among m Map tasks [11]. Let J_M, J_R, and J_{CP} denote the total time cost of three phases, respectively, and T be the total execution time of an MRJ. Then $T \leq J_M + J_{CP} + J_R$ holds due to the overlapping between J_M and J_{CP}.

For each Map task, it performs disk I/O and data processing. Since disk I/O is the dominant cost, we can estimate the time cost for single Map task based on disk I/O. Disk I/O contains two parts: sequential reading and data spilling. Then the time cost for single Map task t_M is

$$t_M = \left(C_1 + p \times \alpha\right) \times \frac{S_I}{m} \qquad (9.1)$$

where:

C_1 is a constant factor regarding disk I/O capability

p is a random variable denoting the cost of spilling intermediate data

For a given system configuration, p subjects to the intermediate data size; it increases while spilled data size grows. α denotes the output ratio of a Map task, which is query specific and can be computed with the selectivity estimation. Assume m' is the current number of Map tasks running in parallel in the system. J_M can be computed as follows:

$$J_M = t_M \times \frac{m}{m'} \qquad (9.2)$$

For J_{CP}, let t_{CP} be the time cost for copying the output of single Map task to n Reduce tasks, it includes the cost of data copying over network as well as overhead of serving network protocols. t_{CP} is calculated with the following formula:

$$t_{CP} = C_2 \times \frac{\alpha \times S_I}{n \times m} + q \times n \qquad (9.3)$$

where:

C_2 is a constant number denoting the efficiency of data copying over network

q is a random variable that represents the cost of a Map task serving n connections from n Reduce tasks

Intuitively, there is a rapid growth of q while n gets larger. Thus, J_{CP} can be computed as follows:

$$J_{CP} = \frac{m}{m'} \times t_{CP} \qquad (9.4)$$

For J_R, intuitively it is dominated by the Reduce task that has the biggest size of input. We assume that the key distribution in the input file is random. Let S_r^i denote the input size of Reduce task i. According to the *central limit theorem* [12], we can assume that for $i = 1, \ldots, n$, S_r^i follows a normal distribution $N \sim (\mu, \sigma)$, where μ is determined by $\alpha \times S_I$ and σ subjects to dataset properties, which can be learned from history query logs. Thus, by employing the rule of "three sigmas" [12], we make $S_r^* = \alpha \times S_1 \times n^{-1} + 3\sigma$, the biggest input size to a Reduce task:

$$J_R = (p + \beta \times C_1) \times S_r^* \tag{9.5}$$

where:

β is a query-dependent variable denoting output ratio, which could be precomputed based on the selectivity estimation

Thus, the execution time T of an MRJ is

$$T = \begin{cases} J_M + t_{CP} + J_R & \text{if } t_M \geq t_{CP} \\ t_M + J_{CP} + J_R & \text{if } t_M \leq t_{CP} \end{cases} \tag{9.6}$$

In our cost model, parameters C_1, C_2, p, and q are system dependent and need to be derived from observations on the execution of real jobs, which are elaborated in the experiments section. This model favors MRJs that have I/O cost dominating the execution time. Experiments show that our method can produce a reasonable approximation of the MRJ running time in real practice.

9.2.3 Multiway Join Processing Using MapReduce

Existing efforts toward efficient join query evaluation using MapReduce mainly fall into two categories: The first category is to implement different types of join queries by exploring the partition of (*key,value*) pairs from Map tasks to Reduce tasks without touching the implementation details of the MapReduce framework and the second category is to improve the functionality and efficiency of MapReduce itself to achieve better query evaluation performance. For example, MapReduce Online [13] allows pipelined job interconnections to avoid intermediate result materialization. A parallel contract (PACT) model [14] extends the MapReduce concept for complex relational operations. Our work, as well as the works of Vernica et al. [15] on set similarity join and Okcan et al. [7] on theta-join, falls in the first category. We briefly survey some most related works in this category.

Afrati et al. [6,16] present their novel solution for evaluating multiway equi-join in one MRJ. The essential idea is that, for each join key, they logically partition the Reduce tasks into different groups such that a valid join result can be discovered on at least one Reduce task. Their optimization goal is to minimize the volume of data copying over the network. But the solution only works for the equi-join scenario. Because for equi-join, as long as we make the join attribute the partition key, the *joinable* data records that have the same key value will be delivered to the same Reduce task. However, for theta-join queries, such partition method for (*key,value*) pairs cannot even guarantee the correctness. Moreover, answering complex join queries with one MRJ may not guarantee the best time efficiency in practice. Wu et al. [4] targets at the efficient processing of multiway join queries over massive volume of data. Although they present their work in the context of equi-join, their focus is how to decompose a complex query into multiple MRJs and schedule them to eventually evaluate the query as fast as possible. However, their decomposition is still join attribute oriented. Therefore, the original query is decomposed into multiple pairwise joins and selecting the optimal join order is the main problem. On the contrary, although we also explore the scheduling of MRJs in this work, each MRJ being scheduled can involve multiple relations and multiple join conditions. Therefore, our solution truly tries to explore all possible evaluation plans. Moreover, the work of Wu et al. [4] does not take the limit of processing unit into consideration, which is a critical issue in real practice. Some other works try to explore the general work flow of single MRJ or multiple MRJs to improve the whole throughput performance. Hadoop++ [17] injects optimized user-defined functions (UDFs) into Hadoop to improve query execution performance. RCFile [18] provides a column-wise data storage structure to improve I/O performance in MapReduce-based warehouse systems. MRShare [11] explores the optimization opportunities to share the file scan and partition key distribution among multiple correlated MRJs. YSmart [5] is a source-to-source Structured Query Language (SQL) to MapReduce translator. It proposes a common MapReduce framework to reduce redundant file I/O and duplicated computation among Reduce tasks. Another recent work presented a query optimization solution that can avoid high-cost data repartitioning when executing a complex query plan in the structured computations for optimized parallel execution (SCOPE) system [19].

9.3 CASE STUDY 1: MULTIWAY THETA-JOIN PROCESSING

Data analytic queries in real practices commonly involve multiway join operations, where the join condition can be defined as a binary function θ that belongs to $\{<, \leq, =, \geq, >, <>\}$, which is known as theta-join. Compared with equi-join, it is more general and expressive in relation description and surprisingly handy in data analytic queries. Consider the following application scenario:

Example 1: Assume we have n cities, $\{c_1, c_2, \ldots, c_n\}$, and all the flight information $FI_{i,j}$ between any two cities c_i and c_j. Given a sequence of cities $<c_s, \ldots, c_t>$ and the stay-over time length that must fall in the interval $L_i = [l_1, l_2]$ at each city c_i, find out all the possible travel plans.

This is a practical query that could help travelers plan their trips. For illustration purposes, we simply assume $FI_{i,j}$ is a table containing flight number, departure time (dt), and arrival time (at). The above request can be easily answered with a multiway theta-join operation over $FI_{s,s+1}, \ldots, FI_{t-1,t}$, by specifying the time interval between two successive flights falling into the particular city's stay-over interval requirement. For example, the θ function between $FI_{s,s+1}$ and $FI_{s+1,s+2}$ is $FI_{s,s+1} \cdot at + L_{s+1} \cdot l_1 < FI_{s+1,s+2} \cdot dt < FI_{s,s+1} \cdot at + L_{s+1} \cdot l_2$.

In fact, evaluating multiway theta-joins has always been a challenging problem along with the development of database technology. Early works, such as References 20–23, have elaborated the problem's complexity and their evaluation strategies. However, their solutions do not scale to process the multiway theta-joins over the data of tremendous volumes. In this case study, we show how to utilize the MapReduce solution framework to efficiently evaluate multiway theta-join queries in a shared-nothing environment.

9.3.1 Solution Overview

The problem that we are looking at is as follows: given a number of processing units (that can run Map or Reduce tasks), mapping a multiway theta-join to a number of MapReduce jobs and having them executed in a well-scheduled order such that the total processing time span is minimized. The solution to this challenging problem includes two core techniques. The first technique is that, given a multiway theta-join query, we examine all the possible decomposition plans and estimate the minimum execution time cost for each plan. Especially, we figure out the rules to properly decompose the original multiway theta-join query and study the most efficient solution to evaluate multiple join condition functions using one MapReduce job. The second technique is that, given a limited number of computing units and a pool of possible MapReduce jobs to evaluate the query, we make decisions on job selection to effectively evaluate the query as fast as possible.

To be specific, we present a resource-aware (*key,value*) pair distribution method in Section 9.4 to evaluate the chain-typed multiway theta-join query with one MapReduce job, which guarantees minimized volume of data copying over the network, as well as evenly distributed workload among Reduce tasks. Moreover, we establish the rules to decompose a multiway join query. Using our proposed cost model, we can figure out whether a multiway join query should be evaluated with multiple MapReduce jobs or a single MapReduce job.

9.3.2 Problem Definition

Our solution targets on the MapReduce job identification and scheduling. In other words, we work on the rules to properly decompose the query processing into several MapReduce jobs and have them executed in a well-scheduled fashion such that the minimum evaluation time span is achieved. In this section, we shall first present the terminologies that we use in this chapter, and then give the formal definition of the problem. We show that the problem of finding the optimal query evaluation plan is NP-hard.

Terminology and statement: For the ease of presentation, in the rest of the chapter, we use the notation of "N-join" query to denote a multiway theta-join query. We use MRJ to denote a MapReduce job.

Consider an N-join query Q defined over m relations $R_1, \dots, R_m$ and n specified join conditions $\theta_1, \dots, \theta_n$. As adopted in many other works, such as in Reference 4, we can present Q as a graph, namely, a join graph. For completeness, we define a join graph G_J as follows:

> **DEFINITION 1:** A join graph $G_J = \langle V, E, L \rangle$ is a connected graph with edge labels, where $V = \{v \mid v \in \{R_1, \dots, R_m\}\}$, $E = \{e \mid e = (v_i, v_j) \Leftrightarrow \exists \theta, R_i \bowtie_\theta R_j \in Q\}$, $L = \{l \mid l(e_i) = \theta_i\}$.

Intuitively, G_J is generated by making every relation in Q a vertex and connecting two vertices if there is a join operator between them. The edge is labeled with the corresponding join function θ. To evaluate Q, every θ function, that is, every edge from G_J, needs to be evaluated. However, to evaluate all the edges in G_J, there are an exponential number of plans since any arbitrary number of connecting edges can be evaluated in one MRJ. We propose a join-path graph to cover all the possibilities. For the purpose of clear illustration, we define a no-edge-repeating path between two vertices of G_J in the first place.

> **DEFINITION 2:** A no-edge-repeating path p between two vertices v_i and v_j in G_J is a traversing sequence of connecting edges $\langle e_i, \dots, e_j \rangle$ between v_i and v_j in G_J, in which no edge appears more than once.

DEFINITION 3: *A join-path graph $G_{JP} = \langle V, E', L', W, S \rangle$ is a complete weighted graph with edge labels, where each edge is associated with a weight and scheduling information. Specifically, $V = \{v \mid v \in \{R_1, \ldots, R_m\}\}$, $E' = \{e' \mid e' = (v_i, v_j)$ represents a unique no-edge-repeating path p between v_i and v_j in $G_J\}$, $L' = \{l' \mid l'(e') = l'(v_i, v_j) = \bigcup l(e), e \in p$ between v_i and $v_j\}$, $W = \{w \mid w(e')$ is the minimal cost to evaluate $e'\}$, and $S = \{s \mid s(e')$ is the scheduling to evaluate e' at the cost of $w(e')\}$.*

In the definition, the scheduling information on the edge refers to some user-specified parameters to run an MRJ such that this job is expected to be accomplished as fast as possible. In this work, we consider the number of Reduce tasks assigned to an MRJ as the scheduling parameter, denoted as $R_N(MRJ)$, as it is the only parameter that users need to specify in their programs. The reason we take this parameter into consideration is based on two observations from extensive experiments: (1) It is not guaranteed that the more computing units involved in Reduce tasks, the sooner an MRJ job is accomplished; and (2) given limited computing units, there is resource competition among multiple MRJs.

Intuitively, we enumerate all the possible join combinations in G_{JP}. Note that in the context of join processing, $R_i \bowtie R_k \bowtie R_j$ is the same as $R_j \bowtie R_k \bowtie R_i$; therefore, G_{JP} is an undirected graph. We elaborate Definition 3 with the following example. Given a join graph G_J, shown in Figure 9.3a, a corresponding join-path graph G_{JP} is generated, which is presented in an adjacent matrix format on the right. The numbers enclosed in braces are the involved θ functions on a path. For instance, in the cell corresponding to R_1 and R_2, {3, 4, 6, 5, 2} indicates a no-edge-repeating

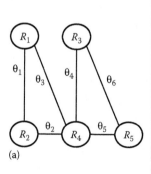

	R_1	R_2	R_3	R_4	R_5
R_1	{1,2,3} $\varepsilon(G_{JP})$	{1} {3,2} {3,4,6,5,2}	{3,4} {3,5,6}	{3} {1,2} {1,2,4,6,5} {3,4,6,5}	{1,2,5} {1,2,4,6} {3,5} {3,4,6,}
R_2	–	{1,3,2} $\varepsilon(G_{JP})$	{2,4} {2,5,6} {1,3,4} {1,3,5,6}	{2} {1,3} {2,4,6,5} {1,3,4,6,5}	{2,5} {2,4,6} {1,3,5} {1,3,4,6}
R_3	–	–	{4,5,6} $\varepsilon(G_{JP})$	{4} {6,5} {4,3,1,2} {6,5,3,1,2}	{6} {4,5} {4,3,1,2,5}
R_4	–	–	–	{4,6,5} {3,1,2} $\varepsilon(G_{JP})$	{5} {4,6} {3,1,2,5} {3,1,2,4,6}
R_5	–	–	–	–	{4,5,6} $\varepsilon(G_{JP})$

(b)

FIGURE 9.3 Example join graph G_J (a) and its corresponding join-path graph G_{JP} (b) presented in an adjacent matrix.

path $\{\theta_3,\theta_4,\theta_6,\theta_5,\theta_2\}$ between R_1 and R_2. In this example, notice that for every node, there exists a closed traversing path (or circuit) that covers all the edges exactly once, namely, the "Eulerian circuit." We use $\varepsilon(G_{JP})$ to denote a "Eulerian circuit" of G_{JP} in the figure. Since we only care what edges are involved in a path, any $\varepsilon(G_{JP})$ would be sufficient. Notice that in the figure, edge weights and scheduling information are not presented. As a matter of fact, this information is incrementally computed during the generation of G_{JP}, which will be illustrated in the later section.

According to the definition of G_{JP}, any edge e' in G_{JP} is a collection of connecting edges in G_J. Thus, e' in fact implies a subgraph of G_J. As we use one MRJ to evaluate e', denoted as $MRJ(e')$, G_{JP}'s edge set represents all the possible MRJs that can be employed to evaluate the original query Q. Let T denote a set of MRJs that are selected from G_{JP}'s edge set. Intuitively, if the MRJs in T cover all the join conditions of the original query, we can answer the query by executing all these MRJs. Formally, we define that T is "sufficient" as follows:

DEFINITION 4: *T*, a collection of MRJs, *is sufficient to evaluate Q iff* $\bigcup e_i' = G_J.E$, *where* $MRJ(e_i') \in T$.

Since it is trivial to check whether T is sufficient, for the rest of this work, we only consider the case that T is sufficient. Thus, given T, we define its execution plan as a specific execution sequence of MRJs, which minimizes the time span of using T to evaluate the original query Q. Formally, we can define our problem as follows:

Problem definition: Given an N-join query Q and k_p processing units, a join-path graph G_{JP} according to Q's join graph G_J is built. We want to select a collection of edges from G_{JP} that correspondingly form a set of MRJs, denoted as T_{opt} such that there exists an execution plan of T_{opt}, which minimizes the query evaluation time. Obviously, there are many different choices of T to evaluate Q. Moreover, given T and limited processing units, different execution plans yield different evaluation time spans. In fact, the determination of MRJ execution order is nontrivial; we give the detailed analysis of the hardness of our problem in the next subsection. As we shall elaborate later, given T and k_p available processing units, we adopt an approximation method to determine in linear time.

Problem hardness: According to the problem definition, we need two steps to find T_{opt}: (1) generate G_{JP} from G_J and (2) select MRJs for T_{opt}. Neither one of these two steps is easy to solve.

For the first step, to construct G_{JP}, we need to enumerate all the no-edge-repeating paths between any pair of vertices in G_J. Assume G_J has the

"Eulerian trail" [24], which is a way to traverse the graph with every edge be visited exactly once. For any pair of vertices v_i and v_j, any different no-edge-repeating path between them is a "subpath" of a Eulerian trail. If we know all the no-edge-repeating paths between any pair of vertices, we can enumerate all the Eulerian trails in polynomial time. Therefore, the complexity of constructing G_{JP} is at least as hard as enumerating all the Eulerian trails of a given graph, which is known to be #-*complete* [25]. Moreover, we find that even G_J does not have a Eulerian trail and the problem complexity is not reduced at all, as we elaborate in the proof of the following theorem.

THEOREM 1: *Generating G_{JP} from a given G_J is a #-complete problem.*

PROOF: If G_J has the Eulerian trail, constructing G_{JP} is #-*complete* (see the discussion above).□

On the contrary, if G_J does not have the Eulerian trail, it implies that there are r vertices having odd degrees, where $r > 2$. Now consider that we add one virtual vertex and connect it with $r - 1$ vertices of odd degrees. Now the graph must have a Eulerian trail. If we can easily construct the join-path graph of the new graph, the original graph G_{JP} can be computed in polynomial time. We elaborate with the following example, as shown in Figure 9.4. Assume v_s is added to the original G_J. By computing the join-path graph of the new graph, we know all the no-edge-repeating paths between v_i and v_j. Then, a no-edge-repeating path between v_i and v_j cannot exist if it has v_s involved. By simply removing all the enumerated paths that go through v_s, we can obtain the G_{JP} of the original G_J. Thus, the dominating cost of constructing G_{JP} is still the enumeration of all Eulerian trails. Therefore, this problem is #-*complete*.

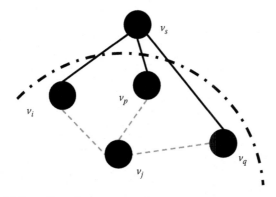

FIGURE 9.4   Adding virtual vertex v_s to G_J.

Although it is difficult to compute the exact G_{JP}, we find that a subgraph of G_{JP}, which contains all the vertices and is denoted as G'_{JP}, could be sufficient to guarantee the optimal query evaluation efficiency. We take the following principle into consideration. Given the same number of processing units, if it takes longer time to evaluate $R_i \bowtie R_j \bowtie R_k$ with one MRJ compared to the total time cost of evaluating $R_i \bowtie R_j$ and $R_j \bowtie R_k$ separately and merging the results, we do not take $R_i \bowtie R_j \bowtie R_k \bowtie R_s$ into consideration. By following this principle, we can avoid enumerating all the possible no-edge-repeating paths between any pair of vertices. As a matter of fact, we can obtain such a sufficient G'_{JP} in polynomial time.

The second step of our solution is to select T_{opt}. Assume the G'_{JP} computed from the first step provides a collection of edges; accordingly, we have a collection of MRJ candidates to evaluate the query. Although each edge in G_{JP} is associated with a weight denoting the minimum time cost to evaluate all the join conditions contained in this edge, it is just an estimated time span on the condition that there are enough processing units. However, when a T is chosen, and the number of processing units is limited, the time cost of using T to answer Q needs to be reestimated. Assume we can find the time cost estimation of T, denoted as $C(T)$. The problem is to find such an optimal T_{opt} from all possible T's, which has the minimum time cost. Apparently, this is a variance of the classic set cover problem, which is known to be NP-hard [26]. Therefore, there are many heuristics and approximation algorithms that can be adopted to solve the selection problem.

As clearly indicated in the problem definition, the solution lies in the construction of G'_{JP} and smartly selects T based on the cost estimation of a group of MRJs. Therefore, for the rest of the chapter, we shall first elaborate our cost models for a single MRJ and a group of MRJs. Then we present our detailed solution for the N-join query evaluation.

9.3.3 Cost Model

We have already presented the execution time cost estimation model for single MRJ in Section 9.2. In this section, we present a generalized analytic study on the execution time of a group of MRJs. In the context of G_{JP} construction and T selection, we study the estimation of $w(e')$, where $e' \in G_{JP} \cdot E$, and $C(T)$, which is the time cost to evaluate T.

There have been some works exploring the optimization opportunity among multiple MRJs running in parallel, such as References 4, 5, and 11, by defining multiple types of correlations among MRJs. For instance, Reference 5 defines "input correlation," "transit correlation," and "job flow correlation," targeting at the shared input scan and intermediate data partition. In fact, their techniques can be directly plugged into our solution framework. Compared to these techniques, the significant difference of our study on the execution model of a set of MRJs is that our work takes the number of available processing units into consideration. Therefore, the optimization problem we study here is orthogonal with the techniques proposed in the existing literatures that we mentioned earlier.

Given T and k_p processing units, we concern about the execution plan that guarantees the minimum task execution time span. However, the determination of MRJ execution order is usually subjected to k_p. For instance, consider the T given in Figure 9.5. MRJ(e_i'), MRJ(e_j'), and MRJ(e_k') can be accomplished in 5, 7, and 9 time units if 4, 4, and 8 Reduce tasks are assigned to them, respectively. Thus, if there are over 16 available processing units, these three MRJs can be scheduled to run in parallel and have no computing resource competition. On the contrary, if there are not enough processing units, parallel execution of multiple MRJs can lead to very poor time efficiency. It is exactly the classic problem of scheduling independent malleable parallel tasks over bounded parallel processors, which is NP-hard [27]. In this work, we adopt the methodology presented in Reference 27. The method guarantees that for any given $\varepsilon > 0$, it takes linear time (in terms of $|T|$, k_p, and ε^{-1}) to compute a scheduling that promises the evaluation time to be at most $(1 + \varepsilon)$ times the optimal one.

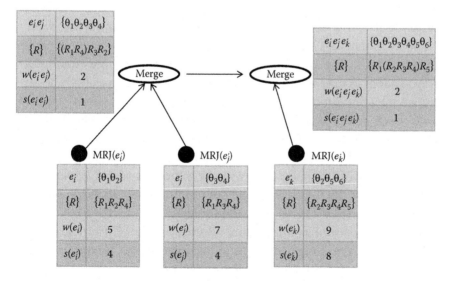

FIGURE 9.5   One execution plan of $T = \{e_i', e_j', e_k'\}$.

Moreover, to evaluate Q with T, not only the MRJs in T must be executed, but a merge step is also needed to generate the final results. Intuitively, if two MRJs share some common input relation, their output can be merged using the common relation as the key. For instance, Figure 9.5 presents one possible execution plan of MRJ(e'_i), MRJ(e'_j), and MRJ(e'_k). Assume there are over 16 available processing units. We execute all three jobs in parallel. Since MRJ(e'_i) and MRJ(e'_j) share the same inputs R_1 and R_4. Therefore, the output of MRJ(e'_i) and MRJ(e'_j) can be merged using the primary keys from both R_1 and R_4. Later on, the output of this step can be further merged with the output of MRJ(e'_k). The total execution time is $9 + 2 = 11$ time units. In the figure, we enclose the merge key with brackets. Note that such a merge operation only has output keys or data identifiers (IDs) involved; therefore, it can be done very efficiently.

9.3.4 Join Algorithm

As discussed in Section 9.3, the key issues of our solution lie in constructing G'_{JP} and selecting T. In Section 9.4, we present an analytic study of the execution schedules of a single MRJ and multiple MRJs. However, we have not yet solved the problem of how to compute a multiway theta-join in one MRJ. Therefore, in this section, we first present our solution to the multiway theta-join processing with one MRJ. Then, we elaborate the construction of G'_{JP} and the selection of T.

9.3.4.1 Multiway Theta-Join Processing with Single MRJ

As discussed in Section 9.2, different from equi-join, we cannot use the join attribute as the hash *key* to answer theta-join in the MapReduce computing framework. The work of Okcan et al. [7] for the first time explores the way to adopt MapReduce to answer a theta-join query. Essentially, it partitions the cross-product result space with rectangle regions of bounded size, which guarantees the output correctness and the workload balance among Reduce tasks. However, their partition method does not have a straightforward extension to solve a multiway theta-join query. Inspired from the work of Okcan et al. [7], we believe that it is a feasible solution to conceptually make the cross-product of multiple relations as the starting point and figure out a better partition strategy.

Based on our problem definition, all the possible MRJ candidates for T is a no-edge-repeating path in the join graph G_J. Thus, we only consider the case of chain joins. Given a chain theta-join query with m different relations involved, we want to derive a (*key,value*)-based solution that guarantees the minimum execution time span. Let S denote the hypercube that comprises the cross-product of all m relations. Let f denote a space partition function that maps S to a set of disjoint components whose union is exactly S. Intuitively, each component represents a Reduce task, which is responsible for checking if any valid join result falls into it. Assume there are k_R Reduce tasks, and the cardinality of relation R is denoted as $|R|$. For each Reduce task, it has to check $\Pi_{i=1}^{m} |R_i|/k_R$ join results. However, it is not true that the more Reduce tasks, the less execution time. As when k_R increases, the volume of data copy over network may grow significantly. For instance, as shown in Figure 9.6, when a Reduce task is added, the network volume increases.

Now we have the two sides of a coin: the number of Reduce tasks k_R and the partition function f. Our solution is described as follows: We first define what an "ideal" partition function is; then, we pick one such function and derive a proper k_R for the given chain theta-join query.

Let $t_{R_i}^j$ denote the jth tuple in relation R_i. Partition function f maps S to a set of k_R components, denoted as $C = \{c_1, c_2, \ldots, c_{k_R}\}$. Let $Cnt(t_{R_i}^j, C)$ denote the total number of times that $t_{R_i}^j$ appears in all the components. We define the *partition score* of f as

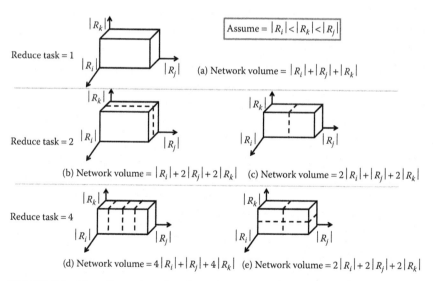

FIGURE 9.6   How the network volume increases when more reduce task(s) are involved. (a) shows the scenario when only one reducer is employed; (b) and (c) show the scenarios when two reducers are employed but with different data partition methods; (d) and (e) show the scenarios when four reducers are employed but with different data partition methods.

$$\text{Score}(f) = \sum_{i=1}^{n} \sum_{j=1}^{|R_i|} \text{Cnt}(t_{R_i}^{j}, C) \tag{9.7}$$

DEFINITION 5: *f is a* perfect partition function *iff for a given S, $\forall k_R$, Score(f) is minimized.*

DEFINITION 6: *For a given S, the class of all perfect partition functions, F, is the* perfect partition class *of S.*

Based on the definition of F, to resolve F for a given S requires the "calculus of variation" [28], which is out of the scope of our current discussion. We shall directly present a partition function f and prove that $f \in F$.

THEOREM 2: *To partition a hypercube S, the* Hilbert space-filling curve *is a perfect partition function f.*

PROOF: The minimum value of score function defined in Equation 9.7 is achieved when the following condition holds:

$$\sum_{u=1}^{|R_i|} \text{Cnt}(t_{R_i}^{u}, C) = \sum_{u=1}^{|R_j|} \text{Cnt}(t_{R_j}^{u}, C) \; \forall 1 \le i, j \le n \tag{9.8}$$

In other words, in a partition component c, assume the number of distinct records from relation R_i is $c(R_i)$. The duplication factor of R_i in this component must be $\Pi_{j=1, j\neq i}^{n} c(R_j)$. Since Hilbert space-filling curve defines a traversing sequence of every cell in the hypercube of $R_1, \ldots, R_n$, if we use a Hilbert curve H as a partition method, then a component c is actually a continuous segment of H. Considering the construction process of H, every dimension is recursively divided by the factor of 2, and such recursive computation occurs the same number of times for all dimensions. Note that H defines a traversing sequence that traverses cells along each dimension fairly, meaning that if H has traversed half of R_i, then H must also have traversed half of R_j, where R_j is any other relation. Thus, given any partition value (equal to the number of Reduce tasks) k_R, a segment of H of length $|H|/k_R$ traverses the same proportion of records from each dimension. Let this proportion be ε. Therefore, the duplication factor for each record from R_i is

$$\prod_{j=1, j\neq i}^{n} \eta \frac{|R_j|}{2^{\eta} \times \varepsilon} \tag{9.9}$$

where:

η is the number of recursions

Note that the derived duplication factor satisfies the condition given in Equation 9.8. Thus, H is a perfect partition function.

After obtaining f, we can further approximate the value of k_R, which achieves the best query evaluation time efficiency. As discussed earlier, k_R affects two parts of our cost model: the network volume and the expected input size to Reduce tasks, both of which are the dominating factors of the execution time cost. Therefore, an approximation of the optimal k_R can be obtained when we try to minimize the following value Δ:

$$\Delta = \lambda \sum_{i=1}^{n} \sum_{j=1}^{|R_i|} \text{Cnt}(t_{R_i}^j, C) + (1-\lambda)\frac{\prod_{i=1}^{m}|R_i|}{k_R} \qquad (9.10)$$

Intuitively, Δ is a linear combination of the two cost factors. Coefficient λ denotes the importance of each cost factor. For instance, if $\lambda < 0.5$, it implies that reducing the workload of each Reduce task brings more cost saving.[*] Note that the first cost factor in Equation 9.10 is also a linear sum function of k_R. Therefore, by making $\Delta' = 0$, we can get $[k_R]$. □

The pseudocode in Algorithm 9.1 describes our solution for evaluating a chain theta-join query in one MRJ. Note that our main focus is the generation of (*key,value*) pairs. One tricky method we employed here, as also be employed in the work of Okcan et al. [7], is randomly assigning an observed tuple t_{R_i}, a global ID in R_i. The reason is that each Map task does not have a global view of the entire relation. Therefore, when a Map task reads a tuple, it cannot tell the exact position of this tuple in the relation.

Algorithm 9.1: Evaluating a chain theta-join query in one MRJ
Data: Query $q = R_1 \bowtie \cdots \bowtie R_m$, $|R_1|\cdots|R_m|$;
Result: Query result
1. Using Hilbert space-filling curve to partition S and compute a proper value of k_R
2. Deciding the mapping: GlobalID(t_{R_i}) → a number of components in C
3. **for** *each Map task* **do**
4.   GlobalID(t_{R_i}) ← Unified random selection in $[1, |R_i|]$
5.   **for** *all components that GlobalID* (t_{R_i}) *maps to* **do**

---

[*] In our experiments, we observe that the value of λ falls in the interval of [0.38,0.46]. We set $\lambda = 0.4$ as a constant.

6.   generate (componentID, t_{R_i})
7. **for** *each Reduce task* **do**
8. **for** *any combination of* $t_{R_1}, \ldots, t_{R_m}$ **do**
9.   **if** *it is a valid result* **then**
10.   Output the result

Constructing G'_{JP}. By applying Algorithm 9.1, we can minimize the time cost to evaluate a chain theta-join query using one MRJ. However, usually a group of MRJs is needed to evaluate multiway theta-joins. Therefore, we now discuss the construction of G'_{JP}, which is a subgraph of the join-path graph G_{JP} and sufficient to serve the evaluation of N-join query Q. As already discussed in Section 9.3.2, computing G_{JP} is a *#-complete* problem, as it requires to enumerate all possible no-edge-repeating paths between any pair of vertices. In fact, only a subset of the entire edge collection in G_{JP} can be further employed in T_{opt}. Therefore, we propose two pruning conditions to effectively reduce the search space in this section.

The first intuition is that to select T_{opt}, the case that many join conditions are covered by multiple MRJs in T_{opt} is not preferred, because each join condition needs to be evaluated only once. However, it does not imply that MRJs in T_{opt} should strictly cover disjoint sets of join conditions. Because sometimes, by including extra join conditions, the output volume of intermediate results can be reduced. Therefore, we exclude $MRJ(e'_i)$ on the only condition that there are other more efficient ways to evaluate all the join conditions that $MRJ(e'_i)$ covers. Formally, we state the pruning condition in Lemma 1.

> **LEMMA 1:** *Edge e'_i should not be considered if there exists a collection of edges ES, and the following conditions are satisfied:* (1) $l'(e'_i) \subseteq \bigcup_{e'_j \in ES} l'(e'_j)$, (2) $w(e'_i) > Max_{e'_j \in ES} w(e'_j)$, *and* (3) $s(e'_i) \geq \sum_{e'_j \in ES} s(e'_j)$.

Lemma 1 is quite straightforward. If an MRJ can be substituted with some other MRJs that cover at least the same number of join conditions and be evaluated more efficiently with less demands on processing units, this MRJ cannot appear in T_{opt}. Because T_{opt} is the optimal collection of MRJs to evaluate the query, containing any substitutable MRJ makes T_{opt} suboptimal. For the second pruning method, we present the following lemma that further reduces the search space:

> **LEMMA 2:** *Given two edges e'_i and e'_j, if e'_i is not considered and $l'(e'_i) \subset l'(e'_j)$, then e'_j should not be considered either.*

PROOF: Since e'_i is not considered, it implies that there is a better solution to cover $l'(e'_i) \cap l'(e'_j)$. And this solution can be employed together with $l'(e'_j) - l'(e'_i)$, which is more efficient than computing $l'(e'_j)$ in one step. Therefore, $l'(e'_j)$ should not be considered.

Note that Lemma 2 is orthogonal to Lemma 1. Since Lemma 1 decides whether an MRJ should be considered as a member of T_{opt}, if the answer is negative, we can employ Lemma 2 to directly prune more undesired MRJs. By employing the two lemmas proposed earlier, we develop an algorithm to construct G'_{JP} efficiently in an incremental manner, as presented in Algorithm 9.2.

Algorithm 9.2: Constructing G'_{JP}

Data: G'_J containing n vertices and m edges, $G'_{JP} = \varnothing$; , a sorted list $WL = \varnothing$;
Result: G'_{JP}

1. **for** $i = 1 : n$ **do**
2.   **for** $j > i$ **do**
3.   **for** $L = 1 : m$ **do**
4.   **if** *there is an L-hop path from R_i to R_j* **then**
5. $e' \leftarrow$ the L-hop path from R_i to R_j
6.   **if** $WL \neq$; **then**
7.   scan WL to find the first group of edges that cover e'
8.   apply Lemma 1 to decide whether to report e' to G'_{JP}
9.   **if** *e' is not reported* **then**
10.   break //Lemma 2 plays the role
11.   insert e' into WL such that WL maintains a sequence of edges in the ascending order of $w(e')$

Since we do not care about the direction of a path, meaning $e'(v_i, v_j) = e'(v_j, v_i)$, we compute the pairwise join paths following a fixed order of vertices (relations). In Algorithm 9.2, we employ the linear scan of a sorted list to help decide whether a path should be reported in G'_{JP}. One tricky part in the algorithm is line 4. A straightforward way is to employ distributed file systems (DFS) search from a given starting vertex. The time complexity is $O(m + n)$. However, it introduces much redundant work for every vertex to perform this task. A better solution is before we run Algorithm 9.2. We first traverse G_J once and record the L-hop neighbor of every vertex. It takes only $O(m + n)$ time complexity. Then, line 4 can be determined in $O(1)$ time. Overall, the worst time complexity of Algorithm 9.2 is $O(n^2 m)$. This happens only when

G_J is a complete graph. In real practice, due to the sparsity of the graph, Algorithm 9.2 is quick enough to generate G_{JP} for a given G_J. As observed in the experiments, G'_{JP} can be generated in the time frame of hundreds of microseconds.

After G'_{JP} is obtained, we select T_{opt} following the methodology presented in Reference 29, which gives $O[\ln(n)]$ approximation ratio of the optimum.

9.3.5 Summary

In this case study, we focus on the efficient evaluation of multiway theta-join queries using MapReduce. The solution includes two parts. First, we study how to conduct a chain-type multiway theta-join using one MapReduce job. We present a Hilbert curve-based space partition method that minimizes data copying volume over network and balances the workload among Reduce tasks. Second, we elaborate a resource-aware scheduling schema that helps the evaluation of complex join queries achieve a near-optimal time efficiency in resource-restricted scenarios.

9.4 CASE STUDY 2: SPARQL QUERY PROCESSING

Along with increasing supports from prevailing search engine projects, such as Rich Snippets from Google and SearchMoney from Yahoo!, as well as the willingness to integrate across-domain knowledge, there emerges a huge volume of public RDF data for management and analysis. For example, the largest RDF dataset [30] available in the Linked Data community [31] has over 3.2 billion triples, and the second largest dataset containing various bio- and gene-related data (Bio2RDF) has over 2.7 billion triples. The increasing demands of massive data-intensive RDF data analysis and the great scalability of cloud platforms have made them the nail and the hammer. Although various efforts have been made to explore the effective analysis of large RDF data on cloud platforms via RDF-specific querying interface, SPARQL [32], the query time efficiency remains a bottleneck to forward cloud-based RDF services in the real world.

In this case study, we shall elaborate how to efficiently evaluate SPARQL queries using the MapReduce computing framework. Followed by a brief introduction on the RDF data model and the SPARQL query, we show the MRJ identification and scheduling strategies with RDF data feature taken into consideration. Moreover, we present the most recent research efforts on distributed SPARQL query evaluation.

9.4.1 RDF and SPARQL Query

9.4.1.1 RDF Data Model

As one of the World Wide Web Consortium (W3C) standards for describing Web resources and metadata, RDF is designed as a flexible representation of schema-relax or even schema-free information for the semantic Web. RDF model can be viewed as a description of the finest granularity of schema-relax relational model. An RDF triple consists of three components: *Subject*, *Predicate* (*Property*), and *Object* (*Value*), which represent that two entities connected by a relationship specified by the *Predicate* or an entity have a certain value on some property. Thus, RDF data can be visualized as a directed graph by treating entities as vertices and relationships as edges between entities. Figure 9.7 shows an example of RDF data describing the authors and their publications. By making *Subjects* and *Objects* the nodes, and *Predicates* the directed edges pointing from the corresponding *Subject* to *Object*, RDF data can be viewed as a directed labeled graph.*

For the clear illustration purpose and terminology completeness, we present two formal definitions of RDF data as follows:

DEFINITION 7: RDF triple function $F \rightarrow \{U,B,L\} \times \{U\} \times \{U,L\}$ defines a Subject–Predicate–Object triple, where U is the set of (URIs),† B is the generated set of syntax to distinguish subjects, a.k.a. blank nodes, and L is the set of string-type literal values. U, B, and L are pairwise disjoint.

DEFINITION 8: RDF graph G is a directed labeled graph. $G = \{V,E,V_l,E_l\}$, where V is the set of graph nodes and E is the set of edges; $\forall v \in V, \exists v_l \in V_l$ binds with v and $V_l \subset \{U \cup B \cup L\}$; $\forall e \in E, \exists e_l \in E_l$ binds with e and $E_l \subset U$.

9.4.1.2 SPARQL Queries

SPARQL is the W3C standard interface for RDF query. It is designed to query RDF data in an SQL-like style. A SPARQL query specifies several *basic query patterns* (a.k.a. BQPs), and the query returns are in fact the desired labels of subgraph(s) that exactly matches with the given BQPs. Like the example given in Figure 9.7, to query the name of an author who has coauthored with "Alice" and has a journal published in 1940, a small

---

* More rigorously, it is a directed multiedge-labeled graph, as some *Subject* may have multiple values of the same *Predicate* [32].
† Unified resource identifier.

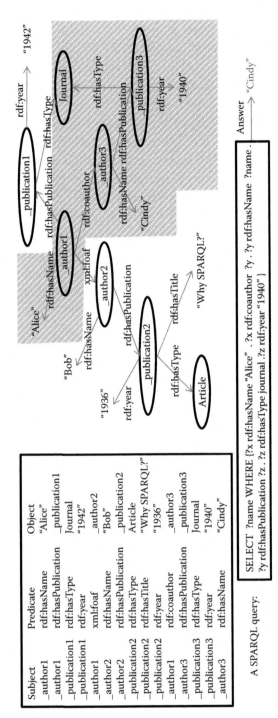

Subject	Predicate	Object
_author1	rdf:hasName	"Alice"
_author1	rdf:hasPublication	_publication1
_publication1	rdf:hasType	Journal
_publication1	rdf:year	"1942"
_author1	xml:foaf	_author2
_author2	rdf:hasName	"Bob"
_author2	rdf:hasPublication	_publication2
_publication2	rdf:hasType	Article
_publication2	rdf:hasTitle	"Why SPARQL?"
_publication2	rdf:year	"1936"
_author1	rdf:coauthor	_author3
_author3	rdf:hasPublication	_publication3
_publication3	rdf:hasType	Journal
_publication3	rdf:year	"1940"
_author3	rdf:hasName	"Cindy"

A SPARQL query:

SELECT ?name WHERE {?x rdf:hasName "Alice" . ?x rdf:coauthor ?y . ?y rdf:hasName ?name .
?y rdf:hasPublication ?z . ?z rdf:hasType journal .?z rdf:year "1940" }

FIGURE 9.7 An illustration example of RDF data and SPARQL query.

subgraph (shown in shadowed area) is identified to be the match of given BQPs. So far, SPARQL(1.1) allows four types of queries to be performed:

- SELECT: returns the desired variable value, as the query example shown in Figure 9.7.

- CONSTRUCT: returns a subgraph of RDF graph G that satisfies all the given BQPs. For example, in Figure 9.7, if we substitute the "SELECT" keyword with "CONSTRUCT," the returned result would be the subgraph covered in the gray shadow.

- ASK: instead of returning the variable value, it is a Boolean function to indicate that a given variable has a value or not.

- DESCRIBE: returns all the associated labels and literal values. Intuitively, it represents some queries such as "SELECT ?p1 ?o ?s ?p2 WHERE {?x ?p1 ?o. ?s ?p2 ?x}."

Studies on the real-world SPARQL queries [32–34] find that over 99% queries are SELECT queries. Therefore, in this work, we only focus on this type of query. So far, SPARQL has evolved to support more advanced functions for flexible queries and the result representation, as well as some simple aggregation functions.

9.4.2 SPARQL Evaluation Using MapReduce

We first summarize the essential ideas of current solutions to this problem. As explained in Figure 9.7, conceptually the SPARQL query evaluation can be considered as a subgraph matching problem. However, to leverage the massive parallelism and scalability of the MapReduce framework, SPARQL queries are usually evaluated in a multiway join fashion. To be specific, considering the variables as the join keys, Map tasks first scan over the dataset to find all the data that satisfy the given BQPs, and then shuffle the data of the same join key to the same Reduce task to examine all the possible valid join results. The state-of-the-art optimization techniques fall into three categories:

1. *Reducing the volume of file scan.* Solutions such as *"Predicate* split" [36,37] and precomputed query forwarding [38] try to evaluate queries only on the computing nodes that hold the desired data. Shared scan [11] is also widely adopted as an effective tool to reduce the file scan cost.

2. *Reducing the I/O cost of intermediate results with* bloom filter [37] *and effective selectivity estimation.* By adopting the selectivity estimation,

multiple MapReduce jobs can be organized and scheduled to achieve the minimum time cost of the query evaluation.

3. *Introducing filters or new hash functions to optimize the performance of MapReduce jobs conducting the join operation.* For example, the work of Afrati and Ullman [6] studies the optimal network shuffling function in case of performing multiway join with one MapReduce job.

9.4.3 Solution Overview

There are a number of challenges to fit RDF query processing, especially join processing, directly into the MapReduce framework. Although attempts were made in existing solutions [39–41], the following problems are not well solved:

- How to map the implied join operations inside a SPARQL query to a number of MapReduce jobs? A MapReduce job can do either *zero-reduce* processing (simple selection), *pairwise* join, or *multiway* join. It is difficult to decide the types for all MapReduce jobs in order to achieve the overall best efficiency.

- How to execute MapReduce jobs efficiently? For a given system and job dependencies, how to make the best use of computing and network resources to maximize the job execution parallelism such that we can achieve the shortest execution time.

- How to organize and manage RDF data on cloud such that MapReduce jobs can scale along with the data volumes involved in different queries?

To solve the above challenges, we present a cost model-based RDF join processing solution on the cloud. To elaborate, we first decompose RDF data into *Predicate* files and organize them according to data contents. Then, we map a SPARQL query directly to a sequence of MapReduce jobs that may employ hybrid join strategies (combination of Map-side join, Reduce-side join, and memory backed join). Finally, based on our cost model of MapReduce jobs for join processing, we present an *All Possible Join* (APJ) tree-based technique to schedule these jobs to be executed in the most extended parallelism style. We also discuss how to extend our solution to handle other types of joins over RDF data.

9.4.4 Problem Definition

For the rest of this case study, we only focus on join operator of a SPARQL query. The "construct" and "optional" semantics are not considered. Therefore, a SPARQL query can be simply modeled as "SELECT *variable(s)* WHERE {BQP(s)}." Intuitively, since we assume that the *Predicate* of BQPs is not a variable [42], the above query can be answered by the following steps: (1) select RDF triplets that satisfy a BQP and (2) join RDF triplets of different BQPs on shared variables. Essentially, we study how to map the original query to one or several join operations implemented by MRJs and schedule the execution of these jobs. To address this problem, we first partition RDF data according to the *Predicate*; each partition is called a *Predicate* file. Then for given BQP$_i$, we denote its corresponding *Predicate* file as PF(BQP$_i$). The statistics of *Subject* and *Object* for each *Predicate* file are also computed to serve the join order selection.

Given a SPARQL query, we can derive a query pattern graph. Each vertex represents a BQP. Two vertices are connected if the two BQPs share the same variable. A formal definition on the query pattern graph is given as follows.

DEFINITION 9: *Graph $G\langle V,E,L_v,L_e\rangle$ is a* query pattern graph, *where $V = \{v|v$ is a BQP\}, $E = \{e|e = (v_i,v_j), v_i \in V, v_j \in V\}, L_v = \{l_v|l_v = (S(v),Sel(var)), v \in V, S(v)$ is the size of PF(v), var is the variable contained in v, Sel(var) is the selectivity of var in PF(v)\} and $L_e = \{l_e|l_e$ is the variable shared by v_i and $v_j, v_i \in V, v_j \in V, (v_i,v_j) \in E\}$.*

For example, considering the following example query [43], the graph on the left of Figure 9.8 presents the derived query pattern graph.

After obtaining the query pattern graph, we can select MRJs by picking connected vertices. Picking BQP$_i$ and BQP$_j$ from G means that an MRJ has been identified, which shall perform a join of *Predicate* files PF(BQP$_i$) and PF(BQP$_j$). Clearly, there are many possible ways to join BQPs, such as join BQPs on single variable or multiple variables, that is, pairwise join or multiway join. Pairwise join perfectly fits into the (*key,value*) semantics of MapReduce computing framework. We just choose the join key as the hashing key, which distributes data from Map to Reduce such that *joinable* data will be sent to the same Reduce task. However, for multiway join, to complete the operation in one phase of MapReduce, default hashing-based data distribution cannot fulfill the request. As studied in Reference 44, a sophisticated (*key,value*) distribution among Reduce tasks needs to be taken.

SELECT distinct *?a ?b ?lat ?long WHERE* {

1 *?a* dbpedia:wikilink dbpediares:actor.

2 *?a* dbpedia:spouse *?b*.

3 *?a* dbpedia:placeOfBirth *?c*.

4 *?b* dbpedia:wikilink dbpediares:actor.

5 *?b* dbpedia: placeOfBirth *?c*.

6 *?d* pos:lat *?lat*.

7 *?c* owl:sameAs *?d*.

8 *?d* pos:long *? long.* }

(a)

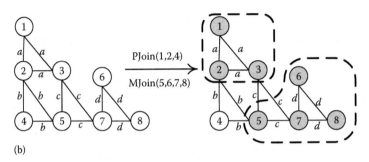

(b)

FIGURE 9.8 Query pattern graph generated from Q7. Part (a) shows the example query, while part (b) shows the query's graph representation in correspondence.

In this chapter, we study an optimal MRJ selection strategy based on a cost model to achieve minimum time span of query processing. For clear illustration, we first classify MRJs into two types: PJoin (Pairwise Join) and MJoin (Multiway Join). We define them as follows:

DEFINITION 10: *Given $G\langle V,E,L_v,L_e\rangle$, PJoin(V') is a join operation on a set of Predicate files PF(v_i), $v_i \in V'$, $V' \subseteq V$, where V' in G are connected by edges labeling with the same variable(s).*

DEFINITION 11: *Given $G\langle V,E,L_v,L_e\rangle$, MJoin(V') is a join operation on a set of Predicate files PF(v_i), $v_i \in V'$, $V' \subseteq V$, where V' in G are connected by edges labeling with more than one variables.*

For example, in Figure 9.8, picking of nodes 1–3 forms a PJoin; whereas picking of nodes 5–8 forms an MJoin. In fact, for a vertex that has been picked in G, we mark it as picked. Every time when we select an MRJ, we make sure that at least one unmarked vertex is picked. Thus, after all vertices in G have been marked, we shall have enough MRJs to answer the original query. Apparently, there are many possible combinations of PJoin and MJoin to answer the query, while each combination implies an execution graph defined as follows.

DEFINITION 12: *An execution graph P is a directed graph with form* $P(V,E)$, *where* $V = \{v|v \text{ is an MRJ}\}$, $E = \{e|e = \langle MRJ_i, MRJ_j \rangle, MRJ_i, MRJ_j \in V\}$.

Given P, we say MRJ_i depends on MRJ_j if and only if MRJ_j's output is a direct input of MRJ_i. A directed edge is added from MRJ_j to MRJ_i in P to specify the execution order. On the contrary, two MRJs are considered independent if they do not incident to the same edge in P. According to P, independent MRJs can be executed in parallel as long as their direct inputs are ready. MRJ that has predecessor(s) must wait until its predecessor(s) are finished.

We summarize the general framework of our solution as follows. We map a SPARQL query to several MRJs, each of which is either a PJoin or an MJoin, and execute the MRJs according to the corresponding P. Since there could be many different P's generated according to different selection of *Join* over G, we want to find such a P that guarantees a minimum query processing time. Formally, we define the problem as follows:

Problem definition: Given a platform configuration Δ, RDF data statistics S, and a query pattern graph G obtained from a SPARQL query Q, find a function $F: (\Delta, S, G) \rightarrow P(V,E)$ such that

1. Let MRJ_i's execution time be t_i. Let $\langle i,j \rangle$ denote a path from MRJ_i to MRJ_j in P, and the weight of this path to be $W_{i,j} = \sum_{k \in \langle i,j \rangle} t_k$.

2. The $\text{Max}\{W_{i,j}\}$ is minimized.

In the problem definition, $W_{i,j}$ indicates the possible minimal execution time from MRJ_i to MRJ_j. Therefore, by minimizing the maximum $W_{i,j}$, the overall minimum execution time of P is achieved.

9.4.5 Query Processing

In this section, we shall first demonstrate how we use the cost model to identify MRJs and generate the execution graph P. Then we show the optimization techniques employed to accelerate the query processing.

9.4.5.1 MRJ Identification and Ordering

The identification and scheduling of MRJs are based on the generated query graph G of a SPARQL query. Given G, the challenge lies in identifying MRJs such that the corresponding query execution plan P guarantees a minimum query answering time. Even for queries with a small number of BQPs and variables, it is impractical to enumerate all possible Ps to find out the optimal solution. Therefore, instead of addressing the generation of P directly, we introduce a tree structure, APJ tree, to represent all possible P's for examination.

First, we consider how two vertices v_i and v_j in G are selected together. If v_i and v_j are the nearest neighbors to each other, no other vertices need to be involved. Otherwise, v_i and v_j together with all the vertices residing on at least one connecting path between v_i and v_j need to be selected. For clear illustration purposes, we introduce the Var-BQP entity concept, which shall derive generalized MRJ types (covers both PJoin and MJoin).

> **DEFINITION 13:** $e^{|\{var_i\}|} = (\{var\}: \{BQP\})$ is a Var-BQP entity, where $\{var\}$ represents a set of edges labeled with elements of $\{var\}$ in G. $\{BQP\}$ represents the set of all the vertices incident to $\{var\}$ in G. If $BQP_i \in \{BQP\}$ does not only incident to edges labeled with elements from $\{var\}$, BQP_i is marked as optional in $\{BQP\}$, denoted by capping BQP_i with a wave symbol.

For example, consider a Var-BQP entity "$a: \{1, \tilde{2}, \tilde{3}\}$" from Figure 9.8. Since BQP 2 and 3 are not only incident to edges labeled with "a," they are considered to be optional and capped with the wave symbol. Now we can answer the one-step selection condition of vertices v_i and v_j with the Var-BQP concept. Selecting v_i and v_j to compose an MRJ is equivalent to select a Var-BQP entity e, where $v_i, v_j \in e \cdot \{BQP\}$. As described in Section 3, MRJs are identified by iteratively selecting and marking vertices from G until all vertices are marked. Correspondingly, it is an iterative selection procedure of Var-BQP entities until all BQPs are selected. Thus, we shall first define the join semantic of Var-BQP entities and present steps to build the APJ tree based on the join of Var-BQP entities. Then we prove that all the possible selection of MRJs can be derived from the APJ tree.

> **DEFINITION 14:** Two Var-BQP entities $e_i^{|\{var_i\}|} = (\{var_i\}: \{BQP_i\})$ and $e_j^{|\{var_j\}|} = (\{var_j\}: \{BQP_j\})$ can be joined together if and only if $\{var_i\} \cap \{var_j\} \neq$; or $\{BQP_i\} \cap \{BQP_j\} \neq$;. Join result of two joinable Var-BQP entities is defined as follows:

$$e_i^{|\{var_i\}|} \bowtie e_j^{|\{var_j\}|} = (\{var_i\} \cup \{var_j\}: \{BQP_i\} \cup \{BQP_j\}) \qquad (9.11)$$

Intuitively, if $e_i^{|\{var_i\}|}$ and $e_j^{|\{var_j\}|}$ do not share a common variable, then there is an absence of join key. Furthermore, if they do not share a common BQP, there is no overlapping of source data. Thus, the join of $e_i^{|\{var_i\}|}$ and $e_j^{|\{var_j\}|}$ is logically invalid.

Based on this join semantic of *Var*-BQP entities, we describe a top-down approach to build an APJ tree, as presented in Algorithm 9.3. Assume G has m vertices and n edges with distinct labels (n variables). By traversing G and grouping BQPs on different variables, we can easily obtain $\{e_i^1\}$ such that $|\{e_i^1\}| = n$ and $\left|\bigcup e_i^1 . \{BQP_i\}\right| = m$. If we apply the join semantics defined earlier, we can further obtain $\{e_i^2\}, \ldots, \{e_i^n\}$. By making each entity a node, and drawing directed edges from e_i and e_j to $e_i \bowtie e_j$, we can obtain a tree structure representing APJ semantics among *Var*-BQP entities (a.k.a. APJ tree).

Algorithm 9.3: Bottom-up algorithm for APJ tree generation

Require: Query pattern graph G of m vertices and n distinct edge labels;
  $V \leftarrow \varnothing$, and $E \leftarrow \varnothing$;
Ensure: APJ tree's vertex set V and edge set E
1. Traverse G to find e_i^1 for each label
2. Add each e_i^1 to V
3. **for** $k = 1$ to $n - 1$ **do**
4. **if** $\exists e_i^k$ and e_j^k are joinable **then**
5.   **if** $\exists BPQ_x$ is only optional in e_i^k and e_j^k among all e^k **then**
6.   Make BPQx deterministic in $e_i^k \bowtie e_j^k$
7.   **end if**
8. $V \leftarrow V \bigcup \{e_i^k \bowtie e_j^k\}$
9. $E \leftarrow E \bigcup \{e_i^k \rightarrow e_i^k \bowtie e_j^k, e_j^k \rightarrow e_i^k \bowtie e_j^k\}$
10. **end if**
11. **end for**

For better illustration, we take G given in Figure 9.8 as an example. An APJ tree can be generated as shown in Figure 9.9. Variables capped with up-arrow indicate the join key. For example, $ab\hat{c}d$ is obtained from $a\hat{c}d$ and $b\hat{c}d$. In fact, any possible P can be obtained by conducting a reverse *breadth first* traversing from an e_i^n, which indicates a final MRJ. Therefore, this tree structure implies all possible combinations of MRJs for identification and scheduling. We give the formal lemma and proof as follows:

LEMMA 3: *An APJ tree obtained from query pattern graph G implies all possible query execution plans.*

PROOF: Based on the generation of APJ tree, $\{e_i^n\}$ gives all the possible combinations that could be obtained from G. Each e_i^n for sure contains all the variables and BQPs, which is exactly the final state in a query execution graph P. e_i^n differentiates from e_j^n as they could be obtained

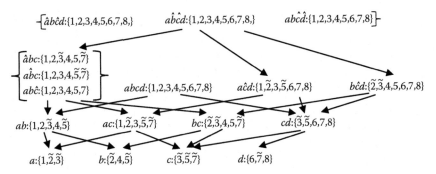

FIGURE 9.9 MRJ identification and ordering of Figure 9.8 example.

from different join plans. Obviously, the join of e_i^k and e_j^k, $1 \le k < n$ (if they are *joinable*), is exactly a PJoin, while e_i^k itself, $1 < k \le n$, is an MJoin.

For each entity e_i^k in the APJ tree, we define its weight as the smaller one of the two cost variables: direct cost $DiC(e_i^k)$ and derived cost $DeC(e_i^k)$. $DiC(e_i^k)$ implies the cost of directly joining $e_i^k \cdot \{BQP_i\}$ with $e_i^k \cdot \{var_i\}$. $DeC(e_i^k)$ implies the accumulative cost of obtaining e_i^k from its ancestors.

LEMMA 4: *The minimum weight of e_i^n indicates the minimum total cost of joining all BQPs to answer the query.*

Lemma 4 can be easily proved by definition. Thus, the problem now becomes finding e_i^n with the minimum weight, which represents the best result for MRJ scheduling. A naive solution is to build up the entire APJ tree and search for the optimal entity. However, we find that it is sufficient to generate and check only part of the APJ tree to obtain the optimal solution. Our top-down search algorithm can effectively prune certain parts of APJ tree that do not contain the optimal solution. Algorithm 9.4 describes searching optimal P during the growth of APJ tree. Since we assume that each BQP only has at most two variables involved, $2m \le n$. It ensures that the complexity of Algorithm 9.4 is no worse than $O(n^2)$.

Algorithm 9.4: MRJ identification for G with m vertices and n edges

Require: Set E for MRJ identification, $E \leftarrow (e_i^1)$ query execution plan $P \leftarrow \emptyset$;; $VT \leftarrow \emptyset$;;

Ensure: P

1. $DeC(e_i^1) \leftarrow DiC(e_i^1)$
2. $VT \leftarrow \bigcup e \cdot BQP, \forall e \in P$
3. $k \leftarrow 1$
4.   **repeat**

5. sort $e_i^k \in E$ on weight in ascending order
6. **while** $\bigcup e_i^k \cdot \{BQP\} = m$ and $\bigcup e_i^k \cdot \{var\} = n$, $e_i^k \in E \backslash \{e_j^k\}$, e_j^k have the heaviest weight in E **do**
7. $$E \leftarrow E \backslash \{e_j^k\}$$
8. **end while**
9. **repeat**
10. for any $e_i^k \in E$
11. **while** $\exists e_j^k \in E \backslash \{e_i^k\}$ that can be joined with e_i^k **do**
12.   **if** $(e_i^k \cdot BQP \cup e_j^k \cdot BQP) \not\subseteq VT$ **then**
13.   **if** $DiC(e_i^k \bowtie e_j^k) \geq DeC(e_i^k \bowtie e_j^k)$ **then**
14. $P \leftarrow P \cup \{e_i^k, e_j^k, e_i^k \bowtie e_j^k\}$
15. **end if**
16.   **if** $DiC(e_i^k \bowtie e_j^k) < DeC(e_i^k \bowtie e_j^k)$ **then**
17. $P \leftarrow P \backslash \{e_i^k, e_j^k\}$
18. $P \leftarrow P \cup \{e_i^k \bowtie e_j^k\}$
19. **end if**
20. **end if**
21. $E \leftarrow E \cup \{e_i^k \bowtie e_j^k\}$
22.   update VT
23. **end while**
24. $E \leftarrow E \backslash \{e_i^k\}$
25.   **until** $\nexists e_i^k \in E$
26. $k \leftarrow k + 1$
27. **until** $k = n$

THEOREM 3: *P computed with Algorithm 9.4 is optimal.*

PROOF: First we prove that Algorithm 9.4 finds the entity e_{opt}^n with the minimal weight. Lines 5–8 in the algorithm guarantees this property. E only contains entities of minimal weight, which are sufficient to cover all the m BQPs and n variables. Thus, when k increases to n, the first e_i^n in E is the entity with the minimum weight.

Now we prove that the computed P is optimal. Since we already found the optimal entity e_{opt}^n, if e_{opt}^n's weight is $DiC(e_{opt}^n)$, P would only contain one MJoin that joins all BQPs in one step (lines 16–19); otherwise, e_{opt}^n's weight is derived from his parents, which would have been added to P (line 13–15). Iteratively, P contains all the necessary entities (equivalent to MRJs) to compute e_{opt}^n, and the longest path weight of P is just e_{opt}^n's weight, which is already the optimal.□

9.4.5.2 Join Strategies and Optimization

With the algorithm presented above, we can get the MRJs and their execution sequence. Note that we have an assumption that cloud computing system can provide as much computing resources as required. In fact, the similar claim was made by Amazon EC2 service [45]. Based on this assumption, we can make MRJs not having ancestor–descendant relationship in P be executed in parallel. Moreover, we adopt hybrid join strategy and bloom filter to improve query efficiency, which are discussed in detail as follows:

Hybrid join strategy. There are three basic strategies for MapReduce join: Reduce-side join (repartition join), in-memory join (broadcast join), and Map-side join (improved repartition join) [46]. Our hybrid strategy works as follows: Reduce-side join is used as the default join strategy. When a *Predicate* file is small enough to be loaded into the memory, we load this file in several Map tasks to perform in-memory join. Map-side join is adopted on the condition that ancestor MRJs' outputs are well partitioned on the same number of reducers.

Bloom filter. We use bloom filter to implement the in-memory join. As described in Section 5, there are a large number of small *Predicate* files. If the query contains a BQP that refers to a small *Predicate* file, we can always read in this file completely into the main memory and generate the bloom filter of *Subject* or *Object* variables, which can be done quite efficiently with one Map task. Since the generated bloom filter file is much smaller, it can be loaded into the memory later on for each MRJ. With the help of bloom filter, large number of irrelevant RDF triples will be filtered out at the Map phase, which greatly reduces the network volume and workload on Reduce tasks. Experiments have proven the effectiveness of this optimization strategy.

9.4.6 Implementations

In this section, we first present our overall system design. Then we elaborate on the data preprocessing techniques, which facilitate the join processing.

9.4.6.1 System Design

We use Hadoop distributed file systems (HDFS) [47] to set up a repository of large-scale RDF dataset. Figure 9.10 presents a system overview of our solution to RDF query processing. The whole system is backed with well-organized RDF data storage on HDFS, which offers block-level management that promises efficient data retrieval. The query engine accepts users' queries and decides

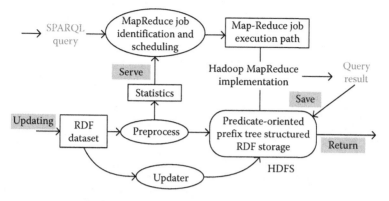

FIGURE 9.10   System design.

the corresponding MRJs and an optimal execution plan for each query. Notice that the SPARQL query engine can be deployed to as many clients as desired. Therefore, the query engine will not become a performance bottleneck.

9.4.6.2 Preprocessing and Updates

The preprocessing of raw RDF data involves four steps: (1) group by *Predicate*, (2) sort on both *Subject* and *Object* for each *Predicate* file (the similar strategy was used in References 48 and 49), (3) blockwise partition for each *Predicate* file, and (4) build a "prefix tree" to manage all the *Predicate* files. Intuitively, the first three steps help us obtain a content-based blockwise index for the entire RDF dataset. The last step is motivated by the fact that many *Predicates* are described within the same *namespace*. By transferring the *namespace* into a prefix tree structure, each *Predicate* file is stored under its prefix directory. Such a data structure promises effective metadata compression and RDF semantics maintenance. Figure 9.11 shows an example of how the top five largest *Predicate* files from the Billion Triple Challenge 2009 dataset are organized in HDFS. *SO* represents an ordered storage according to the alphabet order of *Subject* and vice versa (*OS*).

We take special care of *Predicate* files of small size. For those *Predicate* files whose sizes are less than a threshold, for example, <1% of the block size, we first have them merged and sorted according to the *Predicate* value and then split them into block-sized files. *B*-tree is used to manage these special blocks, which would be sufficiently small to fit into a query engine's main memory.

As mentioned in the previous section, the sizes of *Predicate* files and the selectivities of *Subject* and *Object* for each *Predicate* file will serve MRJ identification and scheduling. In fact, statistics collection is a side product of the data partition process.

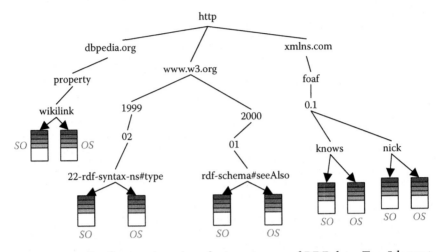

FIGURE 9.11   Predicate-oriented prefix tree storage of RDF data: Top-5 largest *Predicate* files from the Billion Triple Challenge 2009.

When there come new updates, the updated strategy is as follows. First, locate the block for new update by checking its *Predicate*. If the *Predicate* can be interpreted in the prefix tree directory, then locate the directory and further refer to the data blocks from both the *Subject*- and *Object*-sorted list. If the *Predicate* indicates that this triple belongs to a small predicate file, search the B^+ tree to locate a data block that contains triples with the same *Predicate* value. Second, after the data storage block is identified, just append the update to the original records until certain threshold is met. When a data block must split, scanning and sorting are performed to guarantee the range integrity of new split blocks. The update of corresponding block-level index is an in-memory operation, and therefore can be done very efficiently.

9.4.7 Related Work

There are mainly two categories of solutions for RDF management and query processing. One is to use traditional relational database management system (RDBMS) technologies, either stand-alone server or distributed (parallel) computing framework. RDF data are represented as tables in databases. Intensive research interests lies in RDF decomposition (SW-Store [50]) or composition (property table [51]), index building and searching (Hexastore [49], RDF-3X [48]), and query optimization [43]. However, due to the limitation of RDBMS's scalability, the above solutions cannot meet the demands for the management of extremely large-scale RDF data in the coming future. Moreover, the query efficiency claimed in these solutions is generally obtained by taking the advantages provided by the system architecture. Thus, these solutions are closely related with system hard states, which are hard to maintain.

The other category solution is to incorporate NoSQL database to address the scalability and flexibility issues in the first place. Many works, such as References 39–41 and 52–55, adopt the cloud platform to solve the RDF data management problem. However, many of them focus on utilizing high-level definitive languages to create simplified user interface for RDF query processing, which omits all the underlying optimization opportunities and has no guarantees on efficiency.

There are a few works directly conducting RDF query processing within the MapReduce framework. RDFGrid [56] is an open source project, which models RDF data as objects and processes with Hadoop MapReduce. It provides a plug-in parser and simple aggregation processing of RDF data. However, join semantics and strategy still need to be defined by users. A greedy strategy is proposed in Reference 41, which is similar to the solution of Reference 39 and always tries to pick a join that may produce the smallest size of intermediate results. This strategy also has no guarantee on the overall efficiency. Husain et al. [39] presents an improved work in Reference 40 and focuses on effective RDF data storage and querying. Our solution is different from Reference 39 in the following aspects. First, we further partition large predicate files into data blocks containing *Subject* (*Object*) ordered triplet sequence. In addition, we have all data blocks organized with their content metadata attached, which allow for locating and processing efficient block level data. Second, by identifying the optimal MRJ execution plan and employing optimization techniques (hybrid join strategies and bloom filter), our solution promises more efficient query processing. As already given in the experiments, the query processing time of our solution can be 1 order of magnitude less compared to that in Reference 39. Third, we use real datasets for evaluation and reveal how job configuration affects MRJ efficiency, which is absent in Reference 39. Reference 16 studies the general strategy of replicating data from Map to Reduce that guarantees a minimal network volume in solving different kinds of joins with one Map-Reduce procedure. Since this framework leaves no space for pipelining and materialization, its practical effectiveness still needs justification.

9.4.8 Summary

In this case study, we present an efficient evaluation of SPARQL queries over large RDF dataset with MapReduce. By analyzing the dominating cost of an MRJ, we define the transmission scheme from an original SPARQL query to MRJs. We propose a novel APJ tree-based MRJ scheduling technique that guarantees an optimal query processing time. We elaborate our optimization techniques (hybrid join and bloom filter) and further explore how to fit our methodology for other join operators.

REFERENCES

1. Borthakur, D. et al. (2011). Apache Hadoop goes realtime at Facebook. In: *Proceedings of the 2011 ACM SIGMOD International Conference on Management of Data*. June 12–16, ACM, Athens, pp. 1071–1080.
2. Das, S. et al. (2010). G-Store: A scalable data store for transactional multi key access in the cloud. In: *Proceedings of the 1st ACM Symposium on Cloud Computing*. June 10–11, ACM, Indianapolis, IN, pp. 163–174.
3. Iosup, A. et al. (2011). Performance analysis of cloud computing services for many-tasks scientific computing. *IEEE Transactions on Parallel and Distributed Systems* 22(6): 931–945.
4. Wu, S. et al. (2011). Query optimization for massively parallel data processing. In: *Proceedings of the 2nd ACM Symposium on Cloud Computing in conjunction with SOSP*. October 26–28, ACM, Cascais, Portugal.
5. Lee, R. et al. (2011). YSmart: Yet another SQL-to-MapReduce translator. In: *Proceedings of the 31st International Conference on Distributed Computing Systems*. June 20–24, IEEE Computer Society, Minneapolis, MN, pp. 25–36.
6. Afrati, F.N. and Ullman, J.D. (2011). Optimizing multiway joins in a mapreduce environment. *IEEE Transactions on Knowledge and Data Engineering* 23(9): 1282–1298.
7. Okcan, A. et al. (2011). Processing theta-joins using MapReduce. In: *Proceedings of the 2011 ACM SIGMOD International Conference on Management of Data*. June 12–16, ACM, Athens, pp. 949–960.
8. Zhang, X. et al. (2012). Efficient multi-way theta-join processing using MapReduce. *Proceedings of the VLDB Endowment* 5(11): 1184–1195.
9. Dean, J. and Ghemawat, S. (2008). MapReduce: Simplified data processing on large clusters. *Communications of the ACM* 51(1): 107–113.
10. Agrawal, P. et al. (2008). Scheduling shared scans of large data files. *Proceedings of the VLDB Endowment* 1(1): 958–969.
11. Nykiel, T. et al. (2010). MRShare: Sharing across multiple queries in MapReduce. *Proceedings of the VLDB Endowment* 3(1/2): 494–505.
12. Jaynes, E.T. (2003). *Probability Theory: The Logic of Science*. Cambridge: Cambridge University Press.
13. Condie, T. et al. (2010). MapReduce online. In: *Proceedings of the 7th USENIX Conference on Networked Systems Design and Implementation*. April 28–30, USENIX Association, San Jose, CA, pp. 21–21.

14. Battré, D. et al. (2010). Nephele/PACTs: A programming model and execution framework for web-scale analytical processing. In: *Proceedings of the 1st ACM Symposium on Cloud Computing.* June 10–11, ACM, Indianapolis, IN, pp. 119–130.
15. Vernica, R. et al. (2010). Efficient parallel set-similarity joins using MapReduce. In: *Proceedings of the 2010 ACM SIGMOD International Conference on Management of Data.* June 6–10, ACM, Indianapolis, IN.
16. Afrati, F.N. et al. (2010). Optimizing joins in a map-reduce environment. In: *Proceedings of the 13th International Conference on Extending Database Technology.* March 22–26, ACM, Lausanne, Switzerland.
17. Dittrich, J. et al. (2010). Hadoop++: Making a yellow elephant run like a cheetah (without it even noticing). *Proceedings of the VLDB Endowment* 3(1/2): 515–529.
18. He, Y. et al. (2011). RCFile: A fast and space-efficient data placement structure in MapReduce-based warehouse systems. In: *Proceedings of the 2011 IEEE 27th International Conference on Data Engineering.* April 11–16, IEEE Computer Society, Hannover, Germany, pp. 1199–1208.
19. Zhou, J. et al. (2010). Incorporating partitioning and parallel plans into the scope optimizer. In: *2010 IEEE 26th International Conference on Data Engineering.* March 1–6, IEEE, Long Beach, CA, pp. 1060–1071.
20. Chaudhuri, S. and Vardi, M.Y. (1993). Optimization of real conjunctive queries. In: *Proceedings of the 12th ACM SIGACT-SIGMOD-SIGART Symposium on Principles of Database Systems.* May 25–28, ACM Press, Washington, DC, pp. 59–70.
21. Tan, K.L. and Lu, H. (1991). A note on the strategy space of multiway join query optimization problem in parallel systems. *SIGMOD Record* 20(4): 81–82.
22. Yuanyuan, F. and Xifeng, M. (2010). Distributed database system query optimization algorithm research. In: *3rd IEEE International Conference on Computer Science and Information Technology.* July 9–11, IEEE, pp. 657–660.
23. Lee, C. et al. (2001). Optimizing large join queries using a graph-based approach. *IEEE Transactions on Knowledge and Data Engineering* 13(2): 298–315.
24. Gibbons, A. (1985). *Algorithmic Graph Theory.* Cambridge: Cambridge University Press.
25. Brightwell, G. and Winkler, P. (2004). Note on counting Eulerian circuits. *CoRR* cs.CC/0405067, retrieved from http://arxiv.org/abs/cs.CC/0405067
26. Cormen, T.H. et al. (2009). *Introduction to Algorithms* (3rd edn.). Cambridge, MA: MIT Press.
27. Jansen, K. (2004). Scheduling malleable parallel tasks: An asymptotic fully polynomial time approximation scheme. *Algorithmica* 39: 59–81.
28. Gelfand, I. et al. (2000). *Calculus of Variations.* Mineola, NY: Dover Publications.
29. Feige, U. (1998). A threshold of ln *n* for approximating set cover. *Journal of the ACM* 45(4): 634–652.
30. Harth, A. *Billion Triples Challenge 2010 Dataset.* Available at http://km.aifb.kit.edu/projects/btc-2010/.
31. Linked Data, http://linkeddata.org/.

32. World Wide Web Consortium recommendations—SPARQL 1.1 Query Language, http://www.w3.org/TR/sparql11-query/.
33. Arias, M. et al. (2011). An empirical study of real-world SPARQL queries. *CoRR* abs/1103.5043, retrieved from http://arxiv.org/abs/1103.5043
34. Picalausa, F. and Vansummeren, S. (2011). What are real SPARQL queries like? In: *Proceedings of the International Workshop on Semantic Web Information Management.* June 12, ACM, Athens, Greece.
35. Duan, S. et al. (2011). Apples and oranges: A comparison of RDF benchmarks and real RDF datasets. In: *Proceedings of the 2011 ACM SIGMOD International Conference on Management of Data.* June 12–16, ACM, Athens, pp. 145–156.
36. Husain, M.F. et al. (2011). Scalable complex query processing over large semantic web data using cloud. In: *Proceedings of the 2011 IEEE 4th International Conference on Cloud Computing.* IEEE, pp. 187–194.
37. Zhang, X. et al. (2012). Towards efficient join processing over large RDF graph using MapReduce. In: *Proceedings of the 24th International Conference on Scientific and Statistical Database Management.* pp. 250–259.
38. Prasser, F. et al. (2012). Efficient distributed query processing for autonomous RDF databases. In: *Proceedings of the 15th International Conference on Extending Database Technology.* pp. 372–383.
39. Husain, M.F. et al. (2010). Data intensive query processing for large RDF graphs using cloud computing tools. In: *Proceedings of the 2010 IEEE 3rd International Conference on Cloud Computing.* July 5–10, IEEE, Miami, FL.
40. Husain, M.F. et al. (2009). Storage and retrieval of large RDF graph using Hadoop and MapReduce. In: *Proceedings of the 1st International Conference on Cloud Computing.* December 1–4, Springer, Beijing, China.
41. Myung, J. et al. (2010). SPARQL basic graph pattern processing with iterative MapReduce. In: *Proceedings of the 2010 Workshop on Massive Data Analytics over the Cloud in conjunction with WWW.* April 26, Raleigh, NC.
42. Tanimura, Y. et al. (2010). Extensions to the pig data processing platform for scalable RDF data processing using Hadoop. In: *Proceedings of the 26th International Conference on Data Engineering Workshops.* March 1–6, pp. 251–256.
43. Neumann, T. and Weikum, G. (2009). Scalable join processing on very large RDF graphs. In: *Proceedings of the 2009 ACM SIGMOD International Conference on Management of Data.* pp. 627–640.
44. Afrati, F.N. and Ullman, J.D. (2010). Optimizing joins in a map-reduce environment. In: *Proceedings of the 13th International Conference on Extending Database Technology.* pp. 99–110.
45. Amazon Web Services, http://aws.amazon.com/ec2/.
46. Blanas, S. et al. (2010). A comparison of join algorithms for log processing in MapReduce. In: *Proceedings of the 2010 ACM SIGMOD International Conference on Management of Data.* June 6–10, ACM, Indianapolis, IN, pp. 975–986.
47. The Apache Software Foundation. Hadoop. http://hadoop.apache.org/.

48. Thomas, N. and Weikum, G. (2010). The RDF-3x engine for scalable management of RDF data. *The VLDB Journal* 19(1): 91–113.
49. Weiss, C. et al. (2008). Hexastore: Sextuple indexing for semantic web data management. *Proceedings of the VLDB Endowment.* 1(1): 1008–1019.
50. Abadi, D.J. et al. (2009). SW-Store: A vertically partitioned DBMS for semantic web data management. *The VLDB Journal* 18: 385–406.
51. Jena Project, http://jena.sourceforge.net/.
52. Newman, A. et al. (2008). A scale-out RDF molecule store for distributed processing of biomedical data. In: *Proceedings of the 17th International World Wide Web Conference on Semantic Web for Health Care and Life Sciences Workshop.* April 22, Beijing, China.
53. Newman, A. et al. (2008). Scalable semantics—The silver lining of cloud computing. In: *Proceedings of the 2008 4th IEEE International Conference on eScience.* December 7–12, IEEE Computer Society, Indianapolis, IN.
54. Urbani, J. et al. (2009). Scalable distributed reasoning using MapReduce. In: *Proceedings of the 8th International Semantic Web Conference.* October 25–29, Springer, Chantilly, VA.
55. McGlothlin, J.P. et al. (2009). RDFKB: Efficient support for RDF inference queries and knowledge management. In: *Proceedings of the 2009 International Database Engineering and Applications Symposium.*
56. RDFGrid Project, http://rdfgrid.rubyforge.org/.

CHAPTER 10

Development of a Framework for the Desktop Grid Federation of Game Tree Search Applications

I-Chen Wu

National Chiao Tung University
Hsinchu, Taiwan

Lung-Pin Chen

Tunghai University
Taichung, Taiwan

CONTENTS

W E DISCUSS THE DEVELOPMENT of desktop grids of dynamic game tree applications, which are widely used but considered beyond the scope of the previous platforms. The proposed desktop grid platform adopts a push-mode streaming infrastructure to support tightly coupled task control that is vital to the target applications. In addition, the new platform provides a software framework in order to facilitate complex application development on the desktop grids. The users have reported successful results in rapid application development as well as efficient performance for a variety of game tree search applications.

10.1 INTRODUCTION

Desktop grid is a network computing model that can harvest unused computational power from desktop-level computers [2,4,16]. Considering the fact that today's personal computers are more powerful than workstations or even mainframes 20 years ago, this model can offer low-cost and readily available resources by employing a large enough number of workers. Today, more and more research organizations have built desktop grids as a solution for their large-scale e-science projects [1,4,17,23].

Unlike most distributed computing models, desktop grids have remarkable resilience in host connections. The execution of desktop grid applications is coordinated by a central server node, which distributes the task units over widely worker nodes, awaits the execution results, and eventually consolidates the result. The worker nodes can be of different operating systems and are not necessarily connection oriented. For volunteer computing, a server partitions and assigns tasks to the public anonymous participants called volunteers. Since these volunteers are autonomous and can connect or disconnect from time to time, the single-worker response time is not a major concern. Statistically, the resource availability can be maintained at a certain level with a large enough number of volunteers. Another model of desktop grid is to have dedicated workers that are maintained and directly controlled by the organization. Using dedicated computers guarantees both quality and quantity of worker nodes.

Most existing desktop grids are developed, which are intended to host bag-of-task (BoT) applications that contain a large set of task units without explicit precedence relations. These independent tasks can be successfully executed via massive parallelism over widely worker nodes. The server usually does not try to precisely control each single worker for a shortest response time. Instead, it simply uses the polling mechanism to distribute and collect the execution results over workers in a daily or even weekly basis. This model has been demonstrated by a number of successful e-science projects including data mining, parallel simulations, computational biology, and computer imaging.

The nature of loosely coupled communication of BoT application tasks in desktop grids makes resource sharing much easier compared to other network computing models. Based on the notion of reciprocal resource sharing, the emerging desktop grid federation has enabled many overloaded e-science projects for the resource-restricted organizations [1,2]. Some related infrastructures for resource sharing are discussed as follows:

- *Single volunteer desktop grid.* In this approach, all the worker nodes in different organizations are directly connected to a central server [3,4]. The server manages the tasks based on worker credits or membership profile. This approach cannot support the cooperative federation with customized policies.

- *Grid computing community.* A grid computing community [5–7] usually relies on a central resource broker to provide a single point of access for the resources across several organizations. A desktop grid can also require worker nodes via the broker of the grid system.

- *Peer-to-peer (P2P) computing platform.* This platform is developed based on the P2P network and can easily achieve fairness. The key is to transfer the data via the P2P network and to evenly distribute communication and computation load over the entire network.

Nevertheless, the above approaches do not fit the requirements of game tree search applications, which need to generate and prune tasks over loosely coupled worker nodes in a timely manner. To address the above issues, we discuss a software framework of desktop grid federation for enabling the dynamic computation applications. We develop a resource broker that uses two-stage scheduling to ensure fairness resource sharing for the workers in and across the desktop grids. Also, the proposed broker supports push-mode communication that can generate and prune tasks in a timely manner. So, prompt interaction and dynamic job scheduling can be achieved. For example, in case that one move of a board game is found to be winning, the push-back winning message promptly hints the clients to stop jobs under other sibling.

The proposed platform has been used for the research programs of game tree search involving at least five academic organizations. The experience demonstrates that, by using the desktop grid federation with a well-designed broker, a set of resource-restricted organizations can perform large-scale dynamic computation tasks via reciprocal resource sharing.

The remainder of this chapter is organized as follows. Section 10.2 reviews our previous work on computer board game systems. Section 10.3 discusses the requirement of paralleling game tree search applications. Sections 10.4 and 10.5 discuss the design and development of the software framework of the desktop grid system. Finally, Section 10.6 provides the concluding remarks.

10.2 GAME TREE SEARCH APPLICATIONS

10.2.1 Computer Board Games

A typical board game contains two players, *Black* and *White*, which alternately place black and white stones on empty intersections of a Go board (a 19 × 19 board) in each turn. The common computer board games include Connect6 [26], Chess [19,20], Chinese Chess [29], Go [15,18], and Shogi [19]. For example, in Connect6, two players alternately place two black and white stones, respectively, on empty intersections of a Go board (a 19 × 19 board) in each turn. Black plays first and places one stone initially. The player who gets six consecutive stones of his/her own first horizontally, vertically, and diagonally wins.

A *game tree* is a directed graph whose nodes represent the states of the game board and whose edges represent the moves. The computer board games heavily rely on tree search algorithms in several ways. Starting from

TABLE 10.1   Complexities of Computer Board Games

Game	Board Size	State-Space Complexity	Game Tree Complexity	Branching Factor
Tic-tac-toe	9	3	5	4
Chinese Checkers	121	23		
Chess	64	47	123	35
Connect6	361	172	140	46,000
Shogi	81	71	226	92
Go (19×19)	361	171	360	250

a state, the game tree search algorithm is used to evaluate all possible moves and select a move based on certain policy. The challenge of computer board game comes from the fact that the size of state space of the game trees is usually exponential to the input size, as shown in Table 10.1. Thus, an efficient tree search algorithm seeks to prune useless paths and go deep to the possible best moves. A typical strategy for evaluating the best moves is to run a Monte Carlo tree search (MCTS) simulation of the game playing processes [15,18,19].

Among the above board games, Connect6 attracted much attention due to three merits: fairness, simplicity of rules, and game complexity. First, Connect6 is fair in the sense of balancing. For example, each player has one stone more than the other, after finishing a move and before the opponent makes the next move. Second, Connect6 is simple in the sense that no extra rules are imposed. In contrast, prohibition rules of double threes and double fours for Black are imposed in Renju, a professional version of five-in-a-row games [2], for the sake of balancing. Third, Connect6 is complex in the sense of game tree complexity, since the combination of choosing two intersections to place stones is normally much higher than that of choosing one.

10.2.2 Application Components

The game tree search application contains two major modules: a *game record editor* and a *job-level* (or *JL*) *module*. The game record editor is the interface of the computer board games which displays game status and waits for the player commands. The JL module is the component that executes jobs such as searching the best game moves or detecting the end-of-game.

Using the game record editor, players can browse, interpret, and process the game records that are stored in the standard Smart Game Format

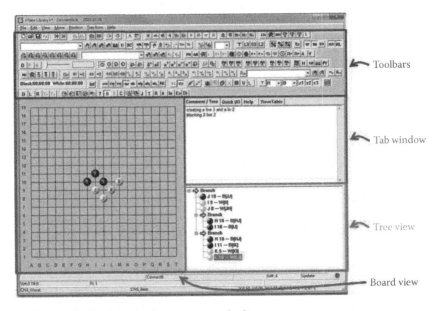

FIGURE 10.1 The layout of a game record editor.

(SGF) [8]. In addition to storing and querying, some game record editors also support the extensive features such as variations in move trees, move comments, threat hints, and plug-ins.

We modified the open-source editor named RenLib to fit our applications such as Connect6 [5], Go, and Chinese Chess. Figure 10.1 shows the layout of our game record editor. The panel in the lower left part of the figure is the board view that shows the current position. The one in the lower right part is the tree view that shows the game record tree. The one in the center right part is the tab window that provides the users with some utilities, for example, comments on a position or a console output for debugging. Toolbars listed on the top are used to provide the users with a variety of functionalities via buttons.

The game editing module uses the model-view-controller (MVC) design pattern and includes two components: model and view corresponding to the same name in MVC. The controller is not encapsulated into a class because we use Microsoft foundation classes (MFC) to construct our software framework that has its own mechanism to map user interface (UI) events to functions.

The JL module accepts the game tree search jobs from the game record editor and dispatches them to the workers for running. The jobs include by giving a start position, finding the best move, expanding all moves, or

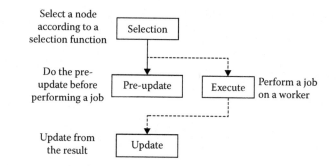

FIGURE 10.2 JL model.

running an MCTS simulation. The execution result can be the best move, all the expanded moves, or the simulation result for updating the tree. The JL model, shown in Figure 10.2, includes four phases: selection, pre-update, execution, and update. The JL module enables template-based application development of the game tree search applications. A JL task can be deployed to a desktop grid federation with dynamic control functionalities that are vital to the game tree applications.

10.3 PARALLELISM OF GAME TREE SEARCH APPLICATIONS

In the board game application, both the editor and the JL module take huge amount of time or uncertain time for executing tasks, making them difficult to be integrated. Thus, it becomes significant to offload the game tree search jobs to other workers.

In Connect6 applications, two approaches or their combination are used to run jobs in parallel. The first approach is simply to trigger jobs in parallel manually. When the users read intriguing positions, they can trigger "M" and "V" operations for these positions via the editor tools shown in Figure 10.1.

The "M" operation is used to invoke NCTU6 to find a best move at a given game state, while the "V" operation is used to verify whether the moves lead to a win (or lose) state. When these operations are triggered, the jobs are generated and then dispatched to remote available workers. If no more workers are available, these jobs wait in a job queue maintained inside the program Connect6Lib itself and will be dispatched later. Using the editor tools, the game developers can control the execution of several operations for different positions concurrently.

The second approach is to use a program to run jobs in parallel automatically. In Reference 9, JL proof number search (JL-PNS) is used to generate moves automatically. JL-PN search is a kind of PN search, where each search tree node is a heavy job, each requiring tens of seconds or more.

PN search is a kind of best-first search algorithm that was successfully used to prove or solve the theoretical values of game positions for many games [9,10], such as Connect Four, Gomoku, Renju, Checkers, Lines of Action, Go, and Shogi. PN search is based on an AND/OR search tree, where each node is associated with PN/disproof number that indicates the minimum number of node expansions (or evaluations) to prove/disprove the node. During each round, the search chooses one node, named the most proving node (MPN), expands it, and then reevaluates the PN/disproof number of the node and its ancestors. An important property about MPN is as follows: If the MPN is proved/disproved, the PN/disproof number of the root decreases by one. Thus, if the root is to be proved/disproved, the PN search will use the MPNs to lead to proving/disproving the root.

Like the most best-first search, PN search has a well-known disadvantage: the requirement of maintaining the whole search tree in the memory. Therefore, many variations [9,10] were proposed to avoid this problem, such as two-level PNS (PN^2), Depth-First Proof-Number Search (Df-PN), Proof-Number* (PN*), and Proof-number and Disproof-number Search (PDS). With the JL-PN search, it becomes possible to maintain the JL-PN search tree inside the client memory without much problem according to our experiences for Connect6.

In JL-PN search for Connect6 (described in more detail in Reference 9), NCTU6 is used to expand OR nodes (or generate nodes from OR nodes), whereas Verifier is used to expand AND nodes. In our experiences, the search tree usually contains no more than 1 million nodes, which can fit process (client) memory well. Assume that it takes 1 minute (60 seconds) to run NCTU6. A volunteer computing system with 60 processors takes about 11 days to build a tree up to 1 million nodes. In such cases, we can manually split one JL-PN search into two.

From the above two approaches, Connect6 applications require both prompt interaction (for the first approach) and dynamic job scheduling (for the second approach). For the former, prompt interaction, the users want to read the returning messages such as the best moves and all the possible defensive moves promptly. As for the latter, highly dynamic job scheduling, in the case that some node is proved, all the subtree nodes should be aborted immediately. Similarly, if the node is disproved, the subtree nodes should be aborted.

The pruning process of tree search algorithm can be even more complex for the random simulation processes. In the JL-PN search [9], it is dynamic to choose the MPNs. From the above observations, it is clearly inappropriate to use the Berkeley Open Infrastructure for Network Computing (BOINC), or some similar middleware systems, which are based on the pull model. The argument here shows the necessity of using the push model.

10.4 SYSTEM DESIGN AND DEVELOPMENT

This section discusses the design and implementation of the computer game desktop grid (CGDG) framework. Section 10.4.1 discusses the organization of users and workers. Section 10.4.2 discusses the broker protocol, including group management, worker management, task management, and connection management.

10.4.1 Organizations

The CGDG system consists of users, workers, and a broker. A user is usually the game record editor mentioned in Section 10.2, which accepts the game player's instructions and initiates the game computation tasks. The tasks are queued in the broker and dispatched to some workers. The JL module is a worker component that executes a tree search task.

The CGDG maintains four types of users, each with a different permission level, as described below:

- System administrator: The administrator with full access to every aspect of system data, including user profiles, organization profiles, broker policies, and job priorities

- Organization administrator: Similar to the system administrator but restricted to the data of an organization.

- Standard user: A registered user that can submit normal tasks via the game record editor.

- Advanced user: A user that is authenticated to submit tasks with high priority.

In order to distinguish the tasks in and across organizations, users and workers are grouped based on their organizations. Figure 10.3 illustrates several users and workers in two organizations.

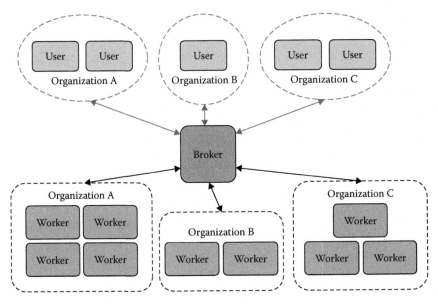

FIGURE 10.3   Organization of users and workers.

10.4.2 Resource Broker

This section presents the resource broker for allocating tasks among workers in several organizations. The system is called a push-based volunteer computing (PVC) system, since all connections among the broker, clients, and workers are all dedicated and are allowed to push jobs or messages immediately. For example, the clients push jobs to the broker that in turn pushes to workers, and the workers push or stream the results back to the broker that in turn pushes or streams them back to the clients immediately. So, prompt interaction and highly dynamic job scheduling can be achieved (Figure 10.4).

10.4.3 Broker Algorithm

This section discusses the design and development of the resource allocation policies used in the resource broker of the desktop grid federation.

In the computing environment of the desktop grid federation, resource competition among organizations is essential. In order to facilitate resource sharing, we adopt the following resource allocation principles. First, the tasks of an organization can be assigned to its own workers with highest priority. Thus, an organization donates its resources only when its task completion rate exceeds that of generation. Among different organizations, tasks are allocated based on *fairness* and *starvation-free* principles.

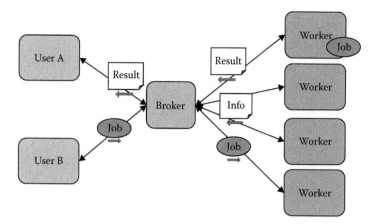

FIGURE 10.4 The push-based volunteer computing system.

In the CGDG federation, the broker records the *credit* for each organization, which is calculated based on the amount of resources that an organization donates to the others. The amount of resources is calculated in terms of CPU cycles, storage space, and network bandwidth. When there are two or more organizations that try to assign tasks via the federation broker, the priority is basically proportional to the credits. Also, in order to prevent starvation, the credits decline over time in a certain ratio.

10.4.4 Broker Protocols

This section discusses the communication protocols between the broker, users, and workers.

10.4.4.1 Group Management: Join and Leave

The group management commands are used for adding and removing users, workers, and broker from the system. In this report, we simply discuss the commands for users; the commands for other roles are similar and are ignored herein.

10.4.4.1.1 NEW_USER, INIT_USER, and REJECT_USER   The group management protocol for users includes three commands: NEW_USER, INIT_USE, and REJECT_USE. A user sends NEW_USER command to the broker to request to join the desktop grid. Upon granted permission, the user then submits his/her account information to login to the system. The broker replies with either a INIT_USER or a REJECT_USER message, according to the authentication result (Figure 10.5).

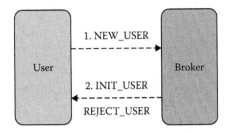

FIGURE 10.5 Communication diagram of user login process.

10.4.4.2 Worker Management
The worker selection commands consist of the instructions used to query the status of a specified worker. The selection conditions can be categorized into two types: the pattern of id/name or the pattern of hardware specification. The conjunction and disjunction of several clauses are also supported.

10.4.4.2.1 WALIVE and WCLOSE   A user periodically sends WALIVE message to the broker to notify the availability of this user. Also, the user sends a WCLOSE message to the broker upon termination of the user application. Figure 10.6 illustrates the WALIVE message between a worker and the broker.

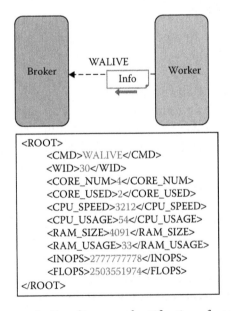

FIGURE 10.6 Communication diagram of notification of worker status.

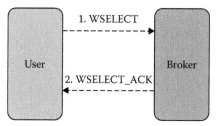

FIGURE 10.7 Communication diagram of worker selection.

10.4.4.2.2 WSELECT The worker selection commands consist of the instructions used to query the status of a specified worker. The selection conditions can be categorized into two types: the pattern of id/name or the pattern of hardware specification. The conjunction and disjunction of several clauses are also supported (Figure 10.7).

An example of querying hardware specification by using union and intersection is demonstrated as follows:

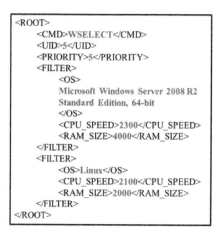

10.4.4.3 Task Management

The task management commands are used for managing the life cycle of the execution of a task, including initiation, cancel, abort, suspend, and wakeup.

The JABORT message is used to abort the execution of a task. Since the task may be hosted in some other organization, this message is first submitted to the broker and then forwarded to the destination worker. The destination worker issues a JABORT_ACK to the broker after successfully aborting the task (Figure 10.8).

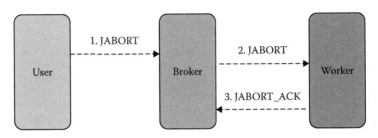

FIGURE 10.8   Communication diagram of job abortion.

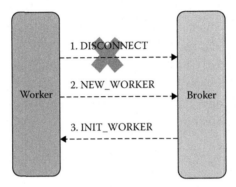

FIGURE 10.9   Communication diagram of reconnecting workers.

Also, the WSLEEP and WAWAKE messages are used to put tasks into sleep mode and active mode, respectively.

10.4.4.4 Connection Management Commands

As mentioned earlier, the CGDG maintains dedicated connections between users and workers for enabling dynamic game tree search tasks. There are user–broker and worker–broker HEART_BEAT messages to identify the network node availability. When the broker detects that some node fails, it will try to automatically reestablish the lost connection via sending NEW_WORKER and INIT_WORKER messages (Figure 10.9).

10.5 TEMPLATE-BASED SOFTWARE DEVELOPMENT OF DESKTOP GRID APPLICATIONS

10.5.1 Software Framework

Although desktop grid federation has enabled many large-scale e-science projects for the resource-restricted organizations, software development and maintenance could be costly. The JL module offers a template of JL search algorithm, which uses a design pattern called template method, and provides the accessibility to the desktop grids. The JL developers can rapidly implement their JL search algorithm by extending this template.

The JL module accepts the game tree search jobs from the game record editor and dispatches them to the workers for running. The jobs include by giving a start position, finding the best move, expanding all moves, or running an MCTS simulation. The execution result can be the best move, all the expanded moves, or the simulation result for updating the tree.

The JL model, as shown in Figure 10.2, includes four phases: *selection, pre-update, execution*, and *update*. First, in the selection phase, select a node according to a selection function based on some search techniques. For example, PN search selects the MPN [11,12], and MCTS selects a node based on the so-called tree policy (defined in References 4 and 13). Note that the search tree is supposed to be unchanged in this phase.

Second, in the pre-update phase, update the tree in advance to prevent from choosing the same node. In this phase, several policies can be used to update the search tree. For example, the flag policy is to set a flag on the selected node so that the flagged nodes will not be selected again.

Third, in the execution phase, perform a job for a position on an idling worker as mentioned earlier, for example, finding the best move from a node n, expanding all moves of n, or running a simulation from n for MCTS.

Fourth, in the update phase, update the search tree according to the job result, for example, generating a node for the best move, generating nodes for all expanded moves, and updating the status on the path to the root.

The template of the JL module includes eight functions that are grouped into three event handlers:

1. Initialization event handler: This event is triggered when a user demands to start the computation of the game tree tasks.

2. Idle worker event handler: This event is triggered when an idle worker is available. When an idle worker is reported, the application, deployed to this worker, is invoked to go through the *selection, pre-update*, and *execution* phases in order.

3. Returning job result event handler. This event is triggered when a job result is returned. When this event is reported, the application is invoked to go through the *update* phase.

TABLE 10.2   Functions to Override in a Game Tree Search Application of the Desktop Grid Federation

```
JobLevelProofNumberSearch::initialize(...)
JobLevelProofNumberSearch::select(...)
JobLevelProofNumberSearch::preupdate(...)
JobLevelProofNumberSearch::dispatch(...)
JobLevelProofNumberSearch::parse(...)
JobLevelProofNumberSearch::update(...)
JobLevelProofNumberSearch::checkFinish(...)
JobLevelProofNumberSearch::finalize(...)
JobLevelProofNumberSearch::PnsPolicy(...)
```

The event handler *initialize* is invoked when a user starts a computation of the game tree tasks. The *select* function is invoked once the worker is available. In the *select* function, the user implements the policies of selecting the next child node to be evaluated in the game tree. After some node is selected, the function *pre-update* is invoked to perform the necessary preparation for execution. Then, the *execute* function performs the main functions of the game tree search task. Finally, the *update* function is invoked when the *execute* function finishes.

To summarize, for developing a game tree search application, the developers simply inherit the base, `BaseJobLevelAlgorithm`, and override the abstract functions as listed in Table 10.2.

10.5.2 Application Cases

The applications that are developed by using our framework are listed in Table 10.3. Circle denotes finished projects and triangle denotes ongoing projects. The line counts of each game and the base module are listed in Table 10.4. This platform has been used for the research programs

TABLE 10.3   Projects That Are Developed by Using the Software Framework

Status	Pure Algorithm	Connect6	Go	Chinese Chess	Mahjong	Tic-Tac-Toe
Pure Editor		O	O	O	O	O
JL-PNS	O	O				
JL-MCTS	O	Δ	O			
JL-SSS [30][a]	Δ			Δ		
AI Competition	O	O		O		

Note: O denotes finished projects and Δ denotes ongoing projects.
[a] Traditional "best-first state space search" approach.

TABLE 10.4 Line Count of the Applications That Are Developed by Using the Software Framework

Original Connect6Lib and JL-PNS	Game Record Editing Module	JL Module	Connect6	Go	Chinese Chess	Mahjong	Tic-Tac-Toe
86688	27854	6658	8215	8843	3535	2192	836

of game tree search involving at least five academic organizations. The users have reported successful results in rapid development of efficient game tree search applications for the desktop grid platform.

10.6 CONCLUSIONS

We develop the desktop grids with the push-mode streaming infrastructure in order to support tightly coupled task control. The push-mode streaming communication can significantly reduce the redundant computations as tree nodes can be generated and pruned in a timely manner. This report depicts the requirements of the dynamic tree search applications and discusses how the JL model can be applied to fit the requirements. The users have reported successful results in rapid development of efficient game tree search applications for the desktop grid platform.

ACKNOWLEDGMENTS

This work was supported in part by the National Science Council of the Republic of China (Taiwan) under Contracts NSC 97-2221-E-009-126-MY3, NSC 99-2221-E-009-102-MY3, NSC 99-2221-E-009-104-MY3, and NSC 101-2221-E-029-04.

REFERENCES

1. SETI@home, available at http://setiathome.ssl.berkeley.edu
2. XtremWeb, available at http://www.xtremweb.net/.
3. Anderson, D.P., "BOINC: A system for public-resource computing and storage," *Proceedings of the 5th IEEE/ACM International Workshop on Grid Computing*, Pittsburgh, PA, November 2004.
4. BOINC, available at http://boinc.berkeley.edu
5. Father of the Grid, available at http://magazine.uchicago.edu/0404/features/index.shtml
6. Foster, I., Kesselman, C., and Tuecke, S., "The anatomy of the grid," *International Journal of Supercomputer Applications*, 23: 187–200, 2001.
7. Taiwan UniGrid, available at http://www.unigrid.org.tw/info.html

8. Wu, I.-C. and Han, S.Y., "The study of the worker in a volunteer computing system for computer games," Institute of Computer Science and Engineering, College of Computer Science, National Chiao Tung University, 2011.

9. Wu, I.-C., Lin, H.-H., Sun, D.-J., Kao, K.-Y., Lin, P.-H., Chan, Y.-C., and Chen, B.-T., "Job-level proof-number search for Connect6," *IEEE Transactions on Computational Intelligence and AI in Games*, 5(1): 44–56, 2013.

10. Allis, L.V., van der Meulen, M., and van den Herik, H.J., "Proof-number search," *Artificial Intelligence*, 66(1): 91–124, 1994.

11. Abramson, B., "Expected-outcome: A general model of static evaluation," *IEEE Transactions on PAMI*, 12: 182–193, 1990.

12. Alexandrov, A.D., Ibel, M., Schauser, K.E., and Scheiman, K.E., "SuperWeb: Research issues in Java-Based global computing," *Proceedings of the Workshop on Java for High performance Scientific and Engineering Computing Simulation and Modelling*. Syracuse University, New York, December 1996.

13. Shoch, J. and Hupp, J., "Computing practices: The 'Worm' programs—Early experience with a distributed computation," *Communications of the ACM*, 25(3): 172–180, 1982.

14. Background Pi, available at http://defcon1.hopto.org/pi/index.php

15. Bruegmann, B., Monte Carlo Go, 1993, available at http://www.althofer.de / bruegmann-montecarlogo.pdf

16. Fedak, G., Germain, C., Neri, V., and Cappello, F., "XtremWeb: A generic global computing system," *Proceedings of the 1st IEEE/ACM International Symposium on Cluster Computing and the Grid: Workshop on Global Computing on Personal Devices*, Brisbane, Australia. IEEE Computer Society Press, Washington, DC, pp. 582–587, 2001.

17. Great Internet Mersenne Prime Search, available at http://www.mersenne.org

18. Gelly, S., Wang, Y., Munos, R., and Teytaud, O., "Modification of UCT with patterns in Monte-Carlo Go," *Technical Report 6062*, INRIA, 2006.

19. van den Herik, H.J., Uiterwijk, J.W.H.M., and Rijswijck, J.V., "Game solved: Now and in the future," *Artificial Intelligence*, 134: 277–311, 2002.

20. Hsu, F.-H., *Behind Deep Blue: Building the Computer That Defeated the World Chess Champion*, Princeton, NJ: Princeton University Press, 2002.

21. Karaul, M., Kedem, Z., and Wyckoff, P., "Charlotte: Metacomputing on the Web," *Proceedings of the 9th International Conference on Parallel and Distributed Computing Systems*, Dijon, France, September 1996.

22. Lin, H.H., Wu, I.C., and Shan, Y.-C., "Solving eight layer Triangular Nim," *Proceedings of the National Computer Symposium*, Taipei, Taiwan, November 2009.

23. Regev, O. and Nisan, N., "The POPCORN market—An online market for computational resources," *Proceedings of the 1st International Conference on Information and Computation Economies*. Charleston, SC. ACM Press, New York, pp. 148–157, October 25–28, 1998.

24. Sarmenta, L.F.G., Volunteer computing. PhD thesis, Massachusetts Institute of Technology, Cambridge, MA, June 2001.

25. Sarmenta, L.F.G., "Bayanihan: Web-based volunteer computing using Java," *Proceedings of the 2nd International Conference on World-Wide Computing and Its Applications*, Tsukuba, Japan, Springer-Verlag, Berlin, pp. 444–461, March 3–4, 1998.
26. Wu, I.C., Huang, D.Y., and Chang, H.C., "Connect6," *ICGA Journal*, 28(4): 234–241, 2005.
27. Wu, I.C. and Chen, C.P., "Desktop grid computing system for Connect6 application," Institute of Computer Science and Engineering, College of Computer Science, National Chiao Tung University, August 2009.
28. Wu, I.C. and Jou, C.Y., "The study and design of the generic application framework and resource allocation management for the desktop grid CGDG," Institute of Computer Science and Engineering College of Computer Science, National Chiao Tung University, 2010.
29. Yen, S.J., Chen, J.C., Yang, T.N., and Hsu, S.C., "Computer Chinese Chess," *ICGA Journal*, 27(1): 3–18, 2004.
30. Stockman, G.C., "A minimax algorithm better than alpha-beta?," *Artificial Intelligence*, 12(2): 179–196, 1979.

CHAPTER 11

Research on the Scene 3D Reconstruction from Internet Photo Collections Based on Cloud Computing

Junfeng Yao and Bin Wu

Xiamen University
Xiamen, People's Republic of China

CONTENTS

11.1 INTRODUCTION

11.1.1 Background

With the great benefit of fast development of the Internet, people can now see the whole world by just sitting in front of a computer. Photos, pictures, and videos become the most important media that helps people open new eyes to the world. As more and more people have a desire to upload their photographs to large image hosting Websites, such as Flickr and Google Images or blogs, to show or store their experience and travels, billions of images can be instantly accessible through image search engines provided by these image Websites. These pictures cover thousands of virtually famous places, which are taken from a multitude of viewpoints, at many different times of day, and under a variety of weather conditions. After typing some key words, such as "the Great Wall," the user can easily get millions of photographs gathered by the image search engine. The resulting picture set is often organized as thumbnails or lists, about 20 pictures appear on the screen and the users click the next or previous bottom to switch to another 20 pictures (Figure 11.1).

Although it is the most normal way to display large image set, there exist two weak points. The first one is that pictures are treated as independent views of events or scenes; although they may be grouped together or labeled in meaningful ways, but still unconnected, the user can hardly attain the structure of the space as a whole. Another weak point is that when the users try to find some specific viewpoints or a detail of a particular object, it can be very hard.

11.1.2 Scene Reconstruction from Internet Photo Collections

As described in the previous section, the vast, rich, and disconnected photo collections become a great challenge to give better user experience. How can we make use of them to effectively communicate the experience of being at a place—to give someone the ability to virtually move around

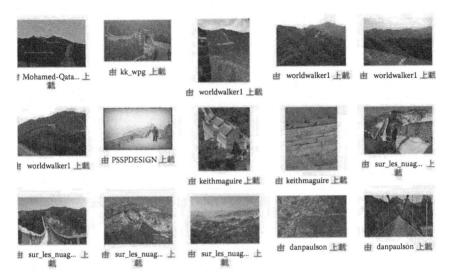

FIGURE 11.1   Search results for the Great Wall from Flickr.

and explore a famous landmark, in short, to convey a real understanding of the scene? Simply displaying them on the screen cannot finish this job. It is worth to say that several commercial software applications have started to present large photo collections in a much more structured way, for instance, Google Street View simulates the experience of walking down the street of major cities by displaying omnidirectional photos taken at intervals along every city street. Such applications, combining the high visual fidelity of photos with simplified 3D navigation control, are helpful for bringing experience as walking around the street. However, this kind of photos requires specific camera hardware, careful attention, and time-consuming postprocessing, which make it unavailable for unorganized image typical of the Internet.

To make full use of unordered photo collections, scene reconstruction [1] is a series of new 3D reconstruction algorithms that operate on large, diverse image collections. These algorithms recover both camera pose and scene geometry and demonstrate, for the first time, that 3D geometry can be reliably recovered from photos downloaded from the Internet using key search.

In Figure 11.2, scene construction takes large collections of photos from the Internet (sample images shown at the top) and automatically

FIGURE 11.2   3D reconstructions from Internet photo collection. (From Keith N. Snavely, Scene reconstruction and visualization from Internet photo collections, Doctoral thesis, University of Washington, pp. 1–67, 2008. With permission.)

reconstructs 3D geometry (bottom). The geometry includes camera information and a point cloud of the scene. In these images of reconstructions, the recovered cameras are shown as black wireframe pyramids, and the scene is rendered as a point cloud.

After scene reconstruction, visualization of 3D photo collections and scenes is a series of computer graphics and interaction techniques based on the built-up camera information and point positions. It provides new ways to browse photo collections and to visualize the world in a 3D way. However, since this part of technique is not the main concern of this thesis, you can find the detailed description in Reference 2.

11.1.3 Challenges in Scene Reconstruction

Although the scene reconstruction pipeline provides a robust and stable way to construct the geometry of the scene, it is still hard to apply to the large photo collections from the Internet because of the hard calculation and memory space. There are several parts of the scene reconstruction pipeline that requires significant computational resources: scale-invariant feature transform (SIFT) [3] feature detection, pairwise feature matching, F-matrix estimation, H-matrix estimation, linking matches into tracks, and incremental structure from motion (SfM) [4]. We will describe all these steps in detail and the time complexity in Section 11.3. In my evaluation, I run the scene reconstruction ("bundler-v0.3-source," the original source code presented in Reference 1) on about 415 photos; it takes about 22 hours to finish all the steps before incremental SfM. It is not hard to imagine that when the number of photos increases to 10,000 or even 100,000, which is not strange in the Internet image collections, the time consuming will become unacceptable and all these delicate works may make no sense.

Running scene reconstruction on a super computer may become a tempting solution to reduce time cost. However, even if we have a computer that is 10 or 20 times faster than a normal computer, it cannot help decrease the time cost into an acceptable range because the heavy amount of calculation is far above the computing power of a single computer. Another point worth to say is that since the system will run for a long time, how to back up the data, how to deal with errors such as shutdown or system fault, and how to recover the data from error status become significantly important. We need to control this risk by buying expensive but trustable hardware and designing the program well.

Parallel computing may become the only solution to deal with large photo collections. However, it requires much more careful design than running on a super computer of the implementation of scene reconstruction. Since every node in a cluster may fail down, the first thing of our system is to detect the status of each node periodically and make response. When a node fails, we need to make sure that we have backup data in another computer and can resume the previous job successfully. Moreover, we must design a program to combine the data from all nodes. All such additional work makes the pipeline of scene reconstruction more complex and hard to control.

If there exists a platform that has already waived us from such additional work, and provided us the powerful but simple parallel computing model in front of us, why shall we not explore the possibility of applying scene reconstruction on this platform—the platform of cloud computing?

11.1.4 Cloud Computing

Cloud computing is Internet-based computing, in which shared resources, software, and information are provided to computers and other devices on demand. In a cloud, multiple computers are connected through network, controlled by one or more master nodes. The platform of cloud provides the key features of reliability, scalability, security, device and location independence, and so on, which will benefit the scene reconstruction pipeline a lot.

In this section, there are two fundamental research goals:

1. Describe the scene reconstruction pipeline in detail.

2. Discuss our research on applying scene reconstruction on cloud computing platform.

11.2 SCENE RECONSTRUCTION PIPELINE

The scene reconstruction pipeline takes an unordered collection of images (for instance, from the Internet search or a personal collection) and produces 3D camera and scene geometry. In particular, for each input photo, the pipeline determines the location from which the photo was taken and the direction in which the camera was pointed, and recovers the 3D coordinates of a sparse set of points in the scene.

The basic principles behind recovering geometry from a set of images are fairly simple. Humans implicitly use multiview geometry to sense depth with binocular vision. If we see the same point in the world in both eyes, we can implicitly "triangulate" that point to determine its rough distance. Similarly, given two photographs of the same scene, a list of pixels in images A and B, and the relative poses of the cameras used to capture the images, the 3D position of the matching pixels can be calculated. However, even though we can get the corresponding pixels by comparing a pair of images, the geometry of cameras that took these two images often keeps unknown. In another word, only if we determine that the camera poses according to the matching pixels as well, can we construct the structure of the scene geometry successfully. Fortunately, the correspondences place constraints on the physical configuration of the two cameras. Thus, given enough point matches between two images, the geometry of the system becomes constrained enough that we can determine the two-view geometry, after which we can estimate the 3D point positions using triangulation. This procedure is also known as SfM. In general, SfM can deal with an arbitrary number of images and correspondences, and estimate camera and point geometry simultaneously.

11.2.1 SIFT Algorithm

SIFT is an algorithm in computer vision to detect and describe local features in images. To begin with, SIFT extracts features from a set of reference images and stores them into a database. An object is recognized in a new image by individually comparing each feature from the new image to this database and finding candidate matching features based on the Euclidean distance of their feature vectors. Since SIFT feature descriptor is invariant to scale, orientation, and affine distortion, and partially invariant to illumination changes, this method can robustly identify objects even among clutter and under partial occlusion. It has wide applications, such as object recognition, robotic mapping and navigation, image stitching, 3D modeling, gesture recognition, video tracking, and match moving.

In scene reconstruction, the pipeline uses SIFT to detect the key features of images for the initial matching. There are five main steps in SIFT listed below:

1. Scale-space extrema detection

2. Accurate keypoint localization

3. Orientation assignment

4. Compute local image descriptor

5. Keypoint matching

11.2.1.1 Scale-Space Extrema Detection

To apply SIFT, transformation of the format of input image is necessary. In general, we change the input image into gray scale. Also, to make full use of the input, the image can be expanded to create more sample points than were presented in the original.

The most important goal of SIFT is to achieve scale invariance. In order to attain scale invariant, theoretically, we need to search stable features in all possible scales. But in practice, since it is impossible to get all scales, here we sample the scale space for reasonable frequency to attain scale-invariant feature detection. In SIFT, difference of Gaussian (DoG) filter is used to build up the scale space [3]. There are two reasons for applying DoG: (1) it is an efficient filter and (2) it provides a close approximation to scale-normalized Laplacian of Gaussian. The normalization of the Laplacian with the factor σ^2 is required for true invariance. In SIFT, the author proved that DoG scale space will be almost the same stable as that of Laplacian Gaussian (Figure 11.3).

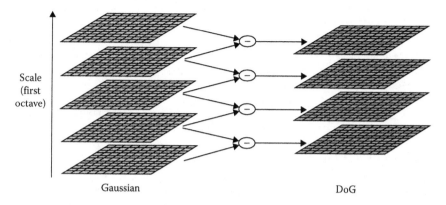

Gaussian   DoG

FIGURE 11.3   DoG and Gaussian pyramids. DoG, difference of Gaussian. (From David G. Lowe, *International Journal of Computer Vision*, 60, 91–110, 2004. With permission.)

It is very simple to get DoG filter; we only need to subtract images convoluted by Gaussian low-pass filter in different scales. Also, since we have preprocessed the input image to prevent significant aliasing, we need to make sure that the blurring factor σ is above threshold T.

The final step of extrema detection is to compare each pixel with its neighbors: 8 neighbors in its current scale level and 18 neighbors in its nearest neighbors. Only if the pixel is bigger than all of its neighbors, it will be chosen as a keypoint candidate.

11.2.1.2 Accurate Keypoint Localization

In the previous step, we have found out the possible feature points. Then, we must eliminate extrema that are along the edges but are unstable to small amounts of noise because of poorly determined position. According to the scale-space value, first we can remove the points that are near to the boundary. Second, we eliminate poorly defined peaks in the DoG function which have a large principal curvature across the edge but a small one in the perpendicular direction.

11.2.1.3 Orientation Assignment

By assigning a consistent orientation to each keypoint based on local image properties, this keypoint descriptor can be represented relative to this orientation, and therefore achieve invariance to image rotation. In SIFT, the orientation histogram is used to determine the major orientation.

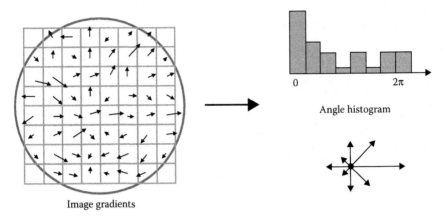

Image gradients

Angle histogram

FIGURE 11.4 The orientation assignment.

First, we compute the histogram in a window area whose center is the keypoint computed before and then we divide it into 36 bins covering the 360° range of orientations. Each sample added to the histogram is weighted by Gaussian function. The angle of the longest bin referred is the orientation of the keypoint. Therefore, to improve the robust, if the other bins are over 80% value of the longest bin, they will be treated as candidates (Figure 11.4).

11.2.1.4 Local Image Descriptor

This step of SIFT is to generate the local descriptor that is highly distinctive yet and is as invariant as possible to remaining variations, such as change in illumination or 3D viewpoint. Here SIFT still uses orientation histogram to store the local descriptor. To begin with, SIFT computes each pixel's orientation included in 16 × 16 window centered as keypoint using the same function in the orientation assignment step with a much fewer degree bins of 8. Then it subdivides the window into 4 × 4 subwindows and recompute the weights of each orientation and produce a 4 × 4 × 8 descriptor (Figure 11.5).

11.2.1.5 Keypoint Matching

Approximate nearest neighbor (ANN) searching algorithm [6] is a library written in C++, which supports data structures and algorithms for both exact and ANN searching in arbitrarily high dimensions. In SIFT, to

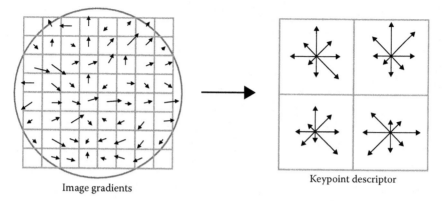

Image gradients

Keypoint descriptor

FIGURE 11.5   Local image descriptor.

determine an inlier of matched keypoints, we should find the nearest neighbor and the second nearest neighbor of each point. The nearest neighbor is defined as the keypoint with minimum Euclidean distance for the invariant descriptor vector. For robustness, SIFT uses the ratio of the nearest neighbor to the second nearest neighbor: $d_1/d_2 \leq 0.6$ to determine whether the matches are inliers. However, computing the exact nearest neighbors in dimensions much higher than 8 seems to be a very difficult task. Few methods seem to be significantly better than a brute force computation of all distances. To improve the efficiency, by computing the nearest neighbors approximately using ANN, it is possible to achieve significantly faster running times with relatively small actual errors.

11.2.2 RANSAC Paradigm

In this section, we introduce the RANSAC paradigm, which is capable of smoothing data that contain a significant percentage of gross errors. SIFT, the local feature detector, always makes mistakes, as described in Section 11.2.1, although it is one of the best local feature detectors. There are two kinds of mistakes: classification errors and measurement errors. Classification errors occur when a feature detector incorrectly identifies a portion of an image as an occurrence of a feature, whereas measurement errors occur when the feature detector correctly identifies the feature, but slightly miscalculates one of its parameters (e.g., its location in the image). To get rid of noise data from images and refine matching pair of keypoints, the RANSAC paradigm is used to compute the epipolar geometry constraint of fundamental matrix and set up the homography matrix.

The RANSAC procedure is opposite to that of conventional smoothing techniques: Rather than using as much of the data as possible to obtain an initial solution and then attempting to eliminate the invalid data points, RANSAC uses as small an initial dataset as feasible and enlarges this set with consistent data when possible. For instance, given the task of fitting an arc of a circle to a set of 2D points, the RANSAC approach would be to select a set of three points (since three points are required to determine a circle), compute the center and radius of the implied circle, and count the number of points that are close enough to that circle to suggest their compatibility with it (i.e., their deviations are small enough to be measurement errors). If there are enough compatible points, RANSAC would employ a smoothing technique, such as least squares, to compute an improved estimate for the parameters of the circle now that a set of mutually consistent points has been identified.

The formal RANSAC paradigm procedure is stated as follows:

1. Given a model that requires a minimum of n data points to instantiate its free parameters, and a set of data points P such that the number of points in P is greater than $n[\#(P) \geq n]$ randomly selects a subset S1 of n data points from P and instantiates the model. Use the instantiated model M1 to determine the subset S1* of points in P that are within some error tolerance of M1. The set S1* is called the consensus set of S1.

2. If #(S1*) is greater than some threshold T, which is a function of the estimate of the number of gross errors in P, use S1* to compute (possibly using least squares) a new model M1*.

3. If #(S1*) is less than t, randomly select a new subset S2 and repeat the above process. If, after some predetermined number of trials, no consensus set with t or more members has been found, either solve the model with the largest consensus set found or terminate in failure.

The RANSAC paradigm contains three unspecified parameters: (1) the error tolerance used to determine whether a point is compatible with a model, (2) the number of subsets to try, and (3) the threshold T, which is the number of compatible points used to imply that the correct model has been found. In Section 11.2.3, we will discuss how RANSAC is applied to eliminate spurious matching key pairs.

11.2.3 Geometric Consistency Test

We now have a set of putative matching image pair (I, J), and for each matching image pair, a set of individual feature matches. Because the matching procedure is imperfect: Many of these matches—both image matches and individual feature matches—will often be spurious. Fortunately, it is possible to eliminate many spurious matches using a geometric consistency test. This test is based on the fact that, no matter what the actual shape of the scene is, there is a fundamental constraint between two perspective images of the static scene by the possible configurations of the two cameras and their corresponding epipolar geometry. In this section, we will first introduce the principle idea of projective geometry and derive the formula of computing the fundamental matrix F and the homography matrix H. Then we will discuss the procedure of computing F and H using RANSAC paradigm.

11.2.3.1 Projective Geometry and Transform

In Euclidean geometry of IR^2, the parallel lines will never meet in a single point. In order to escape from the exception, the projective geometry is proposed. In the projective plane IP^2, one may state without qualification that two distinct lines meet in a single point and two distinct points lie on a single line.

11.2.3.1.1 The 2D Projective Plane   A line in the plane is represented by an equation such as $ax + by + c = 0$, with different choice of a, b, and c giving rise to different lines. Thus, a line may naturally be represented by the vector $(a, b, c)^T$. Also, since $(ka)x + (kb)y + kc = 0$, the vector $(a, b, c)^T$ is the same as $k(a, b, c)^T$. After defining the lines, a point $x = (x, y)^T$ lies on the line $l = (a, b, c)^T$ if and only if $ax + by + c = 0$. This can be represented by $(x, y, 1)(a, b, c)^T = (x, y, 1) \times l = 0$. In this way, points are presented by homogeneous vectors. An arbitrary homogeneous vector representative of a point is of the form: $x = (x, y, z)^T$. So we have the result as follows:

1. The point x lies on the line l if and only if $x^T \times l$.

2. The line l through two points x and x' is $l = x \times x'$.

11.2.3.1.2 Ideal Points and the Line at Infinity   Consider two lines $ax + by + c = 0$ and $ax + by + c' = 0$. These are presented by vectors $l = (a,b,c)^T$ and $l' = (a,b,c')^T$ for which the first two coordinates are the same. The intersection is $l \times l = (c' - c)(b, -a, 0)^T$, and ignoring the scale factor $(c' - c)$, the point is $(b, -a, 0)^T$. Now we may find that the inhomogeneous representation of this point $(b/0, -a/0)^T$ makes no sense, except to suggest that the point of intersection has infinitely large coordinates. This observation agrees with the usual idea that parallel lines meet at infinity.

Now we may have the definition of ideal point that has the last coordinate $x_3 = 0$. The set of all ideal points may be written as $(x_1, x_2, 0)^T$. Note that this set lies on a single line, the line at infinity, denoted by the vector $l_\infty = (0, 0, 1)^T$, since $(0, 0, 1)(x_1, x_2, 0)^T = 0$.

11.2.3.1.3 Projective Transformations   2D projective geometry is the study of properties of the projective plane IP^2 which are invariant under a group of transformations known as projectivity, which is an invertible mapping h from IP^2 to itself such that points $p1$, $p2$, and $p3$ lie on the same line if and only if $h(p1)$, $h(p2)$, and $h(p3)$ do.

A planar projective transformation is a linear transformation on homogeneous four-vectors represented by a nonsingular 3×3 matrix:

$$\begin{pmatrix} x_1' \\ x_2' \\ x_3' \end{pmatrix} = \begin{bmatrix} h_{11} & h_{12} & h_{13} \\ h_{21} & h_{22} & h_{23} \\ h_{31} & h_{32} & h_{33} \end{bmatrix} \begin{pmatrix} x_1 \\ x_2 \\ x_3 \end{pmatrix} \tag{11.1}$$

Or more briefly, $x' = Hx$.

Note that the matrix H occurring in this equation may be changed by multiplication by an arbitrary nonzero-scale factor without altering the projective transformation. Consequently, we say that H is a *homogeneous matrix*.

11.2.3.1.4 Projective 3D Geometry   Similar to 2D Geometry projective plane, point X is represented in homogeneous coordinates as a four-vector. Specifically, the homogeneous vector $X = (x_1, x_2, x_3, x_4)^T$ with $x_4! = 0$ representing the points at finite position while $x_4 = 0$ representing the points at infinity.

A plane in three-space may be written as

$$\pi_1 X + \pi_2 Y + \pi_3 Z + \pi_4 = 0 \tag{11.2}$$

Clearly this equation is unaffected by multiplication by a nonzero-scale factor and the homogeneous representation of the plane is the four-vector $\pi = (\pi_1, \pi_2, \pi_3, \pi_4)$.

Also we may have $\pi^T X = 0$, which expresses that the point X is on the plane π. It is easy to find that the first three components of the plane correspond to the plane normal of Euclidean geometry. This formula may also be written as

$$\left(\pi_1, \pi_2, \pi_3\right)^T \left(x_1, x_2, x_3\right)^T + d = 0 \tag{11.3}$$

where:
$$d = \pi_4$$

11.2.3.2 Camera Model

A camera is a mapping between the 3D world (object space) and the 3D image. The principal camera of interest here is central projection.

11.2.3.2.1 The Basic Pinhole Model   We consider the central projection of points in space onto a plane. Let the center of projection be the origin of a Euclidean coordinate system, and consider the plane $Z = f$, which is called the image plane or focal plane. In this model, a point $X = (X, Y, Z)^T$ is mapped to the point on the image plane where a line joining the point X to the center of projection meets the image plane. Ignoring the final image coordinate, we see that $(X, Y, Z)^T \rightarrow (fx/z, fy/z)^T$. This is a mapping from Euclidean three-space to two-space. The mapping can be represented by homogeneous vectors as

$$\begin{pmatrix} X \\ Y \\ Z \\ 1 \end{pmatrix} \rightarrow \begin{pmatrix} fX \\ fY \\ Z \end{pmatrix} = \begin{bmatrix} f & & & 0 \\ & f & & 0 \\ & & 1 & 0 \end{bmatrix} \begin{pmatrix} X \\ Y \\ Z \\ 1 \end{pmatrix} \tag{11.4}$$

Or more briefly, $m = K\begin{bmatrix}1|0\end{bmatrix}X$, where $m = (fX, fY, Z)$ and $K = \mathrm{diag}(f, f, 1)$.

Because in Reference 1, the author believes that all the principal points are in the center of image plane, there is no need to fix the principal point.

11.2.3.2.2 Camera Rotation and Translation   In general, points in space will be expressed in terms of a different Euclidean coordinate frame, known as the world coordinate frame. The two coordinate frames are related via a rotation and a translation. If χ is an inhomogeneous three-vector representing the coordinate of a point in the world coordinate frame, and χ_{CAM} represents the same point in the camera coordinate frame, which is written as $\chi_{CAM} = R(\chi - \Gamma)$, where Γ represents the coordinates of the camera center in the world coordinate frame and R is a 3×3 rotation matrix representing the orientation of the camera coordinate frame. The equitation can be set as

$$\chi_{CAM} = \begin{bmatrix} R & -R\Gamma \\ 0 & 1 \end{bmatrix} \begin{pmatrix} X \\ Y \\ Z \\ 1 \end{pmatrix} \tag{11.5}$$

Then we have the relationship between the image plane point and the 3D point.

$$m = PX \tag{11.6}$$

where:
$$P = KR\begin{bmatrix} I | \Gamma \end{bmatrix}$$

11.2.3.3 Epipolar Geometry and Fundamental Matrix

The epipolar geometry is the intrinsic projective geometry between two views. It is independent of scene structure, and only depends on the cameras' internal parameters and relative pose. Then the fundamental matrix F encapsulates this intrinsic geometry. It is a 3×3 matrix of rank 2. For each pair of images, we define the following:

An *epipole e,e'* is the point of intersection of the line joining the camera centers (the baseline) with the image.

An *epipolar plane* π is a plane containing the baseline. There is a one-parameter family of epipolar planes.

An *epipolar line* l_m, l'_m is the intersection of an epipolar with the image plane. All epipolar lines intersect at the epipole.

The fundamental matrix is the algebraic representation of epipolar geometry. In this part, we derive the fundamental matrix from the mapping between a point and its epipolar line, and then specify the properties of the matrix.

11.2.3.3.1 Algebraic Derivation of F   Two cameras whose projective matrices are P and P' and the relevant images are I and I'. m is an image point of I

and the ray back-projected from m1 by P is obtained by solving PX = m. Then we have

$$X(s) = P^+ m + sC \tag{11.7}$$

where:
$$P^+ = P^T (PP^T)^{-1}$$
$$P^+ P = 1$$

The epipolar line of I′ is

$$l'_m = e' \times m' = (P'C) \times [P'X(s)] = (P'C) \times (P'P^+ + SP'C) = [e']_x P'P^+ m \tag{11.8}$$

More briefly, we have $l'_m = Fm$, where $F = [e']_x P'P^+$.
Because m′ is in the line of l'_m, we have $m'Fm = 0$.

11.2.3.4 Computing Fundamental Matrix Using Eight-Point Algorithm

In this section, the equations on F generated by point correspondences between two images and their minimal solution are described. Then an algorithm is then described for automatically obtaining point correspondences so that F may be estimated directly from an image pair.

11.2.3.4.1 Basic Equations   In Section 11.2.3.3, the fundamental matrix is defined by the equation: $m'Fm = 0$. Given sufficiently many point matches $m \leftrightarrow m'$ (at least seven), the equation can be used to compute the unknown matrix F. Specifically, the equation corresponding to a pair of points (x, y, 1) and (x′, y′, 1) is

$$x'xf11 + x'yf12 + x'f13 + y'xf21 + y'yf22$$
$$+ y'f23 + xf31 + yf32 + f33 = 0 \tag{11.9}$$

The nine-vector made up of the entries of F in row-major order is denoted by f. From a set of n point matches, we obtain a set of linear equations of the form:

$$Af = \begin{bmatrix} x_1 & x_1 & x_1 & y_1 & \cdots & x_1 & y_1 & 1 \\ \cdot & & \cdot & & \cdots & \cdot & \cdot & 1 \\ x'_n & x_n & y'_n & y_n & \cdots & x_n & y_n & 1 \end{bmatrix} f = 0 \tag{11.10}$$

This is a homogeneous set of equations, and f can only be determined up to scale. However, if the data are not exact, because of noise in the point coordinates, then the rank may be greater than 8 (in fact equal to 9). In this case, one finds a least-squares solution.

11.2.3.4.2 Least-Squares Solution   Adding constraint to Af,

$$\begin{cases} \min & \|Af\| \\ \text{subject to} & \|f\| \end{cases} = 1 \qquad (11.11)$$

To attain F, we obtain UDV^T by applying singular value decomposition (SVD). The solution $f = V9$, which is the ninth column of matrix V. Finally, since the rank of F is 2, the epipolar lines cannot intersect at the same point. However, in Equation 11.10, the matrix we computed has a rank of 3. Then we have to compute F_F.

11.2.3.4.3 Rank 2 Constraint   Then we have the formula below:

$$\begin{cases} \min & \|F - F_F\| \\ \text{subject to rank} & (F_F) \end{cases} = 2 \qquad (11.12)$$

To compute F_F, we apply SVD to F_F again: $F = U\text{diag}(s_1, s_2, s_3)V^T$ and $F_F = U\text{diag}(s_1, s_2, 0)V^T$.

11.2.3.5 Automatic Computation of F Using RANSAC
Since the matching pairs of key features are spurious, we cannot determine which eight points are reliable and which are not. Then we induce the RANSAC paradigm to help us set up the reliable corresponding matches. This part introduces the pipeline of automatic computation of F using RANSAC.

Pipeline of Automatic Computing F

1. **Interest points:** Compute interest points in each image (the SIFT algorithm).
2. **Putative correspondences:** Compute a set of interest point matches based on proximity and similarity of their intensity neighborhood (the ANN algorithm).
3. **RANSAC robust estimation:** Repeat for N samples, where N is determined as predetermine trial or no better F'' is found.

a. Select a random sample of eight correspondences and compute the fundamental matrix F using eight-point algorithm.

b. Use F to apply to a new eight-point set S' and compute the weight of error matches.

c. If the weight of error matches is smaller than a specific threshold T, then use S' to develop new F.

d. Else choose a new random set S'' to determine F'' and repeat the above steps.

e. Until predetermine trial or no better F'' is found, stop the step.

11.2.3.6 Computing Homography Matrix

In Section 11.2.3.1, we have mentioned that homography matrix H is used for describing the transform of points between two flats. Each point correspondence gives rise to two independent equations in the entries of H. Given a set of four such point correspondences, we obtain a set of equations $Ah = 0$, where A is the matrix of equation coefficients built from the matrix rows A_i contributed from each correspondence, and h is the vector of unknown entries of H. We seek a nonzero solution h, since the obvious solution $h = 0$ is of no interest to us. Since h can only be computed up to scale, we impose the condition $h_j = 1$, for example, $h_9 = 1$, which corresponds to H_{33}, then we can gain the inhomogeneous solution of H as follows:

$$
\begin{bmatrix}
0 & 0 & 0 & -x_i w_i' & -y_i w_i' & -w_i w_i' & x_i y_i' & y_i y_i' \\
x_i w_i' & y_i w' & w_i w' & 0 & 0 & 0 & x_i x_i' & y_i x_i'
\end{bmatrix}
\tilde{h} =
\begin{pmatrix}
-w_i y_i' \\
w_i x_i'
\end{pmatrix}
\tag{11.13}
$$

where:

$\tilde{h}$ is an eight-vector consisting of the first eight components of h

Concatenating the equations from four correspondences then generates a matrix equation of the form $M\tilde{h} = b$, where M has eight columns and b is an eight-vector. Such an equation may be solved for $\tilde{h}$ using standard techniques for solving linear equation (such as Gaussian elimination) in the case where M contains just eight rows, or by least-squares techniques in the case of an overdetermined set of equations.

11.2.3.7 Automatic Computation of a Homography Using RANSAC

Similar to the procedure of computing F, we use RANSAC paradigm to help us select the best corresponding matches. The main steps of automatic computing H are listed below:

Main Steps of Automatic Computing H

1. **Putative correspondences:** Matches fit fundamental matrix *F*.
2. **RANSAC robust estimation:** Repeat for *N* samples, where *N* is determined as predetermine trial or no better *F* is found.
    a. Select a random sample of four correspondences and compute the homography *h* using inhomogeneous solution of *H*.
    b. Use *H* to apply to a new four-point set *F* and compute the weight of error matches.
    c. If the weight of error matches is smaller than a specific threshold *T*, then use *S′* to develop new *F*.
    d. Else choose a new random set *F* to determine *F* and repeat the above steps.
    e. Until predetermine trial or no better *F* is found, stop the step.

11.2.4 Computing Track of Matches

In Section 11.2.3, all matches have been refined, and then we organize the matches into point by finding connected sets of matching features across multiple images. For instance, if feature $f_1 \in F(I_1)$ matches feature $f_1 \in F(I_2)$, and f_2 matches feature $f_1 \in F(I_3)$, these features will be grouped into a track $\{f_1, f_2, f_3\}$. Tracks are found by examining each feature f in each image and performing a breath-first search of the set of features in other images that match f until an entire connected component of features has been explored. These features are then grouped together into a track, and the next unvisited feature is considered, until all features have been visited. Because of spurious matches, inconsistencies can arise in tracks; in particular, a track can contain multiple features from the same image, which violates the assumption that a track corresponds to a single 3D point. In this case, the tracks are identified as inconsistent, and any image that observes a track multiple times has all of their features removed from that track.

11.2.5 Reconstructing the Initial Pair

In scene reconstruction pipeline, estimating the parameters starts from a single pair of cameras, and then more cameras will be added into the construction. However, determining the first pair of cameras should be very critical. Because if the reconstruction of the initial pair gets stuck in the wrong local minimum, the optimization is unlikely to ever recover. The images should have a large number of matches, but also have a large baseline (distance between camera centers), so that the initial two-frame reconstruction can be robustly estimated. Since homography matrix H represents the transformation between two images of a single plane or two images taken at the same location (but possibly with different direction). Thus, if a homography cannot fit to the correspondences between two images, it indicates that the cameras have some distance between them, and that is what we want.

In Section 11.2.3, homography matrix between each pair of matching images is created and the percentage of feature matches that are inliers to the estimated homography is stored. Then a pair of images that have the lowest percentage of inliers, but have at least threshold over matches, will be chosen as initial image pair.

The system estimates the extrinsic parameters for the initial pair using five-point algorithm [7], and then tracks visible in the two images are triangulated, giving an initial set of 3D points.

11.2.5.1 Recovering Camera Geometry

In Section 11.2.3, we have examined the properties of F and image relations for a point correspondence $x \leftrightarrow x'$. We now turn to one of the most significant properties of F that the matrix may be used to determine the camera matrices of the two views.

11.2.5.1.1 The Essential Matrix   To compute the extrinsic parameters of camera, the conception of essential matrix used to extract extrinsic parameters should be introduced first. The essential matrix is the specialization of the fundamental matrix to the case of normalized image. Historically, the essential matrix was introduced before the fundamental matrix, and the fundamental matrix may be thought of as the generalization of the essential matrix in which the (inessential) assumption of calibrated cameras is removed. The essential matrix has fewer degrees of freedom, and additional properties, compared to the fundamental matrix. These properties are described below.

Consider a camera matrix decomposed as $P = K[R|t]$, and let $x = PX$ be a point in the image. If the calibration matrix K is known, then we may apply its inverse to the point x to obtain the point. Then we have $x = [R|t]X$, where x is the image point expressed in normalized coordinates. It may be thought of as the image of the point X with respect to a camera $[R|t]$ having the identity matrix I as calibration matrix.

The camera matrix $K - IP = [R \mid t]$ is called a normalized camera matrix and the effect of the known calibration matrix has been removed. Now, consider a pair of normalized camera matrices $\det(V) > 0$ and $P' = [R \mid t]$. The fundamental matrix corresponding to the pair of normalized cameras is customarily called the essential matrix, and according to Section 11.2.3, it has the form:

$$E = [t]_x R = R \left[R^T t \right]_x \tag{11.14}$$

Snavely [1] uses the five-point algorithm to compute E; however, since we have $E = K'^T FK$, we can easily compute the essential matrix E if the camera is calibrated [$K = \mathrm{diag}(f,f,1)$, and f can be loaded from photos]. Once the essential matrix is known, R, t, and the camera matrices can be recovered from it.

11.2.5.1.2 Recover R and t from E According to the theorem, let the SVD of the essential matrix be $E \sim U\mathrm{diag}(1,1,0)V^T$, where U and V are chosen such that $\det(V) > 0$ and $\det(U) > 0$. Then $t \sim t_u \equiv [u_{13}u_{23}u_{33}]^T$ and R is equal to $R_a \equiv UDV^T$ or $R_b \equiv UD^TV^T$. Any combination of R and t according to the above prescription satisfies the epipolar constraint. To resolve the inherent ambiguities, we assume that the first camera is $P = [I \mid o]$ and t is of unit length. Then the possible second camera matrix is in the four possible solutions: $P_A = [R_a \mid t_u], P_B = [R_a \mid -t_u], P_C = [R_b \mid t_u], P_D = [R_b \mid -t_u]$. One of the four choices corresponds to the true configuration.

The four solutions are illustrated in Figure 11.6, where it is shown that a reconstructed point X will be in front of both cameras in one of these four solutions only. Thus, testing with a single point to determine if it is in front of both cameras is sufficient to decide between the four different solutions for the camera matrix P'.

11.2.5.2 Triangulate 3D Points

In Section 11.2.5.1, we have computed the camera matrices P and P', let x and x' be the two points in the two images that satisfy the epipolar constraint, $x'^T Fx = 0$. This constraint may be interpreted geometrically in terms of the rays in space corresponding to the two image points. In particular, it means that x' lies on the epipolar line Fx. In turn, this means that the two rays back-projected from image points x and x' lie in a common epipolar plane, that is, a plane passing through the two camera centers.

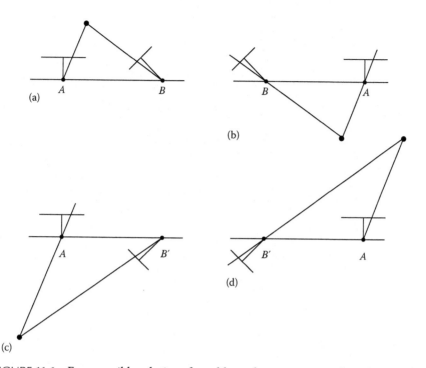

FIGURE 11.6 Four possible solutions for calibrated reconstruction from E. (From Noah Snavely, Steven M. Seitz, and Richard Szeliski, *International Journal of Computer Vision*, 80, 189–210, 2007. With permission.)

Since the two rays lie in a plane, they will intersect at some point, which is the 3D position of the real point (Figure 11.7).

Now let us come to the definition of triangulating 3D points: with the precondition of estimated camera matrices, to estimate a 3D point X that exactly satisfies the supplied camera geometry, so it projects as

$$x = PX, \ x' = P'X \tag{11.15}$$

and the aim is to estimate X from the image measurements x and x'. Notice that the equation $x = PX$, $x' = P'X$ is an equation involving homogeneous vectors; thus, the three-vectors $x = PX$ are not equal; they have the same direction but may differ in magnitude by a nonzero-scale factor. The first equation may be expressed in terms of the vector cross-product as $x \times PX = 0$. This form will enable a simple linear solution for X to be derived. This cross-product results in three equations:

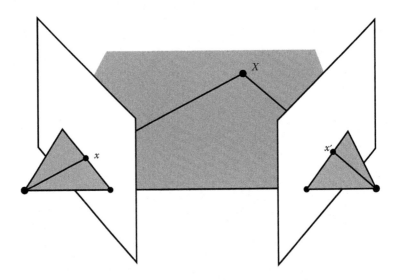

FIGURE 11.7   Triangulation from two image points x and x'.

$$\begin{cases} x\left(p^{3T}X\right)-\left(p^{1T}X\right)=0 \\ y\left(p^{3T}X\right)-\left(p^{2T}X\right)=0 \\ x\left(p^{2T}X\right)-y\left(p^{1T}X\right)=0 \end{cases} \tag{11.16}$$

Where P^{iT} are the rows of P. Only two of the equations are linear independent. Then we combine the two equations derived from $x' \times PX = 0$; an equation of the form $AX = 0$ can then be composed, with:

$$A = \begin{bmatrix} xp^{3T} - p^{1T} \\ yp^{3T} - p^{2T} \\ x'p^{3T} - p'^{1T} \\ y'p^{3T} - p'^{2T} \end{bmatrix} \tag{11.17}$$

This is a redundant set of equations, since the solution is determined only up to scale. Obviously, A has rank 3, and thus has a 1D null space that provides a solution for X. However, since there exists deviation of image points, we may probably fail to get the exactly solution of X. The direct linear transformation (DLT) method [8] is applied: We first obtain the

SVD of A. Then unit singular vector corresponding to the smallest singular value is the solution X.

11.2.6 Adding New 3D Points and Cameras

In Section 11.2.5, we successfully constructed the initial pair of camera matrices and initial 3D points. Then we will add more cameras and points by turn. Cameras that observe the threshold over the number of tracks whose 3D locations have already been estimated will be selected. To initialize the pose of the new camera, for each correspondence $X_i \leftrightarrow x_i$, we derive a relationship:

$$\begin{bmatrix} 0^T & -\omega_i X_i^T & y_i X_i^T \\ \omega_i X_i^T & 0^T & -x_i X_i^T \\ -y_i X_i^T & x_i X_i^T & 0^T \end{bmatrix} \begin{pmatrix} P^1 \\ P^2 \\ P^3 \end{pmatrix} = 0 \qquad (11.18)$$

Where each P^i is a four-vector, the ith row of P, where $i = 1, 2, 3, \ldots$. Only two of the equations are linear independent. Since the matrix P has 12 entries and 11 degrees of freedom, it is necessary to have 11 equations to solve for P. We need to combine six-point correspondence to form the $Ap = 0$, where A is an 11 × 12 matrix in this case. In general, A will have rank 11, and the solution vector p is the 1D right null space of A. Of course, there exists noise in the correspondences; the DLT is applied again to compute the optimal P for the selected six points. Moreover, since we cannot select the best six points to compute the optimal P, we apply RANSAC paradigm to help us select the best P among possible candidates.

Then we add points observed by the new camera into the reconstruction. A point is added if it is observed by at least two cameras, and if triangulating the points gives a well-conditioned estimate of its location. Once the new points have been added, sparse bundle adjustment (SBA) [9] is performed on the entire model.

This procedure of initializing cameras, triangulating points, and SBA is repeated, until no remaining camera observes a sufficient number of points (in Reference 1, at least 20). In general, not all images will be reconstructed. The reconstructed images are not selected by human work, but are determined by the algorithm as it adds images until no more can reliably be added.

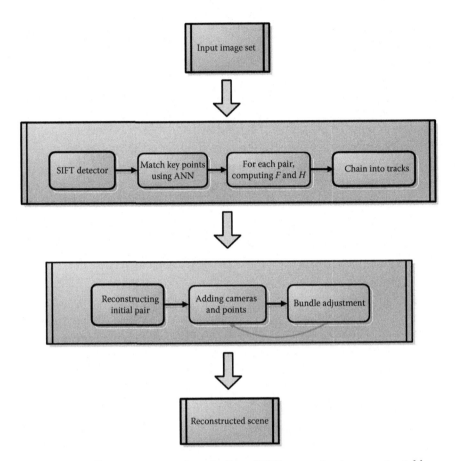

FIGURE 11.8 Scene reconstruction pipeline. ANN, approximate nearest neighbor; SIFT, scale-invariant feature transform.

11.2.7 Time Complexity

In Sections 11.2.1 through 11.2.6, we have introduced the scene reconstruction pipeline in detail. Then let us recall all the steps of the pipeline and analyze the time complexity of each step. The scene reconstruction pipeline is listed in Figure 11.8.

As mentioned in Section 11.1, there are several parts of the scene reconstruction pipeline that require significant computational resources: SIFT feature detection, pairwise feature matching and estimation of F- and H-matrix, linking matches to tracks and incremental SfM. For each part of the algorithm, the time complexity will be listed below with analysis.

SIFT. The feature detection step is linear in the number of input images [$O(n)$]. Running SIFT on an individual image, however, can take a significant amount of time and use a large amount of memory, especially for high-resolution images with complex textures. In my evaluation, the maximum number of SIFT features detected in an image was over 20,000. SIFT ran for 1.1 minutes on that image, on a test machine with Intel Core 2 Duo CPU E8400 and 2 GB memory. SIFT spent an average of 11 seconds processing each image in this collection (and about 1.2 hours of CPU time in total).

Feature matching. The feature matching step does take a significant percentage of the total processing time. This is mostly due to its relatively high complexity. Since each pair of images is considered, it has quadratic time complexity in the number of input image $O(n^2)$. In my evaluation, for the image set of 415 images, each image pair took an average of 2.7 seconds; in total, it took about 20.3 hours to match the entire collection.

F- and H-matrix. Because *F-* and *H*-matrix are only run on the pairs of images that successfully match, they tend to take a much smaller amount of time than the matching itself. Although the worst case is $O(n^2)$; however, for Internet image set, the percentage of image pairs that match is usually fairly small. In my evaluation, for the image set of 415 images, it took about 45 minutes to finish the *F-* and *H*-matrix estimation stage.

Linking up matches to form tracks. In this step, only a breadth-first search on the graph of feature matches is performed, marking each feature as it is visited. Grouping matches into tracks took about 3.3 minutes for my evaluation.

Structure from motion. This step also takes a significant percentage of the total processing time. Since bundle adjustment will run after every camera or point added into the reconstruction, when the image set becomes larger, the pipeline will call the bundle adjustment more times. Because of the high complexity of this step, we will just introduce the practice time cost in my evaluation. In practice, the time required for SfM of an image set of 400 pictures was 33 minutes. Compared to the time spent on feature matching as 20.3 hours, it took much less time than feature matching. However, this step is hard to achieve parallel computing, due to the high algorithm complexity.

11.2.8 Conclusion

In this section, we have introduced the main steps of scene reconstruction and analyze the time complexity of each step. The main work is to understand all the algorithms mentioned in this chapter in some details and to verify the source code, "bundler-source-0.3," to see whether it has matched the algorithms described above.

11.3 SCENE RECONSTRUCTION BASED ON CLOUD COMPUTING

From Section 11.2, we have the fundamental idea of scene reconstruction and the time complexity of each step. Noting that reconstructing scene from an image set of 415 photos took about 21 hours by one computer, we can easily imagine that if we apply this technology to construct the 3D structure of a landscape or a city from 10,000 photos or more, the most challenging problem we have to meet is to find a solution that can speed up the process of reconstruction, otherwise we could finally get our new reconstructed world from 10,000 photos after months or even years of time. Because of the large amount of calculation and the possibility of parallel computing in some steps (i.e., feature matching, SIFT detector, and F- and H-matrix computing), we come up with the idea of utilizing the benefits of parallel computing to help us reduce the wall time cost. Also, since the reconstruction pipeline will become a large job working for day and night, and produce large amount of data, we need a platform to help us deal with the key feature of reliability, scalability, and security, which will benefit the scene reconstruction pipeline a lot. Then we come to the solution we found: scene reconstruction based on cloud computing.

11.3.1 Introduction to Google Cloud Model

When we talk about cloud computing, the first word comes into our mind is Google. Google's cloud computing technique is customized for specific Google network application. According to the characteristic of large-scale internal network data, Google proposed the fundamental architecture of distributed parallel computing cluster, using software control to deal with the problem of node loss that often happens in cluster.

From 2003, Google has continually presented papers that revealed the way it deals with distributed data and the core idea of its cloud computing technique. According to the papers Google presented in recent year, Google's cloud computing infrastructure includes three systems that are independent but closely linked to each other: Google File System (GFS) [10],

MapReduce programming model [11], and large-scale distributed database, the BigTable [12].

In this section, we will introduce the properties of GFS and MapReduce programming model, which are used in our system to show how they fit the requirement of parallel reconstructing scene.

11.3.1.1 Google File System

GFS, which is a scalable distributed file system for large distributed data-intensive applications. It is designed to provide an efficient, reliable access to data using large cluster of commodity hardware. Now let us check whether the system assumption of GFS fits the scene reconstruction pipeline:

1. The system is built from many inexpensive commodity components that often fail.

2. Multi-gigabyte files are the common case and should be managed efficiently, whereas small files must be supported but not optimize for them.

3. The two kinds of reads are large streaming reads and small random reads.

4. The workloads also have many large, sequential writes that append data to file.

5. The system must efficiently implement well-defined semantics for concurrently appending.

Obviously, the conditions 1, 3, 4, and 5 just fit our requirement. However, based on my understanding of scene reconstruction, the pipeline will possess most of the files as small files (photos, key descriptor files, and matches). We can organize them as group files to solve this problem.

11.3.1.1.1 Architecture of GFS   A GFS cluster consists of a single master and multiple chunk servers and is accessed by multiple clients, as shown in Figure 11.9. Notice that the chunk server and the client can be run on the same machine.

For each cluster, the nodes are divided into two types: one master node and a large number of chunk servers. Chunk servers store the data files, with each individual file broken up into fixed size chunks (hence the name)

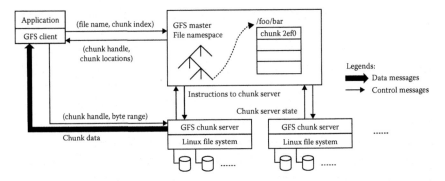

FIGURE 11.9   GFS architecture. GFS, Google file system. (From Sanjay Ghemawat, Howard Gobioff, Shun-Tak Leung, *19th ACM Symposium on Operating Systems Principles*, Lake George, NY, 29–43, 2003. With permission.)

of about 64 MB. Each chunk is replicated several times throughout the network (three in GFS).

The master server does not usually store the actual chunks, but rather all the metadata associated with the chunks, such as the namespace and access control information, the mapping from files to chunks, and the current locations of chunks. It also controls system-wide activities such as chunk lease management, garbage collection of orphaned chunks, and chunk migration between chunk servers. The master periodically communicates with each chunk server to give it instructions and collect its state.

11.3.1.1.2 Single Master   Having a single master of each cluster vastly simplifies our design and enables the master to make sophisticated chunk placement and replication decisions using global knowledge. To avoid it becoming a bottleneck, we must minimize its involvement in reads and writes. In fact, the client asks the master which chunk servers it should contact and then interacts with the chunk servers directly for the operations.

In Figure 11.9, we can find that the clients send the master a request containing the file name and the chunk index. Then the master replies with the corresponding chunk handle and locations of the replicas. The client then sends a request to one of the replicas, most likely the closest one, and further reads of the same chunk require no more client–master interaction.

After a simple introduction of the GFS, we come to the familiar MapReduce model, which our implementation is based on.

11.3.1.2 MapReduce Model

The MapReduce model is a programming model for large-scale distributed data processing. It has the properties as follows:

1. Simple, elegant concept

2. Restricted, yet powerful programming construct

3. Building block for other parallel programming tools

4. Extensible for different applications

Also, an implementation of a system to execute such programs can take the advantages listed below:

1. Parallelism

2. Tolerate failures

3. Hide messy internals from users

4. Provide tuning knobs for different applications

11.3.1.2.1 Basic Programming Model   The computation takes a set of input key–value pairs and produces a set of output key–value pairs. The user of the MapReduce model expresses the computation as two functions: Map and Reduce (Figure 11.10).

The Map function, written by the user, takes an input pair and produces a set of intermediate key–value pairs. The MapReduce groups together all intermediate values associated with the same intermediate key k' and passes them to the Reduce function.

The Reduce function, also written by the user, accepts an intermediate key k' and a set of values for that key. It merges together these values to form a possibly smaller set of values. Typically, just zero or one

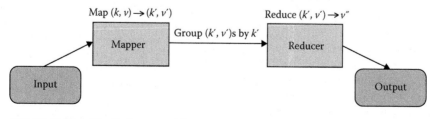

FIGURE 11.10   MapReduce model.

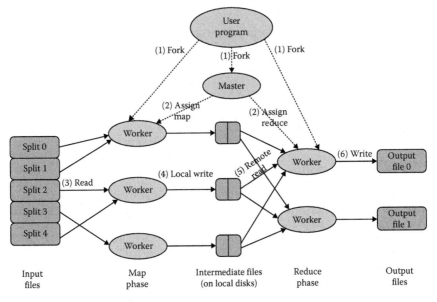

FIGURE 11.11   MapReduce overview.

output value is produced per Reduce invocation. The intermediate values are supplied to the user's Reduce function via an iterator. This allows us to handle lists of values that are too large to fit in memory.

11.3.1.2.2 MapReduce Execution Example   In this section, the overview of MapReduce invocation is described. In Figure 11.11, we can see that

1. User program first forks the map and reduce workers and the master node.

2. Master assigns map and reduce tasks to each worker.

3. Map workers read one of splits, which are split from the input data by the user-defined input format function and execute the Map function.

4. Map workers store the internal key–value pair into the local file system.

5. Reduce workers, which may be different nodes from map workers, will read the internal key–value pair, which have been sorted by any rules, remotely.

6. Reduce workers execute the reduce function, and then atomic write to the GFS.

11.3.2 Open-Source Cloud Framework—Hadoop

Although Google has revealed the design of its cloud computing infra-structure, the implementation of its system still keeps in secret. Then we come to the famous open-source cloud framework—Hadoop. Apache Hadoop is a Java software framework that supports data-intensive dis-tributed applications under a free license. It enables applications to work with thousands of nodes and petabytes of data. Since Hadoop was inspired by Google's MapReduce and GFS paper, the basic knowledge of GFS and MapReduce model introduced in Section 11.3.1 also works well in Hadoop. In Hadoop framework, Hadoop distributed file system (HDFS) is the open-source implementation of GFS and Hadoop also implements the MapReduce model.

Even though the version of Hadoop is 0.20.2, which is still far from 1.0, a wide variety of companies and organizations use Hadoop for production, such as Adobe that uses Hadoop and HBase in several areas from social services to structured data storage and processing for internal use; Amazon Web Service that provides a hosted Hadoop framework running on the Web-scale infra-structure of Amazon Elastic Compute Cloud (Amazon EC2) and Amazon Simple Storage Service (Amazon S3); and Facebook that uses Hadoop to store copies of internal log and dimension data sources and use it as a source for reporting/analytics and machine learning.

In addition, many researchers have extended Hadoop framework for scientific purposes. In Reference 13, a technique is applied to binary image files in order to enable Hadoop to implement image processing techniques on a large scale. Bortnikov [14] shows how Hadoop can be applied for Web-scale computing. He et al. [15] design and implement a MapReduce framework on graphics processors.

In this section, we mainly discuss the Hadoop MapReduce implemen-tation and introduce the MapReduce user interfaces.

11.3.2.1 Overview of Hadoop MapReduce

In Hadoop, a MapReduce job usually splits the input dataset into inde-pendent chunks, which are processed by the map tasks in a completely parallel manner. The framework sorts the outputs of the maps, which are then input to the reduce tasks. Typically, both the input and the output of the job are stored in a file system. The framework takes care of scheduling tasks, monitors them, and reexecutes the failed tasks.

The MapReduce framework consists of a single master JobTracker and one slave TaskTracker per cluster node. The master is responsible for scheduling the jobs' component tasks on the slaves, monitoring them, and reexecuting the failed tasks. The slaves execute the tasks as directed by the master.

Minimally, applications specify the input/output locations and supply map and reduce functions via implementations of appropriate interfaces and/or abstract classes. These, and other job parameters, comprise the job configuration. The Hadoop job client then submits the job (e.g., jar or executable file) and configuration to the JobTracker, which then assumes the responsibility of distributing the software/configuration to the slaves, scheduling tasks and monitoring them, and providing status and diagnostic information to the job client.

11.3.2.2 MapReduce User Interfaces

This section provides a short view on every user-facing aspect of the MapReduce framework. Applications typically implement the mapper and reducer interfaces to provide the map and reduce methods. These form the core of the job.

11.3.2.2.1 Mapper   Mapper maps input key–value pairs to a set of intermediate key–value pairs. The Hadoop MapReduce framework spawns one map task for each InputSplit generated by the InputFormat for the job. Overall, mapper implementations are passed to the JobConf for the job via the JobConfigurable.configure(JobConf) method and override it to initialize themselves. The framework then calls map(WritableComparable, Writable, OutputCollector, Reporter) for each key–value pair in the InputSplit for that task. Applications can then override the Closable.close() method to perform any required cleanup.

11.3.2.2.2 Reducer   Reducer reduces a set of intermediate values that share a key to a smaller set of values. Reducer has three primary phases: shuffle, sort, and reduce.

1. Shuffle: Input to the Reducer is the sorted output of the mappers. In this phase, the framework fetches the relevant partition of the output of all the mappers via the Hypertext Transfer Protocol (HTTP).

2. Sort: The framework groups Reducer inputs by keys (since different mappers may have output the same key) in this stage. The shuffle and sort phases occur simultaneously; while map outputs are being fetched, they are merged.

3. Reduce: In this phase, the reduce (WritableComparable, Iterator, OutputCollector, Reporter) method is called for each <key, (list of values)> pair in the grouped inputs. The output of the reduce task is typically written to the FileSystem via OutputCollector.collect (WritableComparable, Writable).

After the basic introduction to the cloud computing framework, we come to the second core idea of this chapter: running scene reconstruction on Hadoop.

11.3.3 Scene Reconstruction Based on Cloud Computing

This implementation of running scene reconstruction pipeline on Hadoop is based on the source code, "bundler-v0.3-source," with modifications to fit the requirements of MapReduce model. There are three main changes compared to the source code:

1. Redesign the I/O operations as the input/output format for MapReduce.

2. Parallel compute each step using MapReduce model.

3. Design the pipeline for three main MapReduce modules to form the whole scene reconstruction system.

11.3.3.1 I/O Operations Design

11.3.3.1.1 InputFormat   Since Hadoop is traditionally designed for large ASCII text file processing utility, especially for each line as a record. However, to fit the requirements of scene reconstruction, we need to extent the current application programming interface (API) in the Hadoop library to deal with different input requirements beyond the ASCII text files. There are two interfaces that need to be implemented in order to allow Hadoop to work with custom file formats: (1) InputFormat interfaces that split the input files into splits and (2) RecordReader interface that defines the key–value pair type and content. There are several methods that need to be implemented within the two interfaces. The most important method in FileInputFormat is the getSplit function, which determines the way data are split and the content of each split. Notice that split can be part of the input file or a collection of input files. After the input data are split, the RecordReader read one of the splits at a time and generates the key–value pair for mapper. The key–value

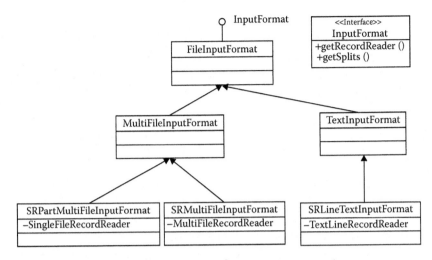

FIGURE 11.12   Specific FileInputFormat for scene reconstruction.

pair is determined by the next method in RecordReader. User can define what portion of the file will be read and sent to the mapper and it is here where user can determine the way key–value pair is extracted from input split.

In our implementation, we have designed three types of FileInputFormat with their relevant RecordReader to help us control the input for mapper, as shown in Figure 11.12.

In the figure, SRPartMultiFileInputFormat and SRMultiFileInputFormat extend MultiFileInputFormat, which is provided by the Hadoop library. Both of them split input files into file collections, at least one file in a collection. To be different, SRPartMultiFileInputFormat just reads part of a file, maybe the first 50 bytes or simply the file name of a file. For instance, when we extract the focal information from a jpg file, we just need to extract the Exif (exchangeable image file format) of a jpg file, and then the RecordReader treats the file name as key and the focal information from Exif as value. SRMultiFileInputFormat is much simple, which splits only a file as a collection and then reads the file from beginning to end. The RecordReader will set the file name as key and the content of the file as value. The SRLineTextInputFormat extends TextInputFormat, which is used for ASCII text operation. Differed from TextInputFormat, SRLineTextInputFormat treats each line as record and attaches additional information such as ID with it. Then RecordReader will define the ID as key and the line content as value.

11.3.3.1.2 OutputFormat   Similar to InputFormat, the user must implement the OutputFormat and RecordWriter interfaces to determine the user-defined output format. In this system, we use two kinds of OutputFormat, which are included in the Hadoop output library: (1) FileOutputFormat writes all the key–value pairs into a single file and (2) MultiOutputFormat writes all the values that have the same key into a file named according to the key.

11.3.3.2 Scene Reconstruction Pipeline

This section discusses our system, including four MapReduce modules and the input data/output result of each module. In Figure 11.13, we can see the whole reconstruction pipeline.

11.3.3.3 Extract Focal Length

This is the first MapReduce module invoked in our pipeline, and here we assume that all the input images have been uploaded to the HDFS. This step is simple: The image set is split into N splits, possibly two times than the map workers. The RecordReader reads the file name and the Exif value of each image and then passes to the mapper. In Map function, the system reads the Exif and extracts the focal length of each image if exists. Then FileOutputFormat collects the name and focal length of each file and invokes RecordWriter to write into a single file for future use.

11.3.3.4 SIFT Detector

Since no source code is provided from SIFT, the only way to apply SIFT on Hadoop is to invoke external file with the appropriate parameters. The command line format probably looks like this:

$sift < $IMAGE_DIR/$PGM_FILE > $KEY_DIR/$KEY_FILE

Here, sift is the executable file name of SIFT and the $IMAGE_DIR/$PGM_FILE indicates the position of the input pgm file in the file system and the $KEY_DIR/$KEY_FILE defines the position of the output key file. If the parameter of ">$KEY_DIR/$KEY_FILE" is not existent, sift will output the key to the standard output.

Because our images are mostly jpg images, we must first transfer jpg images into pgm images. In our implementation, each input jpg file is treated as a split, and RecordReader will read the whole content of jpg file as value and pass to the mapper. Then two external files are invoked in Map function. First, we throw the content of jpg image into an image transform program and then it will generate a pgm format image. Second, we redirect

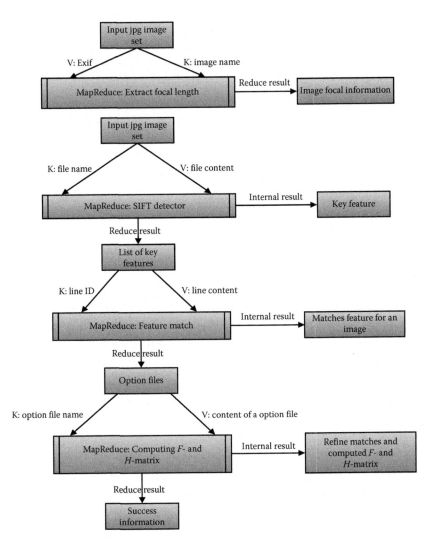

FIGURE 11.13 Scene reconstruction pipeline.

the output of transform program into SIFT, and then compute the key feature of the relevant pgm image. The mapper will generate a image_name.key file in the local file system and upload to the HDFS after mapper. In the reduce phase, a list of the path of key files will be generated.

Another way to run SIFT on Hadoop is to copy the jpg files into local file system and execute the image transform program and SIFT program as well. However, it requires one-time I/O operation than previous implementation. But in my evaluation, it shows little different in efficiency.

11.3.3.5 Feature Match

In this MapReduce module, the key list file generated by SIFT is treated as input file. The FileInputFormat reads each line as value and attributes a line number as key. In the mapper phase, the computed key file is implicated read: first, all key files are copied into local file system and then mapper invokes the external executable file FeatureFullMatch to find matches between one stationary image and the other left images. FeatureFullMatch program is different from provided Bundler-0.3-source, with some modification as: Only image that is appointed by ID will execute match function instead of matching all images in the list. Then FeatureFullMatch tries to find the key file designated in the key list file and generate the match table for that image. After mapper, the generated match table will be uploaded to the HDFS with the name of "ID.matches.init." In the reduce phase, each key–value pair will generate an option file that includes the path of match table and some configuration parameters needed for computing F- and H-matrix, named as "ID.options.txt."

Noting that the latter map job will run much less time than its previous map job because a map job with a relative small ID will have to compare more images than map job with large ID. For instance, to an image set of 415 images, a map job with a ID 1 will try to find matches among left 414 images, whereas a map job with ID 400 only needs to find matches among 15 images. This kind of parallel operation will not influence the wall time cost significantly because image number is much larger than the map workers. So there may exist difference in executed map numbers among data nodes, but almost finish all the map jobs at the same time.

Another weak point of this kind of parallel cooperation is the overhead of I/O and hardware storage, since all map workers will copy the key files from HDFS into local file system for once. However, compared to the large amount of time cost of running this step, the influence will be relatively small.

11.3.3.6 Computing F- and H-Matrix

In this phase, each option file generated by feature match will create a map job. In Map function, similar to feature match, a modified external file, bundler, that fits the property of parallel computing is invoked. Since all the outputs, including refined matches, F-matrix, and inliers of H-matrix, must be gathered to compute tracks and SfM, we must determine what data should be uploaded to the HDFS. Moreover, since the F-matrix, H-matrix, and refined matches as well are parallel computed, as a result, there may exist large number of output files; how to read them in order becomes another concern of our system. We not only rewrite the function of computing F- and H-matrix for parallel computing, but also define new output function to store the necessary information needed in the next step.

To run bundler successfully, the Map function first copies table files containing the matches assigned by the option file and then records the refined matches, the information of F-matrix, and the number of inliers between the images listed in the assigned table file.

11.3.3.7 The Left Part of Scene Reconstruction Pipeline

In scene reconstruction pipeline, two steps are still not transplanted to the Hadoop framework in my implementation: (1) computing tracks and (2) SfM. For the first one, since tracks are founded using a breadth-first search of the sets of features, which is a little hard to parallel, moreover, the wall time cost in this step is very small, about 30 seconds in a total time of 23 hours. However, for the second one, it is over my knowledge range of implementing parallel computing due to its high algorithm complexity. One possible solution proposed in Reference 1 is to divide images into subsets and run two or more incremental SfMs at the same time. But the way to divide image set more reasonable is still hard to figure out. More theory supporters are therefore required to combine the reconstructed scenes into a big one.

11.3.4 Conclusion

In this section, we first introduce the architecture of GFS and the basic MapReduce programming model. With this knowledge, we then describe the implementation of our system running on the Hadoop framework to achieve properties such as parallel computing, reliability, and scalability. Noting that with the help of Hadoop framework, designing a distributed parallel computing system becomes much simpler, users only need to consider how to implement MapReduce model; all the other works can be done by the framework automatically with light configuration.

11.4 HADOOP DEPLOYMENT AND EVALUATION

This section is mainly divided into two parts: The first part emphasizes some keypoints when deploying a Hadoop cluster and the second part shows the experimental result of our system compared to the original scene reconstruction system.

11.4.1 Hadoop Cluster Deployment

In our experiment, we use the version of Hadoop as 0.19.2, which is much more stable than 0.20.2. There exist oceans of materials introducing how to deploy Hadoop cluster on a single node, while some necessary keypoints are missed for deploying Hadoop framework among multiple nodes. In this section, we will give a simple introduction to the installing steps of Hadoop framework for multiple nodes and emphasizes the additional requirement for multiple node cluster. Although Hadoop can run on Windows by installing Cgywin and Linux, because Linux is more efficient and stable, we choose Linux to deploy Hadoop framework.

Operating system: Ubuntu 9.10

Hadoop version: 0.19.2

Preinstall software: JDK6, SSH (Secure shell)

Main installing steps:

1. Configure Java environment variables.

2. Master login every slave without verification using SSH.

3. Modify Hadoop configuration files:

    a. Modify $HADOOP_HOME/conf/hadoop-env.sh.

    b. Modify $HADOOP_HOME/conf/hadoop-site.sh.

4. Format the namenode.

After these four steps, Hadoop can run on a single node successfully; however, in order to run Hadoop on multiple node clusters, some additional but necessary configuration is required.

1. *Close IPv6 address.* In some situations, the datanodes cannot communicate with namenode because the namenode has bounded its port to IPv6 address. We can add the following line into hadoop-env.sh to avoid this situation:

    HADOOP_OPTS=-Djava.net.preferIPv4Stack=true

2. *Define master and slave IP address.* We must configure the /etc/hosts file to help each node understand the IP address of the other node.

All the nodes, including the master and slave nodes, should store the list of all nodes' IP address. All nodes' IP address into/etc/hosts are added as follows:

192.168.1.100 master

192.168.1.101 slave1

192.168.1.102 slave2

192.168.1.103 slave3

192.168.1.104 slave4

192.168.1.105 slave5

3. *Give each node a global hostname.* When reducer tries to read the internal result from mapper, it will find the mapper by its hostname. If we do not give each node a hostname, or two nodes have the same hostname in a cluster, it will report "map result not found" error.

4. *Add slave nodes for master.* When the Hadoop starts, it will launch the datanodes listed in $HADOOP_HOME/conf/slaves. If we want to add new datanodes into the cluster, we must modify this file and add the new datanode's hostname.

After the steps listed above have been configured, the Hadoop framework may run successfully among a cluster, rather than trying to run our system on Hadoop framework.

11.4.2 Evaluation

In this section, we compare the time cost of the original scene reconstruction pipeline and our proposed system. In this evaluation, two Hadoop clusters with different number of datanodes are deployed. The small one is formed by three datanodes and one master, and the big one is formed by seven datanodes and one master. Notice that all the datanodes have the same hardware and operating system. Then we run the original system on one datanode and then run our proposed system among the two clusters.

Three input image collections are applied in the experiment. The first one has an image number of 100, and the second one has an image number of 415. The last image collection has 608 images. Then we first run the relatively two small image collections using the original scene reconstruction system and then run our system on the smaller Hadoop cluster.

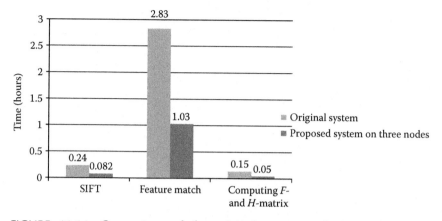

FIGURE 11.14 Comparison of the original system with proposed system (100 images). SIFT, scale-invariant feature transform.

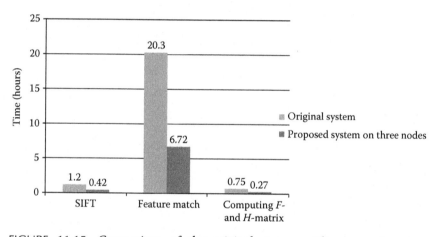

FIGURE 11.15 Comparison of the original system with proposed system (415 images). SIFT, scale-invariant feature transform.

The comparisons of running time cost for each step of scene are shown in Figures 11.14 and 11.15.

From these two figures, we can easily find out that the feature match step takes the most part of the computing time and our proposed system reduces the wall time clock cost significantly. Table 11.1 shows the result of time cost for each system. Our proposed system running on three datanode clusters only spent about a third of time than the original system.

Then we increase the number of datanode cluster and compare the efficiency of increment between different clusters. The first cluster is formed by three datanodes, whereas the second one has seven datanodes. Also, we

TABLE 11.1   Comparison of Total Time Cost between Different Systems

Number of Photos	Original System	Proposed System (Three Nodes)
100	3 hours 16 minutes	1 hours 9 minutes
400	22 hours 15minutes	7 hours 35 minutes

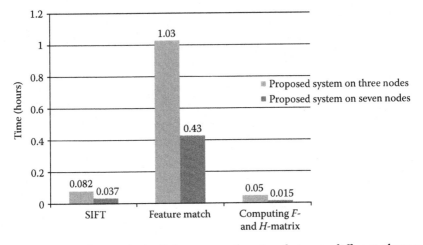

FIGURE 11.16   Comparison of the proposed system between different clusters (100 images). SIFT, scale-invariant feature transform.

add the largest image collection that has 608 images into the experiment. We now reconstruct the three image collections on the two clusters. Figures 11.16 through 11.18 show the time cost for each step of scene reconstruction among different clusters.

From these three experiments, we can see that as the datanode number in a cluster increases, our system will reduce the time cost incrementally. Table 11.2 shows the result of time cost for our system running among different clusters.

11.4.3 Conclusion

The experiment results show that our system performs almost seven times faster than the original system, which is nearly to the node numbers of the Hadoop cluster. It is easy to induce that if we add more nodes into the cluster, the new system will reduce the time cost incrementally.

In general, it is also true that if we run the original system on a super machine that has the computing power similar to the total amount of seven datanodes, it will perform equal or even better than our proposed system. That is because the network blocking and the data interchange may decrease

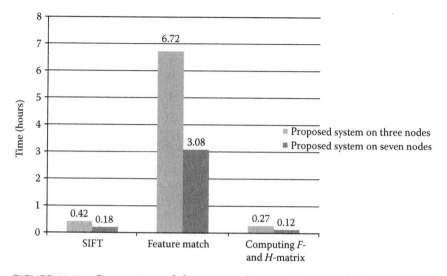

FIGURE 11.17 Comparison of the proposed system between different clusters (415 images). SIFT, scale-invariant feature transform.

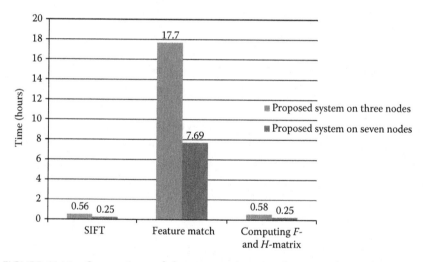

FIGURE 11.18 Comparison of the proposed system between different clusters (608 images). SIFT, scale-invariant feature transform.

TABLE 11.2 Comparison of Total Time Cost between Clusters with Different Node Numbers

Number of Photos	Proposed System (Three Nodes)	Proposed System (Seven Nodes)
100	1 hours 9 minutes	29 minutes
400	7 hours 35 minutes	3 hours 22 minutes
600	18 hours 30 minutes	8 hours 11 minutes

the efficiency of the system. However, since the cost of buying a super machine is more expensive than several normal computers, also when the number of photos increases to 100,000 or more, which is reasonable for image Websites, the super machine will also be hard to bear them. So our system will become a possible solution to deal with large image datasets and reduce the time cost with a relatively low money cost.

11.5 CONCLUSION AND FUTURE WORK

11.5.1 Conclusion

The purpose of this research was to determine if scene reconstruction can be transplanted to Hadoop framework and to see the new properties the Hadoop framework could provide into scene reconstruction. In realizing this goal, the following three keypoints are described in this chapter:

1. Introduce the main steps of scene reconstruction pipeline in some details.

2. Implement the first four steps of scene reconstruction running on Hadoop framework.

3. By means of evaluation, the experimental result shows that scene reconstruction based on cloud computing achieves great success.

Cloud computing with Hadoop could provide relatively inexpensive means process datasets of the magnitude without seriously compromising performance. We believe that our exploration in this field could give some hints for using Hadoop framework to deal with complex algorithms running on large image datasets.

11.5.2 Future Work

Our future work will be the following:

1. Continually transplant the left two steps of scene reconstruction pipeline, computing tracks and SfM to Hadoop framework.

2. Utilize the computing ability of graphic processing unit (GPU) to speed up the scene reconstruction in advance.

3. Explore the possibility of building up a MapReduce framework that can utilize GPU to help accelerate scene reconstruction.

REFERENCES

1. Keith N. Snavely. Scene reconstruction and visualization from Internet photo collections. Doctoral thesis, University of Washington, Seattle, WA, pp. 1–67, 2008.
2. Noah Snavely, Steven M. Seitz, and Richard Szeliski. Photo tourism: Exploring photo collections in 3D. *ACM Transactions on Graphics*, 25(3): 835–846, 2006.
3. David G. Lowe. Distinctive image features from scale-invariant keypoints. *International Journal of Computer Vision*, 60(2): 91–110, 2004.
4. Noah Snavely, Steven M. Seitz, and Richard Szeliski. Modeling the world from Internet photo collections. *International Journal of Computer Vision*, 80(2): 189–210, 2007.
5. Martin Fischler and Robert Bolles. Random sample consensus: A paradigm for model fitting with applications to image analysis and automated cartography. *Communication of ACM*, 24(6): 381–395, 1981.
6. Sunil Arya, David M. Mount, Nathan S. Netanyahu, Ruth Silverman, and Angela Y. Wu. An optimal algorithm for approximate nearest neighbor searching fixed dimensions. *Journal of the ACM*, 45(6): 891–923, 1998.
7. David Nistér. An efficient solution to the five-point relative pose problem. *IEEE Transactions on Pattern Analysis and Machine Intelligence*, 26(6): 756–777, 2004.
8. Richard I. Hartley and Andrew Zisserman. *Multiple View Geometry*. Cambridge: Cambridge University Press, 2004.
9. Manolis Lourakis and Antonis Argyros. The design and implementation of a generic sparse bundle adjustment software package based on the Levenberg–Marquardt algorithm. *Technical Report 340*, Institute of Computer Science, FORTH, Heraklion, Greece, 2004.
10. Sanjay Ghemawat, Howard Gobioff, and Shun-Tak Leung. The Google File System. *Proceedings of the 19th ACM Symposium on Operating Systems Principles*, Lake George, NY, pp. 29–43, 2003.
11. Jeffrey Dean and Sanjay Ghemawat. MapReduce: Simplified data processing on large clusters. *6th Symposium on Operating System Design and Implementation*, San Francisco, CA, pp. 137–150, 2004.
12. Fay Chang, Jeffrey Dean, Sanjay Ghemawat, Wilson C. Hsieh, Deborah A. Wallach, Mike Burrows, Tushar Chandra, Andrew Fikes, and Robert E. Gruber. Bigtable: A distributed storage system for structured data. *Proceedings of the 7th Symposium on Operating System Design and Implementation*, Seattle, WA, pp. 205–218, 2006.
13. Jeff Conner. Customizing input file format for image processing in Hadoop. *Technical Report*, Arizona State University, Mesa, AZ, 2009.
14. Edward Bortnikov. Open-source grid technologies for Web-scale computing. *ACM SIGACT News*, 40(2): 87–93, 2009.
15. Bingsheng He, Wenbin Fang, Qiong Luo, Naga K. Govindaraju, and Tuyong Wang. Mars: A MapReduce framework on graphics processors. *Proceedings of the 17th International Conference on Architectures and Compilation Techniques*, pp. 260–269, 2008.

CHAPTER 12

Pearly User Interfaces for Cloud Computing

First Experience in Health-Care IT

Laure Martins-Baltar, Yann Laurillau, and Gaëlle Calvary

University of Grenoble
Grenoble, France

CONTENTS

12.1 RESEARCH CONTEXT AND PROBLEM

Moving away from a classic and monolithic computing model based on local resources, cloud computing pushes the boundaries to fully rely on online resources, and thus invites inventor to reconsider interaction metaphors for cloud-based environments. As cloud computing combines and leverages the existing technologies already available in data centers (huge storage and computation capacities, virtualized environments based on high-speed data links) (Zhang et al. 2010b), users now have an ubiquitous access to on-demand services in a pay-as-you-go manner (Mell and Grance 2009): "Virtualization makes it possible for cloud computing's key characteristics of multi-tenancy, massive scalability, rapid elasticity and measured service to exist" (Carlin and Curran 2012).

This evolution raises several issues in human–computer interaction (HCI), such as the following: How to cope with offline situations (loss of connectivity, failures) (England et al. 2011; Stuerzlinger 2011; Terrenghi et al. 2010; Vartiainen and Väänänen-Vainio-Mattila 2011)? Which persuasive technologies would foster sustainability and power energy savings (Pan and Blevis 2011)? To which extent would migration from a conventional desktop to a cloud web-based setting fragment user experience (UX) and impact UI consistency (England et al. 2011; Pham 2010)? Are privacy, trust, and data ownership UX issues (Armbrust et al. 2010; England et al. 2011; Odom et al. 2012)? Which plasticity capabilities must UIs have to support the dynamic and ubiquitous provision of services? Which design would appropriately address traceability, control of data, and sharing issues raised by the socialization of cloud services and online activities (Odom et al. 2012; Zhang et al. 2010a)? This work focuses on the two last issues.

This work investigates how to make the desktop metaphor evolve so that it integrates cloud-based services and activities, and thus it supports the convergence of cloud and social computing (Pham 2010). Currently, the classic desktop metaphor is still single-user, data-centered, and designed for the use of local resources. As a first approach, we consider the social dimension of the desktop and agree on representing the relationships between data, people, activities, and services (Väänänen-Vainio-Mattila et al. 2011; Zhang et al. 2010a) so that it promotes the social worth of data. However, the classic hierarchy-based folder metaphor does not suit anymore for online repositories supporting social sharing (Odom et al. 2012; Shami et al. 2011). This chapter proposes the concept of Pearly UIs for cloud and social computing. It first browses the state of the art, and then reports the invention, implementation, and evaluation of Pearly UIs in the context of health care.

12.2 RELATED WORKS

This section covers the convergence of cloud and social computing. We identify three classes of research: the social desktop for the cloud to deal with online resources and the socialization of services; the social navigation and sharing to deal with privacy and traceability issues; and information seeking/refinding to deal with big data and scalability.

12.2.1 Social Desktop for the Cloud

The application-centric model of the traditional desktop is progressively fading away: More and more applications and services are deployed in clouds and made available as Web applications. However, moving to a full Web-based environment breaks down the guaranty of consistent UIs and leads to a fragmented UX (England et al. 2011; Pham 2010).

A possible explanation of this failure (Pham 2010) may be that the current desktop metaphor "heavily reflects" the local nature of resources, "grounded firmly in hierarchy and physical locations" and "evolved very little to support sharing and access control." As a consequence, the "social desktop" was proposed based on the concept of user-created groups. This concept extends the folder metaphor to encompass files, people, and applications. Compared to the usual folder metaphor, a unique instance of an object may be included in different groups, thus providing a lightweight means to allow sharing and access: As a group is associated with users, access is implicitly granted to the members of the group, allowing file sharing.

Similarly, CloudRoom (Terrenghi et al. 2011) is a new desktop metaphor for the cloud, focusing on storage and data retrieval issues. It partitions the desktop into three separate areas (planes in a 3D space) to organize data: (1) a long-term and persistent storage, (2) a timeline overview, and

(3) a temporary storage for work-in-progress activities. It allows session sharing with contacts.

Before the advent of cloud and social computing, when e-mail, voice mail, or instant messaging were considered as prevalent communication tools supporting social networking, Whittaker et al. (2004) first pointed out the limits of the current desktop to support social interfaces. ContactMap is a social software application primarily designed to support communication-oriented tasks. The social desktop is the central element of ContactMap: Similar to shared workspaces, it allows to structure and to visually represent social information as groups of contacts.

In computer-supported cooperative work (CSCW), Voida et al. (2008) advocated moving away from a document- and application-centric desktop to an activity-centric desktop, encompassing collaborative activities, and thus the social dimension. In particular, the Giornata interface includes a contact palette allowing users to manage contacts (individuals or groups) and providing a lightweight means for file sharing.

Grudin (2010) observed that CSCW is slowly moving toward CSCW as "collaboration, social computing, and work," which, from a user-centered point of view, is a foundation of cloud computing.

12.2.2 Social Sharing and Browsing

"Online social networks have become indispensable tools for information sharing" (Kairam et al. 2012). Still, it is difficult for users to target specific parts of their network. Google+'s Circles are similar to Pham's user-created groups: user's social network is structured into circles allowing selective information sharing with parts of his/her social network. However, there is a lack of awareness to trace shared information from circles to circles.

Shami et al. (2011) focus on social file sharing in enterprise, considering the social worth of files. Social metadata are added to files as extra attributes. Such metadata allow a nonhierarchical file organization. In addition, metadata facilitate pivot browsing (i.e., parameterized browsing based on metadata). To do so, the authors have developed the Cattail file sharing system. The system is able to reveal social activity around files using a time-based stream of recent events. Access is supported through three levels of sharing: private, confidential, and public.

Several works have explored metadata to promote different and more efficient file organizations. In particular, Dourish (2003) introduced the concept of placeless document, a paradigm based on document properties that cover both external (e.g., creation date) and internal properties (e.g., it is a photo of me and my son).

12.2.3 Social Information Seeking and Refinding

Information seeking is another facet of the convergence of cloud and social networking. For instance, the social worth of data is also considered to improve Web search engines (Muralidharan et al. 2012): Web search is augmented with social annotations. The goal is "to make relevant social activity explicitly visible to searchers through social annotations." Such an approach is similar to social metadata. In particular, social annotations are another contextual key to facilitate and improve information refinding (i.e., orienteering). For instance, for local file storage, Sawyer et al. (2012) have developed a system that detects people and groups present at the time a piece of information is used or created. Therefore, files are tagged with information (i.e., social orbits) about the present physical context of the social interactions. Thus, social orbits are similar to Pham's user-created groups.

From this broad state of the art, let us conclude that, although in HCI research about cloud computing is still in its infancy, very recent works show a growing interest in this topic. It appears that big data, social networks for data sharing, communication, and collaborative activities are becoming central and firmly linked. However, cloud services are still underexplored. Therefore, we propose the "pearl" metaphor to present to users socially augmented entities (SAEs) as well as the services available in the cloud.

12.3 CASE STUDY: HEALTH-CARE INFORMATION TECHNOLOGY

Health is particularly interesting for information technology (IT). With the evolution of practices and legislation, medical practitioners increasingly need tools for the production of images and the follow-up of medical records. Gastroenterology, the domain under study in this work, strongly relies on endoscopic images for decision making, evaluation of practices as well as for education and research. A medical challenge is to export endoscopic images outside the operating room, in the cloud for instance.

To understand medical practices and to identify users' needs, a thorough study of the field has been conducted in three phases: (1) meetings with doctors and secretaries, (2) analysis and modeling of their needs, and (3) validation of this work with different actors. This study resulted in several models of the patient care process in gastroenterology [17 use cases,

10 task models, and 29 Unified Modeling Language (UML) diagrams]. The models have been validated by medical practitioners, and thus make a strong know-how explicit in the field.

The models revealed not only the importance and variety of medical data, but also a crucial need for medical software applications that better support the social dimension of medical activities: sharing medical data, easily communicating with colleagues to get advices, better capitalizing medical knowledge, and better supporting medical decision making. More precisely, medical practitioners expand four socio-professional networks: (1) health workers, including colleagues, experts, and friends; (2) academics and students; (3) health workers involving follow-up of inpatients; and (4) institutions. However, in practice, information sharing is still informal (e.g., by phone), in particular for medicolegal reasons, depriving practitioners from peer-based decisions, and giving them the feeling of being alone.

Based on these findings, we first improved the usability of the software (named Syseo) used by the practitioners (giving rise to Syseo*; Figure 12.1), and then retargeted it for the cloud (giving rise to the PearlyDesktop running prototype; Figure 12.2). The redesign of Syseo* was driven by two requirements: the improvement of health-care quality and the ability to

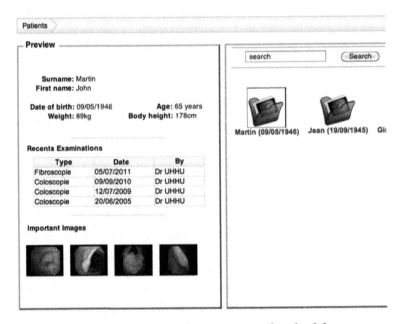

FIGURE 12.1   Syseo*, a data-centered management of medical data.

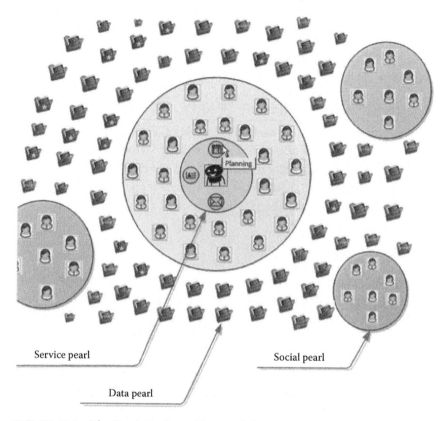

Service pearl

Social pearl

Data pearl

FIGURE 12.2   The PearlyDesktop: User- and data-centric views.

trace and evaluate professional practices. Scenarios (Rosson and Carrol 2002) written with experts in gastroenterology were used to support the design process. Three gastroenterologists and one developer of medical information systems validated the Syseo* prototype.

As shown in Figure 12.1, the UI of Syseo* is data centered, that is, centered on medical records. A medical record is a collection of data about a patient, notably endoscopic images. In gastroenterology, endoscopic images are key, at the heart of practices, and diagnostic and therapeutic approaches.

However, Syseo* supports the social dimension along three function-alities: the sharing of medical data either privately between two doc-tors or publicly for capitalization (e.g., teaching), the request of expert advices, and the management of the practitioner's professional network. Information is stored on the cloud, making it possible for medical prac-titioners to manage medical data and professional relationships within a unique application. Online cloud-based services are envisioned for improving the quality of medical care: sharing confidential medical data among practitioners, requesting advices from experts to improve diagnoses, and taking benefit from online services, such as endoscopic image analysis or 3D reconstruction.

12.4 THE PEARL METAPHOR: PEARLY UI

The pearl metaphor is based on two principles: (1) in terms of abstraction, modeling SAEs instead of classical entities only (actors, data, and tasks) and (2) in terms of presentation, using the sets of actors and data (i.e., the pearls) to visualize social relationships. This metaphor is generic, applicable to several fields such as cloud-based e-mail services. In this chapter, it is applied to health care.

12.4.1 Abstraction of Pearls: SAEs

Lahire (2010) claims that sociality is not restricted to social interactions between groups. He defines sociality as a relationship between two human beings. A document, and thus a data, as a communication trace between two human beings, may represent such a relationship (Pédauque 2003). Therefore, a data has a social status. Based on these observations, we propose to transpose the social relationships from the real world to the digital world. This gives rise to a taxonomy of SAEs (Figure 12.3) based on core entities (data, actor, and task as modeled in HCI and CSCW; Van Welie et al. 1998) and their intra versus inter relationships.

Data cover the information that is manipulated by an actor while achieving a task. In our case study, data can be a medical record, an endoscopic

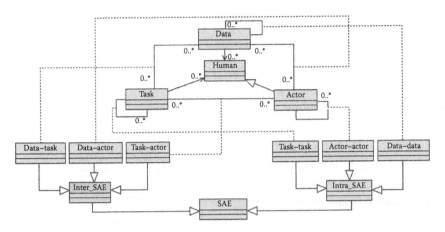

FIGURE 12.3 Taxonomy of SAEs.

image, and so on. Cloud services, such as image analysis or 3D reconstruction, could be applied to them.

Actors denote the stakeholders involved in the system: They manipulate data and perform tasks. In our case study, actors are doctors, secretaries, and medical students. Cloud services, such as medical workflow management, would apply to them.

Tasks describe the goals actors are intended to achieve. They are usually decomposed into subtasks. In our case study, the main task is the patient follow-up. This means for the doctor to perform tests, capture images, communicate with colleagues, and so on. Cloud services, such as best practice recommendation, would be applicable to them.

Relationships between these entities constitute extra information that enriches these entities and create SAEs. Relationship is represented in Figure 12.3 as a line between two entities with a notation indicating the multiplicity of instances of each entities. Relationships may be intra or inter. "Intra" makes reference to relationships between entities of the same type:

- Actor–Actor: Social proximity between actors (e.g., frequent collaboration between two doctors, the patient–doctor relationship) may enhance the Actor entity, and thus may be considered as an SAE. Mailing services based on cloud would apply to these SAEs.

- Data–Data: Data may be socially augmented instead of just being logically linked together (hierarchy, database, etc.). The decoration of a relationship with its type is a good example (e.g., genealogical relationship between a father's and a son's medical records for the prediction of hereditary pathology). Cloud services, such as family history research, would apply to these SAEs.

- Task–Task: Enhancing the tasks with experts' practices is an example of this category. This extra information could be presented to the doctor to serve as an advice. Cloud services such as expert systems would apply to these SAEs.

"Inter" makes reference to relationships that involve entities of different types:

- Data–Actor: Both are socially linked. Actors produce or use data to achieve their goals, giving them a social status (private, confidential, and public). Conversely, data may embed social information about these different actors. In our case study, a medical record (data) is associated with a patient (actor) and, at least, with a practitioner (actor) who takes care of this patient. A medical record can also be associated with a student (actor). This relationship may also in addition give rise to indirect relationships between actors as these actors (patient and student) do not necessarily know each other, but share data. Cloud services such as sharing information (confidential or anonymous) or correlating medical records would apply to these SAEs.

- Data–Task: Tasks are socially linked to data as the social status of data may influence the task execution. For example, updating (task) a medical record (data) by another practitioner, as the referring doctor, is allowed if this medical record is shared. Conversely, performing tasks implicitly augments data with social data [e.g., the production (task) of an endoscopic image (data) added to the medical record available to the patient and the doctor; sharing (task) an image (data) for an expertise]. Services such as traceability of medical activity (last exam, last appointment, last update, etc.) available in the cloud would apply to these SAEs.

- Task–Actor: Actors may achieve the same task in different social contexts. For instance, decision making can be done during a multi-disciplinary meeting involving several specialists or during an examination with just the surgeon. Cloud services such as requesting an advice or sharing information would apply to these SAEs.

This taxonomy provides a powerful support for identifying and selecting the relevant socially augmented data for cloud-based applications. SAE inspired the proposition of Pearly UIs.

Figure 12.2 represents the socio-professional network of the user (at the center). Data pearls are represented by an icon symbolizing a folder and gravitate around these social pearls. Figure 12.4 represents a selected data pearl (at the center). Other data pearls are distributed around the selected data pearl according to their "intra" relationship (i.e., correlation between two data pearls).

12.4.2 Presentation of Pearls: Toward Pearly UIs

The pearl metaphor revisits the classical desktop to support interactive visualization of socially augmented entities.

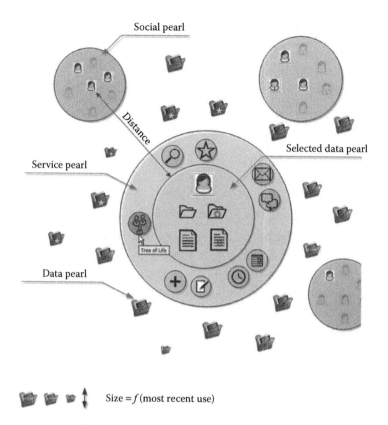

FIGURE 12.4 The PearlyDesktop: User- and service-centric views.

Three kinds of pearls are identified (Figures 12.2 and 12.4):

- Data: They are collections of data about people or groups of people. They embed a part of their history. They can be shared, stored, annotated, and so on, and may be influenced by the context.

- Social relationships: These are communities (friendly, professional, familial, etc.) created by the user.

- Services: The services that apply to data (respectively to actors) are displayed as pearls around the data (respectively the actors) pearl.

Data history is represented by the size of the icon: The bigger the icon is, the more recent the data are.

The "inter" relationship is represented by several ways:

- The spatial proximity of a data pearl (Figure 12.2) with regard to two actors (or communities) indicates its level of sharing.

- Only actors who have a relationship with the data pearl are highlighted in the different social pearls (Figure 12.4).

- The services offered in the service pearls depend on data (respectively on actors) and their relationships.

12.4.3 PearlyDesktop: The Running Prototype

We chose to first represent the relationships between actors using a network-based visualization as it is suitable for social networks (Henry and Fekete 2010). Figure 12.5 represents the socio-professional network of Dr. Roger (at the center). It depicts different kinds of socially augmented data: direct and indirect social relationships between actors (lines), as well as professional communities related to Dr. Roger (pearls labelled "colleagues," "institutions," etc.). The distance between icons indicates the social proximity between actors. For instance, doctors' community (pearl around Dr. Roger) is the closest to Dr. Roger.

In our case study, data pearls are medical records. Their social status is visible on the icon. It may be the following (Figure 12.5):

- Private: Only a patient and his/her referent doctor have access to it.

- Confidential: Sharing is restricted (e.g., to family doctor or patient's relatives).

- Public: The record is anonymously shared, for instance, for pedagogical use.

As shown in Figure 12.6, links between records and actors are labeled to indicate the current status of medical activities.

Three categories are considered:

- Most recent tasks or next tasks to achieve (e.g., last or next examination).

- Sharing status: As shown in Figure 12.6, the state of an expertise request or the state of a medical record for a hospital admission is indicated.

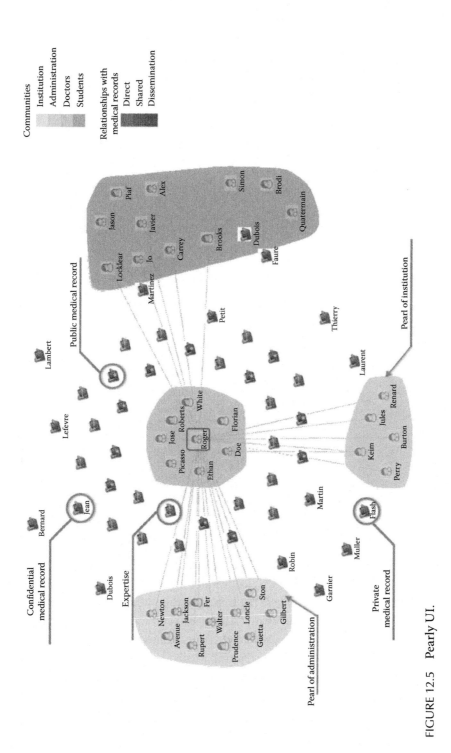

FIGURE 12.5 Pearly UI.

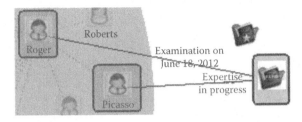

FIGURE 12.6   An inter actor–data relationship.

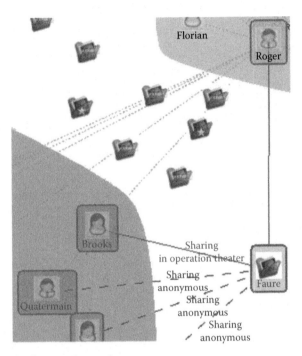

FIGURE 12.7   Indirect relationships.

- Traceability: direct or indirect sharing. For instance, sharing a private medical record with a colleague, a student, or the administration is direct. Conversely, the dissemination of anonymous medical records (e.g., between two students) is indirect. Such relationships between actors are represented using dashed lines. For instance, Figure 12.7 shows that Mr. Faure's medical record is directly shared between Dr. Roger and Brooks (line between Brooks and Dr. Roger). However, this record is indirectly and anonymously shared between Dr. Roger (via Brooks) and three other students: purple dashed lines are displayed between Mr. Faure's medical record and students (e.g., Simon) who have access to this record. Such visualization is powerful for displaying the dissemination network. This is crucial in health care where security is key.

When clicking on an entity (actor or medical record), details are provided at the right side of the UI. Using check buttons at the bottom of the UI, the user may filter information depending on the social status of the medical record or the state of expertise.

12.5 FIELD STUDY

Based on an iterative user-centered approach, we conducted two successive qualitative experiments to evaluate the pearl metaphor. For this purpose, we implemented the PearlyDesktop as a Java prototype based on the Prefuse (Heer et al. 2005) and Vizster (Heer and Boyd 2005) software libraries devoted to interactive visualization.

12.5.1 Protocol

Four experts were recruited: three doctors (participants P1, P3, and P4) to validate the characterization of socially augmented data and the relevance of their presentation, and one specialist of medical information systems for gastroenterology (participant P2). Fifty medical records were entered into the database.

Each interview lasted about an hour and was divided into three parts: (1) presentation of the prototype, (2) playing scenarios, and (3) filling a qualitative questionnaire. All interviews were recorded and then transcribed.

The first part consisted in presenting the main features of the UI. We chose to represent the fictive professional network of Dr. Roger filled with fictive medical records.

For the second part, participants had to perform three different kinds of tasks identified as representative:

1. Search for an expertise

2. Search for a private medical record shared with a student

3. Control the dissemination of a sensitive medical record

At the end of this part, users were invited to provide comments about the prototype (usability, utility, etc.).

At the end of the interview, in order to identify and understand more precisely their vision of socially augmented data, participants had to fill in a qualitative questionnaire. This questionnaire is divided into four sections, articulated on the following: (1) actors, (2) data, (3) tasks, and (4) relationships between these concepts.

12.5.2 Findings

In this section, we summarize the main results related to participants' perception of social data and their pearl-based representation.

12.5.2.1 Pearly UI: A Social Desktop for the Cloud

Gastroenterologists were eager to play with PearlyDesktop. As ephemeral (e.g., related to students) or persistent (e.g., colleagues) medical networks are important aspects of their activity, participants appreciated the network-based visual representation (P1, P2, P3, and P4). "The structure of the interface is fine: the visualization at the center and the details displayed in the panel on the right. Legends and filters are meaningful and easy to use. They provide a better understanding and are useful to filter at a finer grain" (P2). "We can view our closest contacts and shared folders. [etc.] It's interesting to see the referent doctor of a medical record or the specialist to contact for an advice" (P1).

Participants suggested improvements such as the visualization of the "relationships within families to better detect hereditary diseases" (P2, P3, and P4) or of "similarities between medical records to find the right dosage of a medicine drug or to improve medical decision-making" (P4). They also mentioned scalability as critical: How to represent a huge set of data without disrupting the UX? In addition, during the interview, several participants mentioned the importance of the temporal worth of data: access frequency to a medical record, creation date of medical records, or doctor's schedule. They also suggested improvements including an "alphabetical sorting of medical records" (P2), or "a grouping of contacts [within a community] as concentric circles around the center" (P1). Another improvement is to make a clear distinction between "active medical records and medical records of dead patients or patients that will never come back" (P2).

12.5.2.2 Social Sharing and Browsing

Participants very much appreciated the possibility to share data, and thus to support a better communication among stakeholders: "situations of information sharing are numerous: between practitioners about a patient, with laboratories, with the hospital for admissions, [etc.]" (P4). Lack of sharing today is a real problem: "we do not know who see whom" (P4). "We do not have a full access to information about a patient which is crucial to know, for instance, what the different medical treatments are, or to know if the process care for a patient is well managed: our colleagues have difficulties to share data" (P3 and P4). This issue may be explained by "the fear of losing control of their data" (P3), especially with the shift to cloud-based solutions.

Thus, the pearl metaphor was found useful to support sharing and communication activities, thanks to a visualization that merges a view on medical data with the professional network. Participants also pointed out the ability of the metaphor to support traceability. It offers "a better visibility of the medical activity" (P2), and allows "controlling the dissemination of data which gives a feeling of safety" (P3). "For confidential medical records shared with a student, it is essential to always know what happens to this medical record. With another gastroenterologist, the responsibility is shared" (P3). However, if a student "decides to share this medical record with anyone, the doctor is responsible of" (P3).

Browsing medical records is another issue. A gastroenterologist manages "about 90 medical records every month, about 50 records every day" (P4). During his/her entire career, a practitioner usually collects "between 15,000 and 30,000 medical records" (P2 and P4). Adding the social dimension to medical data may facilitate the browsing. Indeed, participants underlined the fear of browsing a huge amount of medical data, in particular when asked for an "advice by phone" (P1 and P3).

12.5.2.3 Social Information Seeking and Refinding
Participants appreciated the organization of the workspace according to the communities.

However, participants also pointed out some issues: "This representation is fine because the pearl is small [to find a medical record]. If the pearl is small, we glance through the different medical records, but if the pearl is larger, how to find easily a medical record?" (P1). Currently, doctors do "not have time to waste for medical record searching" (P1).

It therefore seems necessary to allow users searching using more attributes similarly to pivot browsing: "by age range, by type of examination or pathology, by place of examination" (P3). "When we search for a patient by its name, it's because there is no relevant criteria for this medical record, or pathology has not been informed" (P1). "We do not necessarily remember patient names, but rather a time, a date or a place" (P3). For these situations, our metaphor appears meaningful, as the social worth of data constitutes additional criteria. This idea seems very suitable for "young doctors or doctors managing a large number of patients" (P3).

12.6 DISCUSSION AND IMPLICATIONS

Participants' feedbacks are very positive. Still, these raise issues that must be addressed by future research.

12.6.1 Social Desktop for the Cloud: Pearly UI

The experimentations point out the limits of data-centered metaphors: Information is fragmented and distributed across different services— e-mail services to manage contacts (actors), health-care information systems to manage medical records (data), and so on. Our PearlyDesktop prototype appears as a promising answer to (partly) satisfy the needs of health-care professionals.

Similar to Pham's user-created groups (Pham 2010), we promote an enlarged folder metaphor to integrate the social worth of data. Although Pham's proposal is based on a visualization that only focuses on relationships among actors, our approach merges two views: data and actors. In addition, depending on the social status of actors and data, and the social relationships among actors, the view on data may be restricted or enlarged. For instance, while access to medical records would be restricted for students, a referent doctor may have a full access to a medical record managed by a gastroenterologist.

Compared to CloudRoom (Terrenghi et al. 2010) that relies on different and disconnected views to access data, we foster a metaphor that reveals the context of data such as the social orbits of Sawyer et al. (2012; e.g., physical location related to data such as an operating room or a medical office). The temporal dimension constitutes another means to reveal such a context. However, we have to revisit the interaction to comply with Schneiderman's mantra: "Overview, zoom and filter, details-on-demand": "We should see data, cluster, relationships and gaps of data. [...] The user zooms in on what he wants to see, filter it does not want and click on the item for which he wants information" (Schneiderman 2010). Currently, we investigate how to combine this user-centered representation with a timeline, to allow users to zoom and filter, that preserves the representation of the social context of data.

12.6.2 Social Sharing and Browsing

Thinking cloud is far from being easy for users. They are not familiar with this way of organizing data (Mell and Grance 2009; Odom et al. 2012). Obviously, Dourish's concept of placeless document (Dourish 2003) is fully relevant: Neither absolute paths nor hierarchical organizations make sense anymore.

The health community is strongly constrained by the need of medical confidentiality. Despite the diversity of solution in the medical field, there is a fear of losing data control, which limits the sharing between medical professionals. The communities (i.e., pearls of actors) allow an easier and faster sharing, such as circles in Google+ (Kairam et al. 2012). We are taking the pearls concept a step further by providing a graph representation. This approach allows the user to become aware of these past and future exchanges, and gives a feeling of trust to the users.

Currently, data are presented as a list that doctors do not browse. They search only by keywords. The fear of browsing a huge amount of medical data also appears with this visual representation. We propose to enhance data with the social dimension to support pivot browsing, as proposed by Dourish. Socially augmented will make it possible to parameterize the browsing.

12.6.3 Social Information Seeking and Refinding

During these experiments, the visual representation as well as filters appeared as sufficient for the addressed research scenarios. Surprisingly, the users did not use the search entry. Similar to the social orbits (Sawyer et al. 2012), this reorganization of the workspace according to communities facilitates information seeking.

As proposed by Muralidharan (2012), the social worth of data is used to make relevant the medical activity on record (i.e., patient appointments with a specialist). To indicate the status of activities (the last exchange, the last appointment, etc.), we proposed to label relationships.

In future, we plan to integrate the context into our taxonomy so that it ensures a situated interaction. By context, we mean all the environmental information (e.g., social, physical) that may influence the user while performing his/her task.

Context may be a means for filtering information as well as pushing the right information at the right time. This opens research on user interface plasticity.

12.7 CONCLUSION

This chapter presents a new metaphor targeted for the cloud. The contributions are twofold: a taxonomy of SAE and the concept of pearl for empowering users with these SAE. We consider the social dimension as an approach to provide the first answer to the issues raised by cloud computing. For instance, highlighting the social status of entity constitutes a means to represent sharing of data and traceability. Our application domain, gastroenterology, illustrates this: As underlined by the medical practitioners we have met, this feature is highly relevant about medical confidentiality.

Early feedbacks from medical practitioners encourage pursuing. As underlined, there are several issues to address in order to improve the metaphor and therefore our prototype.

REFERENCES

Armbrust, M., Fox, A., Griffith, R., Joseph, A.D., Katz, R., Konwinski, A., and G. Lee. 2010. A view of cloud computing. *Communications of the ACM* 53(4): 50–58.

Carlin, S. and K. Curran. 2012. Cloud computing technologies. *Computing and Services Science* 1(2): 59–65.

Dourish, P. 2003. The appropriation of interactive technologies: Some lessons from placeless documents. *Computer Supported Cooperative Work* 12(4): 465–490.

England, D., Randles, M., and A. Taleb-Bendiab. 2011. Designing interaction for the cloud. In *CHI'11 Extended Abstracts on Human Factors in Computing Systems*. May 7–12, Vancouver, BC, ACM Press, New York, pp. 2453–2456.

Grudin, J. 2010. CSCW: Time passed, tempest, and time past. *Interactions* 17(4): 38–40.

Heer, J. and D. Boyd. 2005. Vizster: Visualizing online social networks. In *Proceedings of the 2005 IEEE Symposium on Information Visualization*. October 23–25, Minneapolis, MN, ACM Press, New York, pp. 32–39.

Heer, J., Card, S., and J.A. Landay. 2005. Prefuse: A toolkit for interactive information visualization. In *Proceedings of the SIGCHI Conference on Human Factors in Computing Systems*. April 2–7, Portland, OR, ACM Press, New York, pp. 421–430.

Henry, N. and J.D. Fekete. 2010. Novel visualizations and interactions for social networks. In *Handbook of Social Network Technologies and Applications*, ed. Borko Furht. New York: Springer. pp. 611–636.

Kairam, S., Brzozowski, M.J., Huffaker, D., and Chi, E.H. 2012. Talking in circles: Selective sharing in Google+. In *Proceedings of the SIGCHI Conference on Human Factors in Computing Systems*. May 5–10, Austin, TX, ACM Press, New York, pp. 1065–1074.

Lahire, B. 2010. *The Plural Actor.* Cambridge: Polity.

Mell, P. and T. Grance. 2009. The NIST definition of cloud computing. *National Institute of Standards and Technology* 53(6): 50.

Muralidharan, A., Gyongyi, Z., and Chi, E.H. 2012. Social annotations in Web search. In *Proceedings of the SIGCHI Conference on Human Factors in Computing Systems.* May 5–10, Austin, TX, ACM Press, New York, pp. 1085–1094.

Odom, W., Sellen, A., Harper, R., and E. Thereska. 2012. Lost in translation: Understanding the possession of digital things in the cloud. In *Proceedings of the SIGCHI Conference on Human Factors in Computing Systems.* May 5–10, Austin, TX, ACM Press, New York, pp. 781–790.

Pan, Y. and E. Blevis. 2011. The cloud. *Interactions* 18(1): 13–16.

Pédauque, R. 2003. Document: Forme, signe et médium, les reformulations du numérique. *Archive Ouverte en Sciences de l'Information et de la Communication.*

Pham, H. 2010. User interface models for the cloud. In *Proceedings of the 23rd Annual ACM Symposium on User Interface Software and Technology.* October 3–6, ACM Press, New York, pp. 359–362.

Rosson, M.B. and J.M. Carroll. 2002. *Usability Engineering: Scenario-Based Development of Human–Computer Interaction.* San Francisco, CA: Morgan Kaufman Publishers.

Sawyer, B., Quek, F., Wong, W.C., Motani, M., Yew, S.L.C., and M. Pérez-Quiñones. 2012. Information re-finding through physical-social contexts. In *Proceedings of the Workshop of Computer Supported Collaborative Work 2012 on Personal Information Management.* February 11–15, Seattle, WA, ACM Press, New York.

Schneiderman, B. 2010. *Information Visualization for Knowledge Discovery.* San Francisco, CA: Morgan Kaufmann Publishers.

Shami, N.S., Muller, M., and D. Millen. 2011. Browse and discover: Social file sharing in the enterprise. In *Proceedings of the Computer Supported Collaborative Work Conference.* March 19–23, Hangzhou, ACM Press, New York, pp. 295–304.

Stuerzlinger, W. 2011. On- and off-line user interfaces for collaborative cloud services. In *CHI'11 Extended Abstracts on Human Factors in Computing Systems.* ACM Press, New York.

Terrenghi, L., Serralheiro, K., Lang, T., and M. Richartz. 2010. Cloudroom: A conceptual model for managing data in space and time. In *CHI'10 Extended Abstracts on Human Factors in Computing Systems.* April 10–15, Atlanta, GA, ACM Press, New York, pp. 3277–3282.

Väänänen-Vainio-Mattila, K., Kaasinen, E., and V. Roto. 2011. User experience in the cloud: Towards a research agenda. In *CHI'11 Extended Abstracts on Human Factors in Computing Systems.* May 7–12, Vancouver, BC, ACM Press, New York.

Vartiainen, E. and K. Väänänen-Vainio-Mattila. 2011. User experience of mobile photo sharing in the cloud. In *Proceedings of the 9th International Conference on Mobile and Ubiquitous Multimedia.* December 7–9, Beijing, ACM Press, New York.

Van Welie, M., Van der Veer, G.C., and A. Eliëns. 1998. An ontology for task world models. In *Proceedings of the Design, Specification, and Verification of Interactive Systems Conference.* June 3–5, Abingdon, Springer, Heidelberg, pp. 57–70.

Voida, S., Mynatt, E.D., and W.K. Edwards. 2008. Re-framing the desktop interface around the activities of knowledge work. In *Proceedings of the 21st Annual ACM Symposium on User Interface Software and Technology.* October 19–22, Monterey, CA, ACM Press, New York, pp. 211–220.

Whittaker, S., Jones, Q., Nardi, B., Creech, M., Terveen, L., Isaacs, E., and J. Hainsworth. 2004. ContactMap: Organizing communication in a social desktop. *Transactions on Computer–Human Interaction* 11(4): 445–471.

Zhang, C., Wang, M., and R. Harper. 2010a. Cloud mouse: A new way to interact with the cloud. In *Proceedings of the International Conference on Multimodal Interfaces and the Workshop on Machine Learning for Multimodal Interaction.* ACM Press, New York.

Zhang, Q., Cheng, L., and R. Boutaba. 2010b. Cloud computing: State-of-the-art and research challenges. *Internet Services and Applications* 1(1): 7–18.

CHAPTER 13

Standardized Multimedia Data in Health-Care Applications

Pulkit Mehndiratta, Hemjyotasna Parashar, and Shelly Sachdeva

Jaypee Institute of Information Technology
Noida, India

Subhash Bhalla

University of Aizu
Fukushima, Japan

CONTENTS

13.1 INTRODUCTION

Considering the recent developments in network technology, distribution of digital multimedia content through the Internet occurs on a very large scale. Electronic health record (EHR) databases are among the important archives in this vast ocean of data. EHRs are paperless solution to a disconnected health-care world that runs on a chain of paper files. The EHRs are expected to be interoperable repositories of patient data that exist within the data management and decision support system [1]. EHR databases are real-time databases, the state of which keeps on changing with time. It is a complex task for medical expert to retrieve relevant information from the database (in a short period of time). In health-care domain, information sharing and information reachability are related to the safety of patients. A standardized format for the content of a patient's clinical record helps to promote the integration of care among various health-care providers. Medical practices also require sharing of data by many agencies over long periods of time. Thus, the structure and content of the life-long EHRs require standardization efforts for improving interoperability.

Although the patient's paper-based medical records can be scanned and transferred between the providers in a standard image format, it will not fulfill the functional requirements of the EHRs, that is, the image data cannot form the electronically exchanged medical data that can support analysis and decision making. There is a large variety of the potential functions supported by the databases and content of EHRs. The sharing of information between the providers introduces new questions and challenges concerning the data to be exchanged and their format. Text-based EHRs contain large amount of data and provide opportunities for analysis (both quantitative and qualitative). However, programming skills are required. Similarly, the use of standardized diagnostic images creates a scope of storing digital multimedia content and other media as well. It will simplify the data sharing and improve interoperability.

13.1.1 Standardized EHR Databases

Health-care domain generates large quantities of data from various laboratories, wards, operating theaters, primary care organizations, and wearable and wireless devices [2]. Its primary purpose is to support an efficient and quality-integrated health care. It contains information that is retrospective, concurrent, and prospective. EHRs need to be standardized for information exchange between different clinical organizations. The existing standards aim to achieve data independence along with semantic interoperability.

The semantic interoperable EHR standards such as Health Level 7 (HL7) [3], openEHR [4], and CEN/ISO 13606 [5] deal with communication, data representation, and meaningful use. Many medical experts avoid using the standardized EHR system for diagnose, analysis, and decision making. There are several reasons behind this hindrance such as high cost, system complexity, lack of standardization, and different naming convention.

EHRs must provide support for multimedia device and for integrating images, along with alphanumeric, and other forms of data. Currently, multimedia data are not part of any standard EHR functionality [6]. Multimedia data should be integral to standardized EHRs because they can capture information that cannot be easily summarized into the text. Such multimedia records are also crucial for providing best advice to patients to improve the health-care services and reduce the human errors. Structured text data cannot efficiently present the whole history and current status of patients, and need more time during analysis and decision making compared to visual data and multimedia data. These limitations of structured text data can potentially be overcome by storing nontextual types of data, such as multimedia data. Diagnostic reports such as CT scan, X-ray, waveforms, and audio–video file need to be included in EHRs. Although multimedia data are very useful for better clinical judgments, these types of data can be stored in digital form with good clinical interpretations. It can be widely integrated with the existing standardized EHRs. Apart from technical complexity of incorporation of multimedia into EHRs, there are few more challenges for creating multimedia EHR systems [7].

13.1.2 Multimedia Data and Standardized EHRs

Multimedia data represent various types of media contents (text, audio, video, animation, and images). These are often used together. The data contents in the text format are huge when considered in regard to EHRs leading to difficulties in communication and analysis. The overall size can possibly be reduced by storing nontextual type of data (multimedia data).

The addition of these data types will enable the medical experts to explore and create more accurate and efficient analysis. Consider an integration of digital media (images, audio, and videos) and conventional text-based data. Currently, the images retrieved for diagnostic purpose are available on papers. Multimedia EHRs may require data such as nondiagnostic images, data with an audio component, or a video data. Lowe [8] proposes to include the images along with physiological signals into text-based records.

The Institute of Medicine (IOM) recommends the use of multimedia EHRs containing the various media possibilities. While the tools for natural language processing, digital imaging, voice recognition have evolved, a lot of effort needs to be made for the inclusion of standardized multimedia data in EHR services or systems.

13.1.3 Integration of Imaging Standards

In order to image-enable EHRs, the technology should support industry-level standards, such as Digital Imaging and Communications in Medicine (DICOM) [9], HL7 [3], and Integrating the Health Enterprise (IHE) [10]. Figure 13.1 shows the approaches for integration of various imaging standards:

1. DICOM: It permits storing images from multiple modalities and systems, by accepting and cleansing all variations of former DICOM standards, and providing EHRs with standardized DICOM formats.

2. HL7: It can be used for messaging for ordering images, sending image results, and updating patient demographic information.

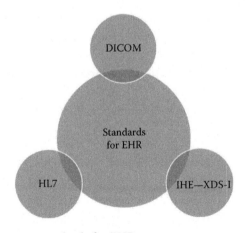

FIGURE 13.1   Imaging standards for EHRs.

3. IHE: It supports Consistent Time Integration Profile (CT), Cross-Enterprise Document Sharing for Imaging (XDS-I), and Audit Trail and Node Authentication (ATNA) for meeting various integration requirements.

XDS-I lays the basic framework for deploying medical images in the EHR. The deployment of XDS-I as the framework for sharing images within the EHR is taking place in many countries including Canada, the United States, Japan, and several European countries.

13.1.3.1 Imaging Standard: DICOM

DICOM [9] is a standard for storing, handling, printing, and transmitting information in medical imaging. It is also known as the National Electrical Manufacturers Association (NEMA) standard PS3.x and as ISO standard 12052:2006 "Health Informatics—DICOM including workflow and data management." Over the years, many versions and updates have been released and the current versions PS3.1x and PS3.x are in compliance with the latest standards and guidelines stated by the IOM. DICOM differs from some data formats. It groups the information in datasets. This means that a file of a chest X-ray image, for example, actually contains the patient ID within the file, so that the image can never be separated from this information by accidents or mistake. This is similar to the way that image formats such as JPEG can also have embedded tags to identify and otherwise describe the image.

DICOM provides a set of protocols for devices for proper network communication and exchange of information during the communication. DICOM gives consistency in multimedia storage services, format, and report structures. Using DICOM standard, clinical structured reports can be generated in which both multimedia and alphanumeric data are integrated. Clinical experts can view the whole history as well as all the multimedia data (such as X-ray, CT scan, waveforms, and other images and interpretations). These types of structured reports contain all the information of patients, interpretations with evidence, and links to the other similar cases. Traditionally, the imaging reports tend to be dictated by a radiologist who interprets the images; reports are then subsequently transcribed into an electronic document by a typist and verified by the radiologist. Structured reports enable efficient radiology workflow, improve patient care, optimize reimbursement, and enhance the radiology ergonomic working conditions [11].

Structured reports are sharable among different clinical experts and can be stored and viewed by any Picture Archiving and Communication System (PACS). These are the broad systems that can visualize and do processing of the images that they manage and archive. The PACS needs to import images for visualization, which leads to consideration of persistency management and information consistency. In order to achieve image import, the system performs image streaming with the help of the Web Access to DICOM Persistent Objects (WADO).

ISO WADO defines a Web-based service that can be used to retrieve DICOM objects (images, waveforms, and reports) via Hypertext Transfer Protocol (HTTP) or HTTP Secure (HTTPS) from a Web server. DICOM Structured Report (DICOM SR) is a general model for encoding medical reports in a structured manner in DICOM's tag-based format. It allows the existing DICOM infrastructure network services, such as storage or query/retrieve, to be used to archive and to communicate, encrypt, and digitally sign structured reports with relatively small changes to the existing systems [12]. Siemens has launched a new PACS called as syngo. plaza [13]. It is the new agile PACS solution for the clinical routine, where 2D, 3D, and 4D reading comes together at one place. EndoSoft application [14] contains an integrated DICOM-compliant solution that generates and exports images from the EndoSoft endoscopy software to a PACS system.

Some of the key benefits of DICOM are as follows:

1. Better communication with the physician

2. More accurate coding of diagnosis and fewer rejections

3. Faster turnaround (the creation of the radiology report is achieved according to an interpretation process)

4. Minimization of the typing work

5. Data are archived, transferred, and managed with the images.

6. References to evidences used for interpretation

7. References to prior reports and similar case history

8. Whole track records can be maintained for patients by maintaining the references of the same type of previous report.

9. Structured report is the stand-alone complete information object.

The DICOM standard is related to the field of medical informatics. Within that field, it addresses the exchange of digital information between medical imaging equipment and other systems. Due to the fact that such equipment may interoperate with other medical devices, the scope of this standard needs to overlap with other areas of medical informatics.

13.1.3.1.1 DICOM and Interoperability   In digitized form, the individual patient's medical record can be stored, retrieved, and shared over a network through enhancement in information technology. Thus, EHRs should be standardized, incorporating semantic interoperability. Semantic interoperability refers to the ability of computer systems to transmit data with unambiguous, shared meaning. The DICOM standard facilitates the interoperability of devices claiming conformance by

1. Addressing the semantics of commands and associated data. Thus, devices must have standards on how they will react to commands and associated data, not just the information that is to be moved between devices.

2. Providing semantics of the file services, file formats, and information directories necessary for off-line communication.

3. Explicitly defining the conformance requirements of implementations of standard. A conformance statement must specify enough information to determine the functions for which interoperability can be expected with another device claiming conformance.

4. Making use of existing international standards wherever applicable, and itself conforms to establish documentation guidelines for international standards.

The DICOM standard facilitates the interoperability of systems claiming conformance in a multivendor environment, but does not, by itself, guarantee interoperability.

13.1.3.1.2 DICOM and Security   The DICOM standard does not address any security issues directly but focus on appropriate security policies, which are necessary for a higher level of security. It only provides mechanisms that could be used to implement security policies with regard to interchange of DICOM objects between application entities. For example, a security policy may dictate some level of access control. This provides the technological means for the application entities involved to exchange sufficient information to implement access control policies. The DICOM standard assumes that the application entities involved in a DICOM interchange can implement appropriate security policies. Essentially, each application entity must ensure that their own local environment is secure before even attempting secure communications with other application entities. The standard assumes that application entities can securely identify local users of the application entity, using the users' roles or licenses. It also assumes that entities have means to determine whether the "owners" (i.e., patients or institutions) of information have authorized particular users, or classes of users to access information.

This standard also assumes that an application entity using Transport Layer Security (TLS) has secure access to or can securely obtain X.509 key certificates for the users of the application entities.

Table 13.1 depicts the scope of the EHR standards with regard to two basic properties of an EHR: the EHR content structure and the access services (communication protocol) [12]. It also gives a brief comparison between the security measures taken by ISO WADO and DICOM structured reporting (SR) standards.

For the content-related features such as the persistent documents, multimedia content, and content that can easily be processed, distribution rules, visualization, and digital signatures (for providing security) are taken into consideration. As indicated in Table 13.1, the standard ISO WADO does not have support for content structure. Thus, none of the above-mentioned features are supported by it. The DICOM SR has support for content structure but does not support the visualization for content. It also does not have any specified rules for distribution for the content.

Similarly, considering the comparison based on access services such as querying, retrieving, and submitting EHR content and content format agnostics, both the standards have support for these. At the same time, none of them has complete support for these. Considering retrieval and storage of the EHR content, ISO WADO has the functionality to support, but it lacks support for the querying and submission of the EHR content. The DICOM SR has support for querying, retrieval, and submission as well as storage supports. Both these standards lack in the formatting agnostics of the EHR content.

In case of the EHR data, the security of the data plays an important role. But, as stated earlier, the DICOM standard does not take security into consideration directly. It gives sufficient storage and flexibility to the system to implement it. Both the above-mentioned standards use TLS. These

TABLE 13.1   Comparison of Two DICOM Standards

	ISO WADO[a]	DICOM SR
Scope of EHR[b] Standards		
EHR content structure	No	Yes
EHR access services	Yes	Yes
Content Structure of Standards		
EHR contains persistent documents.	No	Yes
EHR can contain multimedia data.	No	Yes
EHR document can contain references to multimedia data.	No	Yes
EHR structures content suitable for processing.	No	Yes
EHR supports archetypes/templates.	No	Yes
EHR specifies distribution rules.	No	No
EHR standard covers visualization.	No	No
EHR supports digital signatures on persistent documents.	No	Yes
Analysis of EHR Standards' Access Services		
Service for querying EHR content	No	Yes
Service for retrieving EHR content	Yes	Yes
Service for submitting EHR content	No	Yes
Document-centric storage retrieval	Yes	Yes
Content format agnostic	No	No
Security Features of Standards		
Protocol supports transport-level encryption.	Yes	Yes
Protocol allows to transmit user credentials.	Yes	Yes
Protocol enforces access rules.	No	No

[a] ISO WADO, ISO Web Access to DICOM Persistent Objects.
[b] EHR, electronic health record.

have support for sharing and transmitting the user credentials, but both the systems lack in features for providing access control for the associated data or the EHR content.

13.1.3.1.3 Mapping of Data among Imaging Standards   Cohen et al. [15] describe the conversion of data retrieved from PACS systems through DICOM to HL7 standard [Extensible Markup Language (XML) documents]. This enables the EHR systems to answer queries such as "Get all chest images of patients between the age of 20–30, which have blood type 'A' and are allergic to pine trees." The integration of data from multiple sources makes this approach capable of delivering such answers. International organizations are already developing XML-based standards for the health-care domain, such as HL7 clinical document architecture (CDA). The EHR system will include an extensive indexing and query mechanism, much like the Web. This will help to convert the data collected from one EHR system to be linked and interpreted by another EHR system.

13.1.3.1.4 Evaluation of DICOM Standard   DICOM has many advantages and can play a critical role in health-care industry, when it comes to digital imaging and multimedia. But DICOM as well as all the other available EHR applications lacks in support for other multimedia content such as audio and video files and various contents with waveform. In addition, one major disadvantage of DICOM standard is that it does not allow optional fields. Some image objects are often incomplete because some fields are left blank and others may have incorrect data. Another disadvantage of DICOM is limited point-to-point design for data forwarding. Any data-receiving device always becomes the end point of the communication. It cannot be instructed to relay the received data elsewhere. The rest of this chapter is organized as follows: Section 13.2 discusses about image data. Section 13.3 describes the status of EHRs in cloud computing environment. Section 13.4 presents the issues related to multimedia data in EHRs such as bandwidth requirements. Section 13.5 represents summary and conclusion.

13.2 IMAGE DATA AS PART OF EHRs

Some of the studies [16,17] have demonstrated the need to share images, seamlessly, in order to improve patients' care and disease management, and reduce unnecessary procedures. Noumeir and Renaud [16] have tried to come up with a new architecture of a Web application for testing the interoperability in health care; the proposed software provides functionality to test groups involved in sharing images between different institutions, whereas Noumeir and Pambrun [17] described how the JPEG 2000 Interactive Protocol (JPIP) can be used to deliver medical images from EHRs to the workstation directly without importing rather than streaming them. This also eliminates the problem of persistency and consistency associated with PACS.

13.2.1 Image Data for Medical Needs

Images are widely used in medical treatments from X-rays, dermatology photographs, pathology laboratory tests, and CT scans. Specialists consider multimedia images to be important. Also, each specialist (from different medical fields) uses different techniques to obtain, observe, and understand images. This is of extreme importance and relevance to have digital multimedia images related to any image-intensive specialties. Few examples, concerning how digital multimedia content can improve the quality of treatment for the patients, are considered as follows:

1. Cardiology: Cardiac imaging is of prime importance in medical field to understand disease and to form a diagnosis and treatment routine. Images are required [in operation theaters and in intensive care units (ICUs)] in real time. However, the challenge is to integrate EHRs and cardiac imaging in such a comprehensive manner that the whole patient data are available to the medical practitioner at the point of care and provide help to form clinical decision support.

2. Neurology: In clinical setup, medical practitioners may need different set of images that vary in complexities. Neurosurgeons may also need access to the images taken in the past to get to the root of the disease. Thus, it is desirable to combine image data with clinical data for diagnosis and treatment purpose.

3. Gynecology: This practice considers patients that move from one place to another. Thus, the images are required to be transferrable between the medical practitioners to provide high quality of treatment and consistency. Such an access is also required to have distant consultation through telemedicine.

13.2.2 Considering Multimedia EHRs

In many countries, the EHR services have been implemented with different level of integration of images in them. EHR solutions such as OpenVista [18] and CareCloud [19] in the United States are notable examples. These can operate as self-contained systems. OpenVista is Medsphere's comprehensive EHR solution, which is a single solution. It provides the continuum of acute, ambulatory, and long-term care environments as well as multifacility, multispecialty health-care organizations. A comprehensive architectural framework and modular functional design make this robust system extremely flexible in terms of tailoring a custom EHR solution for each facility. In addition, these EHR solutions have the ability to impose standards and methods for capturing, storing, transferring, and accessing images across the whole system.

13.2.2.1 Adoption of DICOM Standard in EHRs

Radiology information systems (RISs) are employed to manage diagnosis reports generated for reviewing medical diagnostic images. These depend on PACS to manage medical diagnostic images. The protocols for storing and communicating such data are specified by standards such as DICOM. The DICOM standard defines data structures and services for the vendor-independent exchange of medical images and related information [20]. Unlike most other EHR standards, DICOM uses a binary encoding with hierarchical lists of data elements identified by numerical tags and a complex DICOM-specific application level network protocol. Thus, to make the EHR standards interoperable and also to lay the foundation of multimedia EHRs, the DICOM standards for images have been adopted. Using these international standards, EHRs can be implemented to transmit images and image-based reports among the various providers. A recent example is by Oracle [21,22]. Oracle has come up with an enhanced version of DICOM standards. It has named it as "Oracle Multimedia DICOM." The architecture [22] of this multimedia DICOM has two perspectives: the database tier (Oracle database) and the client tier (thick clients). The oracle database holds the DICOM content in tables. The content stored in a column of a table can include DICOM data such as X-rays, ultrasound images, and magnetic resonance images. In the client tier, the ability to access Oracle Database DICOM (ORDDicom) objects in the database is supported through Oracle Multimedia DICOM Java API. Oracle Multimedia DICOM also supports automatic extraction of any or all 2028 standard DICOM metadata attributes [21] as well as any selected private attributes search and business intelligence applications. Built-in functions convert DICOM content to Web-friendly formats such as JPEG, GIF, MPEG, and AVI that can generate new DICOM format images from legacy formats. They have tried to enhance the existing DICOM standard by making few additional changes to the architecture and even to the storage of data, to make query processing and information retrieval more agile and fast.

13.2.2.2 Multimedia Unique Identifier: Standardized Effort

For integration of multimedia data with EHR systems, one of the key challenges is multimedia unique identifier (MUI) of the multimedia data. Multimedia data are generated from different vendors for different purposes, for different departments by different devices for patients. Research is needed to find out the standardized efforts for generating unique identifier for multimedia data of a single patient. Reports based on multimedia data such as

X-ray, CT scan, heartbeat sounds, waveforms, electrocardiography (ECG), and magnetic resonance imaging are generated by different devices with different formats and standard of user interface for the same and different patients. Most of the time devices use the day–date–time format to generate unique identifier. The clinical staff manually enters the patient name, so the uniqueness of unique identifier is not unique all the time and manual entry may increase the probability of human errors. Sometimes, the same tests are being done again for checking improvement in patient's health. Therefore, unique identifier plays a key role in the retrieval process. MUI must be generated in such a manner that it becomes easy for clinical expert to find out all the data regarding a patient (test reports, multimedia data, text data, EHR data, and interpretations of previous clinical experts). MUI should also support groups by query processing where clinical experts find out all the related history of similar health issues of different patients.

13.3 EHRs ON CLOUD COMPUTING SYSTEMS

Cloud computing paradigm is one of the popular health IT infrastructures for facilitating EHR sharing and integration. An ecosystem will evolve that constantly generates and exchanges insights and brings appropriate and important insights into health-care decisions. One of the key benefits will be the ability to exchange data between disparate systems; such a capability is needed by health-care industry. For example, cloud can support various health-care organizations to share EHR (doctor's prescriptions, test reports, and results), which are stored across different systems around the globe as shown in Figure 13.2.

The electronic record sharing between different electronic medical record (EMRs) systems is difficult. The interoperation and sharing among different EMRs have been extremely slow. Heavy investment and poor usability are the biggest obstacles, which hinder the adoption of health IT, especially EHR systems. Cloud computing provides an IT platform to cut down the cost of EHR systems in terms of both ownership and IT maintenance burdens for many medical practices. Cloud computing environment can provide better opportunities and openings to clinicians, hospitals, and various other health-care-related organizations. These units can come to a consensus to exchange health-care multimedia information among themselves, which will help doctors and physicians. It will also help the patients by providing better treatment facilities along with accurate and fast decision-making facilities.

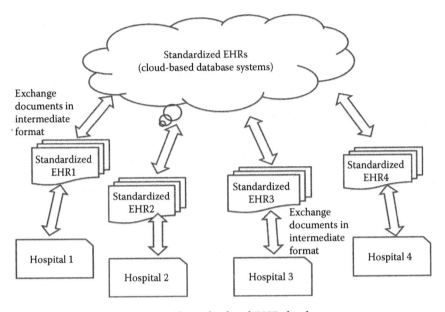

FIGURE 13.2   The utilization of standardized EHR database.

It has been widely accepted and recognized that cloud computing and open health-care standards can generate cutting-edge technology to streamline the health-care system in monitoring patients and managing disease or collaboration and analysis of data. But a fundamental and most important step for the success of tapping health care into the cloud is the in-depth knowledge and the effective application of security and privacy in cloud computing [23,24].

Various Web applications such as Flickr [25] (taken over by Yahoo as replacement of Yahoo pictures) and Facebook [26] present new challenges for processing, storing, and delivering the user-generated content (i.e., multimedia content). A cloud-based multimedia platform can perform the heavy lifting for massive amounts of multimedia storage and processing in the spirit of cloud computing environment. One such architecture has been proposed by Kovachev and Klamma [27], named as cloud multimedia platform. The architecture has been developed in such a manner that it can easily handle massive amounts of multimedia data.

13.4 ISSUES RELATED TO MULTIMEDIA DATA IN EHRs

The slow acceptance of EHRs and multimedia technology in health care is due to many reasons such as adding older records into EHR system, storage and long-term preservation, synchronization of data, hardware limitations, initial cost, semantic interoperability, security and privacy, bandwidth requirements for information exchange and transfer, and usability. These issues need to be addressed and resolved as soon as possible to take full advantage of multimedia EHR systems.

13.4.1 Bandwidth Requirements

Bandwidth is a critical issue when dealing with large amount of multimedia rich data. It may be available in the urban and well-served cities, but in developing countries most of the population belongs to rural and underserved communities. Thus, it is very difficult to meet the bandwidth requirement. Few of the measures that may solve this problem are as follows:

1. Image compression can reduce the size of images and other multimedia content, but it will reduce the quality of the data.

2. Streaming the images live whenever required (just-in-time streaming) can reduce the burden on the network.

3. Prioritizing the order for the optimized presentation may be utilized.

4. Performing rendering of the data by huge visualization engines on the server side and giving the direct access can be utilized.

The above-mentioned measures reduce load on the available bandwidth and help in smooth functioning of multimedia-integrated EHR services.

13.4.2 Archiving and Storage of Multimedia Data

Another important issue in multimedia databases is the management of storage subsystem. There are certain characteristics of multimedia data objects that make their storage an unusual problem. Multimedia objects can be very large. Indeed, in most real-world systems, we may expect multimedia objects to be a few orders of magnitude larger on the average than other objects (typically text files and binary files). Along with that, they have hard real-time timing constraints during display. Multimedia data may require a high degree of temporal data management. The HL7, openEHR, and European Committee for Standardization (CEN) standards support the EHR repositories. The medical personnel needs frequent interaction with these archives. Further, the patient, care professional, care provider, funding organizations, policy makers, legislators, researchers, and analysts also need to access these archives. The problems about information exchange arise, as these archives have different data storage representations. Similarly, the information exchange (among the various archival storages) also poses difficulties.

These problems can be overcome by using some simple techniques, such as the following:

1. Multimedia data are mostly archival and tends to be often accessed in the read mode. This means that modifications of multimedia data objects, in many applications, are relatively few. This allows designers to read optimize the data.

2. Multimedia data also tend to show strong temporal locality. This allows easier management and optimization of cache buffers.

Although these optimization methods are not universal, some applications may still benefit from them.

13.4.3 MUIs for Different Needs

Different specialties use image data in different forms, ranging from dental X-rays, dermatology photographs, and pathology slides to computerized tomography scans for oncologists and magnetic resonance images for cardiologists. Each specialty has different requirements for viewing and interpreting images. For cardiologists, multiple real-time images are required as to correctly identify the disease and then medicate it. In case of neurosurgeons, intraoperative images are needed, along with the images taken in the past for the reference purpose. Similarly, obstetric and gynecologic physicians deal with a patient population that is mobile. For these, images must be taken and stored in such a manner that they can be transportable between providers and annotated in order to have consistent readings and interpretations. Primary care physicians also have their own requirements and areas on which they focus. Radiologists have different requirements such as frequent viewing, electronic ordering of procedures, facilitation of report generation, and support for rapid notification of the ordering physician in case of time-sensitive critical findings. This widespread use of the images in health care requires portable datasets with metadata linked to the images that can be easily viewed and interpreted by other members of health-care team.

13.5 SUMMARY AND CONCLUSIONS: TOWARD MULTIMEDIA EHRs

Over the past few years, the information and communication technology (ICT) has evolved standardized EHRs. EHR applications are the medium to provide an interactive path for storing and retrieving information of one or more patients. Multimedia content has become an integral part of data. But multimedia content in EHRs and in hospitals lacks the capability to integrate images and multimedia data as a component of the enterprise information systems. Clinical decision support systems that employ multimedia data to support comprehensive patient management are scarce. The meaningful use of EHRs embraces multimedia data in all of its criteria, from the definition of laboratory data through to the ability of consumers to upload complex datasets from health-monitoring devices. Image-enabling EHRs offer additional benefits such as increasing referrals and productivity, optimizing operational efficiency, improving patient safety, enhancing the quality of care, and reducing costs.

Essential requirements for multimedia success include DICOM standards and optimum bandwidth to exchange images along with archiving and storage of data, which will help in satisfying the needs of the multiple users to better extent. Multimedia data will be a critical component of EHRs. The features extracted from images, as well as the image data themselves, become part of an expanded database that can be used for decision support, predictive modeling, and research purposes. By implementing EHR applications with multimedia support and integration, medical experts can access and aggregate patients' data more quickly and efficiently. This will not only improve the productivity of health care but also reduce the risk of medical errors. The adoptions of multimedia content in health care will not only free medical practices from the burden of supporting IT systems but also be able to improve and solve many collaborative information issues in health-care organization as well as cost optimization.

Interoperability is a critical step in supporting scalable health-care solutions and can bring many benefits to medical users. The adoption of standards and guidelines has been a move toward interoperability. The lack of standards and technological integration is a key barrier to scaling multimedia health care. Health-care systems in both developed and developing countries continue to struggle to realize the full potential of multimedia-integrated EHRs, and more generally technology, in part due to limited interoperability. The problem of security and privacy hinders the deployment of integrated multimedia electronic health-care systems. Image-enabling EHRs offer additional benefits such as increasing referrals and productivity, optimizing operational efficiency, improving patient safety, enhancing the quality of care, and reducing costs. The full-scale development of the health-care application with integrated multimedia support will result in better reachability of health-care services to the remote corners of the world. It will not only help the developing nations but also be useful for the developed countries.

REFERENCES

1. Karen M. Bell, HHS National Alliance for Health Information Technology (NAHIT). Report to the officer of the National Coordinator for Health Information Technology on defining key health information technology terms. USA, pp. 1–40, 2008.
2. M. Simonov, L. Sammartino, M. Ancona, S. Pini, W. Cazzola, and M. Frasio, "Information knowledge and interoperability for healthcare domain." In *Proceedings of the 1st International Conference on Automated Production of Cross Media Content for Multi-Channel Distribution*, November 30–December 2, pp. 35–42, Florence, Italy, 2005.
3. The Health Level Seven (HL7), http://www.hl7.org/about/index.cfm?ref=nav
4. The openEHR standard, http://www.openehr.org/what_is_openehr
5. The CEN/ISO EN13606 standard, http://www.en13606.org/the-ceniso-en13606-standard
6. B. Seto and C. Friedman, "Moving toward multimedia electronic health records: How do we get there?" *Journal of the American Medical Informatics Association*, 19(4): 503–505, 2012.
7. N. Yeung, "Multimedia features in electronic health records: An analysis of vendor websites and physicians perceptions." Master of Information, University of Toronto, Toronto, ON, 2011.
8. H.J. Lowe, "Multimedia electronic medical health systems." *Academic Medicine*, 74(2): 146–152, 1999.
9. D.M. Heathcock and K. Lahm, "Digital imaging and communication in medicine: DICOM standard." Kennesaw State University, Kennesaw, GA, IS 4490: Health Informatics, 2011.
10. Integrating the Healthcare Enterprise, http://www.ihe.net/.
11. R. Noumeir, "Benefits of the DICOM structured report." *Journal of Digital Imaging*, 19(4): 295–306, 2006.
12. M. Eichelberg, T. Aden, J. Riesmeier, A. Dogac, and G.B. Laleci, "A survey and analysis of electronic healthcare record standards." *ACM Computing Survey*, 37(4): 277–315, 2005.
13. syngo.plaza, http://www.siemens.com/syngo.plaza/.
14. Endoscope Company, http://www.endosoft.com

15. S. Cohen, F. Gilboa, and U. Shani, "PACS and electronic health records." In *Proceedings of SPIE 4685, Medical Imaging*, 2002.
16. R. Noumeir and B. Renaud, "IHE cross-enterprise document sharing for imaging: Interoperability testing software." *Source Code for Biology and Medicine*, 5(1): 1–15, 2010.
17. R. Noumeir and J.F. Pambrun, "Images within the electronic health record." In *Proceedings of the International Conference on Image Processing*, pp. 1761–1764, November 7–10, IEEE Press, Cairo, 2009.
18. The OpenVista EHR Solution, http://www.medsphere.com/solutions/openvista-for-the-enterprise
19. CareCloud, http://www.carecloud.com/.
20. R. Hussein, U. Engelmann, A. Schroeter, and H.P. Meinzer, "DICOM structured reporting: Part 1. Overview and characteristics." *Radiographics*, 24(3): 91–96, 2004.
21. Oracle Multimedia DICOM, Technical report by Oracle for DICOM standards, http://www.oracle.com/us/industries/healthcare/058478.pdf
22. Oracle Multimedia DICOM concepts, Developers guide to implement multimedia DICOM standard, http://docs.oracle.com/cd/B28359_01/appdev.111/b28416/ch_cncpt.htm
23. R. Zhang and L. Liu. "Security models and requirements for healthcare application clouds." In *Proceedings of the 3rd IEEE International Conference on Cloud Computing*, July 5–10, Miami, FL, IEEE, pp. 268–275, 2010.
24. H. Takabi, J. Joshi, and G. Ahn. "Security and privacy challenges in cloud computing environments." *IEEE Security and Privacy*, 8(6): 24–31, 2010.
25. Flickr, http://www.flickr.com/.
26. Facebook, https://www.facebook.com/.
27. D. Kovachev and R. Klamma, "A cloud multimedia platform." In *Proceedings of the 11th International Workshop of the Multimedia Metadata Community on Interoperable Social Multimedia Applications*, Vol-583, May 19–20, Barcelona, Spain, pp. 61–64, 2010.

CHAPTER 14

Digital Rights Management in the Cloud

Paolo Balboni and Claudio Partesotti

ICT Legal Consulting
Milan, Italy

CONTENTS

14.1 INTRODUCTION

Entertainment, gaming, digital content, and streaming services are increasingly provided by means of cloud computing technologies.* The benefits of such deployment model are as numerous as the legal challenges that it poses. In fact, cloud technology generally improves availability, usability, integration, and portability of digital media services while decreasing related costs. However, it creates more complex value chains, involving multiple jurisdictions. The relevant legal aspects are numerous: intellectual property, personal data protection, information society

---

* For example, iTunes, Grooveshark, Spotify, and UltraViolet offer music and/or audio–visual content to stream it over the Internet, download it for off-line listening/viewing, or play it.

service providers' liability, law enforcement agencies' access to content and information stored in the cloud, digital media forensic, and so on.

Given the limited space of this chapter, we decided to focus on intellectual property and data protection. Intellectual property is the most prominent legal challenge in the provision of digital media and personal data protection is the biggest issue related to cloud technology. Being European lawyers, we will take a European point of view to the matters. As the European system is one of the world's most regulated and complex legal systems with respect to both intellectual property and personal data protection, readers may find the European perspective particularly valuable. For the sake of our analyses, it is important to immediately and clearly identify three main categories of subjects involved in the provision of digital media services by means of cloud technology: the right holder(s), the cloud service provider(s) (CSP), and the user(s). By right holder, we identify the subject that makes the digital media service available. It may be the direct intellectual property right (IPR) holder (e.g., a gaming company that produces and makes available games) or it may have the relevant rights to (sub-)license third-party intellectual property (e.g., Apple that makes third-party audio–visual and music content available through its iTunes Store). By CSP, we refer to the supplier of cloud technology that enables a delivery model based on the Internet, where digital media are provided to computers and other devices on demand. By user, we mean the subject who accesses via the Internet digital media on demand. We will develop our analysis from the right holder's point of view. This seems to be the most interesting perspective to us as the right holder is at the center of direct relationships with the CSP, the user, and, possibly, the third-party IPR owner(s).

"[S]implify copyright clearance, management and cross-boarder licensing" is the way forward for the next era of digital right management (DRM) identified by the European Commission in the Digital Agenda for Europe.* The European cloud strategy reaffirms such action stressing that it will "enhance Europe's capacity to exploit the exciting new opportunities of cloud computing for both producers and consumers of digital

---

* Pillar: Digital Single Market. Action 1. http://ec.europa.eu/information_society/newsroom/cf/fiche-dae.cfm. Under this Action, a Directive on Collective Rights Management COM(2012) 372 final and a Directive on Orphan Works COM(2011) 289 final have been proposed; and the Directive on Re-Use of Public Sector Information COM(2011) 877 final has been reviewed.

content"* in the cloud. We believe that cloud for digital media can only work if CSPs and right holders agree on license terms that allow users to access their account from multiple devices and from different territories. More flexibility for the whole digital media environment is needed.

Logically, before extensively dealing with new business models and license terms (Section 14.3), we need to address the personal data protection issue. In fact, the provision of digital media by means of cloud computing technologies is typically made possible through a preliminary user registration form. In other words, users need to first create an account in order to start enjoying the service according to the relevant licence–service agreement.

14.2 PERSONAL DATA PROTECTION ISSUES

Users provide their personal data to the right holder to create an account and to enter into a relevant license agreement in order to access the digital media. The right holder is the collector of the users' data, which will then be automatically shared with the other relevant subjects [i.e., CSP(s) and, possibly, with the third-party IPR owner(s)] in order to provide users with the requested service.[†] Legally speaking, the right holder acts as the data controller,[‡] CSP(s), and third-party IPR owner(s), if any, who will typically be the data processors,[§] and the users are the data subjects.[¶] At the time of the collection of the personal data, the right holder must provide the user with information about

> (a) the identity of the controller [Right Holder] (...); (b) the purposes of the processing for which the data are intended; (c) any further information such as—the recipients or categories of recipients of the data—whether replies to the questions are obligatory or voluntary, as well as the possible consequences of failure to

---

[*] COM(2012) 529 final, Communication from the Commission to the European Parliament, the Council, the European Economic and Social Committee and the Committee of the regions. Unleashing the potential of cloud computing in Europe, p. 6. http://eur-lex.europa.eu/LexUriServ/LexUriServ.do?uri=SWD:2012:0271:FIN:EN:DOC

[†] Users' personal data may also be shared with the subjects or with different subjects (e.g., advertising networks) for other scopes than the provision of the requested service (e.g., for marketing and advertising purposes).

[‡] Article 2(d) Directive 95/46/EC of the European Parliament and the Council of 24 October 1995 on the protection of individuals with regard to the processing of personal data and on the free movement of such data, *Official Journal L 281*, November 23, 1995, p. 0031–0050. http://eur-lex.europa.eu/LexUriServ/LexUriServ.do?uri=CELEX:31995L0046:en:HTML

[§] Article 2(e) Directive 95/46/EC.

[¶] Article 2(a) Directive 95/46/EC.

reply—the existence of the right of access to and the right to rectify the data concerning him.*

However, the crucial point here is that the right holder will in fact be accountable for any data processing activities carried out by the data controllers.† If the relationship between the right holder and the third-party IPR owner(s) does not pose peculiar issues, it is worth taking a closer look at the one between the right holder and the CSP(s).

There has been a lot of discussions and concerns about the generally unclear (inadequate) legal framework for cloud computing, especially when it comes down to personal data protection. Surely, the legal framework leaves ample room for interpretation. Therefore, it is extremely important to clearly lay down parties' duties and obligations in appropriate data processing agreements. In this respect, Article 29 Working Party (WP) Opinion 5/2012 on Cloud Computing [1] specifically addresses the point of putting in place appropriate contractual safeguards. Moreover, in the last year, there has been a lot of new developments in Europe from the personal data protection regulatory point of view. It can be argued that we are now facing a second generation of cloud data processing agreements. The sources of such fundamental changes are to be traced back to the publication at the EU and Member States' levels of the following official documents: Article 29 WP Opinion 05/2012 on Cloud Computing; CNIL's recommendations for companies using cloud computing services [2]; Italian DPA Cloud Computing: il Vademecum del Garante [3]; data protection in the cloud by the Irish Data Protection Commissioner (DPC) [4]; ICO guidance on the use of cloud computing [5]; and last but not least European Cloud Strategy. All these documents need to be read in close connection with the European Commission's proposal for a General Data Protection Regulation (GDPR), which was published on January 25, 2012 [6]. GDPR is currently going through a revision process in the European Parliament and expected to be passed some time in 2014. From these official documents that were

---

* Article 10 Directive 95/46/EC.
† Article 17 Directive 95/46/EC. See also Article 29 Working Party Opinion 3/2010 on the principle of accountability—WP 173 (July 13, 2010). http://ec.europa.eu/justice/policies/privacy/docs/wpdocs/2010/wp173_en.pdf. Furthermore, the "principle of accountability" is one of the main pillars of the European Commission proposal for a regulation of the European Parliament and the Council on the protection of individuals with regard to the processing of personal data and on the free movement of such data (GDPR) COM(2012) 11 final. http://eur-lex.europa.eu/LexUriServ/LexUriServ.do?uri=CELEX:52012PC0011:en:NOT. Moreover, the "principle of accountability" is the common element of most of the personal data protection regulations across the world.

recently published and the actual and the forthcoming applicable personal data protection legislation, we can draw a checklist of information that the right holder needs to check with the CSP before entering into a cloud service contract. More precisely, the right holder should request the CSP(s) to

1. Share information about its identity and the contact details of the data protection officer or a "data protection contact person."

2. Describe in what ways the users' personal data will be processed, the locations in which the data may be stored or processed, the subcontractors that may be involved in the processing, and whether the service requires installation of software on users' systems.

3. Specify whether/how data transfer outside the European Economic Area (EEA)—to countries without "adequate" level of data protection—takes place and on which legal ground (e.g., model contracts, binding corporate rules—Safe Harbor principles alone have not been recognized as an adequate means of transfer in Article 29 WP Opinion 5/2012).

4. Indicate the data security measures in place, with special reference to availability of data, integrity, confidentiality, transparency, isolation (purpose limitation), and "intervenability."*

5. Describe how the right holder can monitor CSP's data protection and data security levels and whether there is a possibility to run audits for the right holder or trusted third parties.

6. Disclose personal data breach notification policy.

7. Provide information on data portability and migration assistance.

8. Disclose data retention, restitution, and deletion policies.

9. Prove accountability by showing policies and procedures CSP has in place to ensure and demonstrate compliance throughout the CSP value chain (e.g., subcontractors).

10. Ensure cooperation with the right holder to be in compliance with data protection laws, for example, to assure the exercise of users' data protection rights.

---

* Article 29 Working Party, Opinion 5/2012 on cloud computing, p. 16.

11. Provide information on how law enforcement request to access personal data is managed.

12. Clearly describe the remedies available to the right holder in case of CSP breaching the contract.

Only the right holder that has obtained all of the above information from the CSP will be able to choose the right business partner, provide all the necessary information to the users, and keep the personal data protection compliance risk under control.*

14.3 SPECIFIC FOCUS ON THE APPLICATION OF CLOUD COMPUTING TO DIGITAL MEDIA

The evolution of the digital market (also before the existence of cloud computing) has pivoted around three main aspects that are intertwined: the technological framework, the business model framework in the distribution of digital content,† and the legal framework.

1. *The technological framework.* The development of the Internet and new technologies had a deep impact on the demand and distribution of digital contents. Technology facilitates the availability of work at anytime and anywhere, and indefinite number of perfect copies with little or no marginal costs; the increasing availability of the broadband Internet connection and end-to-end Internet architectures make it easier to upload, download, distribute, and sample preexisting and digital contents. In addition, the fragmentation of digital contents (e.g., to purchase only one or more songs of a music album, rather than the entire album, and

---

* It is noteworthy to highlight the work done by Cloud Security Alliance for the production of the Privacy Level Agreement (PLA) outline for the sale of cloud services in the European Union. https://cloudsecurityalliance.org/research/pla/. Moreover, in May 2013, the European Commission has set up an expert working group to draft a data protection code of conduct for cloud service providers, which falls under Key Action 2: Safe and fair contract terms and conditions of the European cloud strategy.

† The impact of new technologies (including cloud computing) on IPRs has been traditionally examined in respect of copyrightable works (music, audio–visual works, literary works, software) on which this chapter is focused as well. It is worth noting that the same concerns apply as well in respect of other intangible assets (patents, trademarks, trade secrets). Indeed, the fragmentation of the legal framework in EU countries and the consequent yearning for a single digital market can be respectively overcome and achieved also through the implementation of a unitary patent system in Europe and a modernization of the trademark system at the EU and national level: see COM(2011) 287 final, Communication from the European Commission, a single market for intellectual property rights. Boosting creativity and innovation provides economic growth, high-quality jobs, and first-class products and services in Europe.

create a custom-ordered iTunes playlists) and the diffusion of portable devices influence consumption of digital contents by consumers.

Technology has boosted the distribution of digital contents, allowing a broader circulation of copyrightable works, whether legally or illegally. Indeed, although an increased offer of legally available digital contents is per se a desirable outcome for copyright holders, technological evolution has also significantly increased copyright infringements by means of peer to peer (P2P), illegal streaming, download, and upload. New technologies represent at the same time an opportunity to maximize circulation by the right holders and lawful access by users to creative content (e.g., by means of DRM, whether or not associated with alternative distribution models) and a potential threat that increases the creation and distribution of illegal/counterfeit copies of intellectual property assets by placing illegal content in the jurisdictions, which have a more favorable regime for infringers.[*]

2. *The distribution models.* From the IPR standpoint, the development of new technologies (including but not limited to cloud computing) had a deep impact on the efficiency of the traditional distribution models of content. In a traditional digital scenario, everything is downloaded and stored on a single, personal device; and the right holders exercise their rights to exploit their IPRs by maximizing the control over the number of samples of their works.

In the new technological framework, the traditional market model (and related legal framework) based on the remuneration of the author's work by means of a control over the number of copies distributed in a given territory has rapidly shown its limits: right holders progressively lose control over the distribution of their works and cannot therefore receive a full remuneration for their works. In addition, traditional distribution models are usually national in scope because the content (audio–visual works, in particular) has often been considered as much as a cultural as an economic product, strongly linked to national contexts and cultural preferences.[†]

---

[*] Reference is traditionally made to the scenario where the user uploads/downloads illegal copies of a work (e.g., individually or through P2P platforms), in breach of a third party's exclusive rights. However, illegal exploitation of IPRs might as well be suffered by the CSP in respect to its IPRs which are involved in the supply of cloud services (e.g., the software which manages the cloud, the trademarks identifying the service).

[†] COM(2011) 427 final, Green Paper on the online distribution of audiovisual works in the European Union: opportunities and challenges towards a digital single market, p. 2.

The limits of the traditional distribution model have been further articulated by the development of cloud computing services, which trigger some legal issues: on the one end, technology sets up higher expectations in users, who can store information (pictures, e-mail, etc.) and use software (social networks, streamed video and music, and games) "when and where they need it (e.g., on desktop computers, laptops, tablets and smartphones),"* and is therefore much more reluctant to accept legal and/or technical limitations to their fruition of digital content caused by geographical, technical (e.g., interoperability†), or marketing‡ reasons. On the other end, services may be made available to users in a particular jurisdiction, yet that user's data may be stored and processed at an unknown variety of locations in the same or other jurisdictions, thus making it very difficult to verify the implementation of adequate security measures. In a nutshell, users "store and access data located on external computers that the User does not own, does not control and cannot locate" [7].§

3. *A fragmented legal framework.* The inadequacy of the traditional IPR distribution models is also mirrored in the increasing inadequacy of the traditional legal systems based on (1) the principle of territoriality, as noted above, and (2) the strict control by the right holders over individual uses and the efforts to minimize the number of "units" accessed without payment. Despite its huge potential, the digital single market remains small and highly fragmented: "Internet Europe is still a patchwork of different laws, rules, standards and practices," with little or no interoperability.¶ The current fragmentation of the digital

---

* COM(2012) 529 final, Communication from the Commission to the European Parliament, the Council, the European Economic and Social Committee and the Committee of the Regions. Unleashing the potential of cloud computing in Europe, p. 3.
† Users' demand of interoperability in the cloud computing space is also highlighted in the 2013 BSA Global Cloud Computing Scorecard, which "ranks 24 countries accounting for 80% of the global ICT market based on seven policy categories that measure the countries' preparedness to support the growth of cloud computing": http://cloudscorecard.bsa.org/2013/. The 2013 BSA Global Cloud Computing Scorecard "finds marked improvements in the policy environment for cloud computing in several countries around the world."
‡ Music users would find nowadays anachronistic to be obliged to purchase an entire music when they have the possibility to purchase only one or more songs of the same and create their own custom-ordered iTunes playlists.
§ Such uncertainty has a significant adverse impact on the effectiveness of IPRs' enforcement strategies.
¶ COM(2011) 942 final, Communication from the Commission to the European Parliament, the Council, the Economic and Social Committee and the Committee of Regions, A coherent framework for building trust in the digital single market for e-commerce and online services, p. 2.

market makes it difficult for users in all Member States "to have legal access to a wide range of products and services, offered over the largest possible geographical area."* While "increased demand for online access to cultural contents (e.g., music, films, books) does not recognise borders or national restrictions and neither do the online services used to access them,"† the legal offer of copyrightable works is still subject to different rates and limitations in each Member State. Another hindrance for cross-border transactions is manifested in the diversity of value added tax (VAT) regimes applicable to digital contents, which somehow explains the limited development of the digital books market.

The European Commissions' intended purpose is to achieve a single market for IPRs and to set up a legal framework for building trust in the digital single market for e-commerce and online services.‡

The Commission firmly believes in the necessity to foster alternative distribution models of digital contents to (1) promote the right holders' interests and incentivize the creation and distribution of new copyrightable works in a digital environment and (2) incentivize the access to digital works at the lowest possible costs and the distribution of user-generated content, even if such incentives might indirectly increase digital piracy.

This target requires finding a delicate balance between the different (and somehow conflicting) interests of the right holders (to receive an adequate remuneration for their artistic and/or entrepreneurial efforts), intermediaries (to receive adequate remuneration and be guaranteed clear and unambiguous principles regarding their liability), and users (in terms of freedom of access to digital contents).

The achievement of these targets is challenging, since to a certain extent they are conflicting. The development of cloud computing services further increases the necessity to reach such a balance between the protection

---

* COM(2011) 942 final, Communication from the Commission to the European Parliament, the Council, the Economic and Social Committee and the Committee of Regions, A coherent framework for building trust in the digital single market for e-commerce and online services, p. 5.
† Media release on the new Commission proposal for a Directive of the European Parliament and of the Council on collective management of copyright and related rights and multiterritorial licensing of rights in musical works for online uses in the internal market.
‡ For example, see the above-mentioned COM(2011) 287 and COM(2011) 942 final, Communications to the Commission.

of a fundamental right, such as intellectual property,* with other rights having the same dignity (privacy,† freedom of expression, freedom of speech, right to Internet access). This has also been recently underlined by the Business Software Alliance in its 2013 BSA Global Cloud Computing Scorecard: "Cloud computing readiness can be measured by considering how it addresses (i) data privacy; (ii) security (storing data and running applications on cloud systems); (iii) cybercrime; and (iv) IPRs."‡

To achieve a digital single market, the European Commission has identified five main obstacles,§ including the necessity to enhance legal, cross-border offer of online products and services. A new legal scenario should address these issues and recognize the new business models created or reshaped by the digital market.

The analysis of the reactions by the business and legal framework to the issues triggered by new technologies have been somewhat contradictory. At an EU level, the importance of increasing online distribution and the benefits brought by new distribution models have been often affirmed.¶

Either the legal scenario was not yet "ready to deal with the Internet" this was the case of the 1994 Agreement on Trade Related Aspects of Intellectual Property Rights – TRIPS since at that time it was still not entirely foreseeable impact it may have had on the market for copyrightable goods; or, in other cases [1996 World Intellectual Property Organization (WIPO) Treaties** and the Directive 2001/29/EC of the copyright and related rights in the information

---

* Even before the implementation of the EU directives, the dignity of intellectual property as a fundamental right has been recognized *inter alia* by the Universal Declaration of Human Rights adopted by the United Nations General Assembly on December 10, 1948; the International Covenant on Civil and Political Rights adopted by the United Nations General Assembly on December 16, 1966; and the Charter of Fundamental Rights of the European Union.

† "The protection of the right to intellectual property is indeed enshrined in Article 17(2) of the Charter of Fundamental Rights of the European Union ('the Charter'). There is, however, nothing whatsoever in the wording of that provision or in the Court's case-law to suggest that that right is inviolable and must for that reason be absolutely protected": ECJ, C 360/10, *SABAM v. Netlog*, par.43.

‡ 2013 BSA Global Cloud Computing Scorecard, p. 4.

§ COM(2011) 942 final, Communication from the commission to the European Parliament, the Council, the Economic and Social Committee and the Committee of Regions, A coherent framework for building trust in the digital single market for e-commerce and online services, p. 4.

¶ Reference is made, by way of example, to the Lisbon Strategy; the 1995 Green Paper on Copyright and Related Rights in the Information Society, COM(95) 382 final; the eEurope 2002 and 2005 action plans; the i2010 eGovernment action plan; the Communication from the Commission to the European Parliament, the Council, the European Economic and Social Committee, and the Committee of the Regions on a European agenda for culture in a globalizing world, COM(2007) 242; Europe 2020's growth strategy; and the EU's Digital Agenda for Europe.

** WIPO Copyright Treaty; and WIPO Performances and Phonograms Treaty, both dated December 20, 1996.

society], the "Internet versus copyright" issue was addressed maintaining a traditional approach in favor of the right holders, that is, a "copy-control model trying to replicate physical scarcity of supply online [7]." Protection of online works was achieved by means of (1) the extension of copyright protection into the digital environment (the right holder acquires the right to make available its works to the public copyright works or any other subject matter by way of online distribution); (2) the implementation of technological measures (DRM); and (3) the prohibition of DRM circumvention by third parties.*

Technological measures supported the protection of copyright and tries to remedy the lack of control by the right holders through a combination of technical protection (machine-readable code lines), contractual protection (users must adhere to a standard contract), and licensing model (to discipline the use of DRM by manufacturers of devices that support such DRM standard). DRM measures influence how and when users can access to the digital contents and are subject to interoperability issues, which may require users to purchase more devices to access the same content. This may limit the exploitation of digital content by users and not without raising copyright† and privacy issues.‡ However,

> the development of new technologies shows that the best way to maximize value on the internet is not to control individual uses [...] In

---

* Directive 2001/29/EC of the European Parliament and of the Council of May 22, 2001, on the harmonization of certain aspects of copyright and related rights in the information society (so called "Information Society Directive"). In the preamble of this Directive, it was highlighted that "A rigorous, effective system for the protection of copyright and related rights is one of the main ways of ensuring that European cultural creativity and production receive the necessary resources and of safeguarding the independence and dignity of artistic creators and performers." The same "classic approach" had been adopted in the United States (see the Digital Millennium Copyright Act and the Sonny Bono Copyright Term Extension Act 1998). More recently, at a multilateral level, the Anti-Counterfeiting Trade Agreement (ACTA) has been signed by Australia, Canada, Japan, Morocco, New Zealand, Singapore, South Korea, and the United States. At a EU level, the treaty was signed but subsequently rejected by the EU Parliament; hence, it will not come into force in the European Union. In addition, there are other national legislative and non-legislative initiatives focused on online copyright infringement.

† For example, DRM measures cannot distinguish whether a copyrighted work is in public domain.

‡ DRM measures can record data related to the use of protected content by the users, including the Internet Protocol (IP)/media access control (MAC) address of the computer. Such data can be used for antipiracy purposes to track users' illegal habits. The use of personal data of Internet users (e.g., by means of spyware in P2P networks) to identify illegal exploitation of copyrighted works and the identity (IP address) of the infringers has been widely debated at both the US and the EU level and highlighted the difficulty to achieve a balance between the exclusive rights of right holders, the freedom of enterprise of ISPs, and the right to privacy of end users. As a paradigm of the difficult balance between copyright and privacy, see in this respect ECJ—*Scarlet Extended SA v Société belge des auteurs, compositeurs et éditeurs SCRL (SABAM)*, C 70/10.

a 21st century cloud … a copyright holder should seek to maximize access (and the number of people who pay, in one form or another) for such access, and not to minimize the number of "units" accessed without payment, because this is not how value is derived [7].

It has been acknowledged at a EU level that "artists, entrepreneurs and citizens should benefit from a borderless, digital single market, where accessing and distributing content is easy and legal, generating more value and more visibility."* The opportunities offered by the Internet and cloud computing to the distribution of digital contents are so significant that the traditional approach of the right holders appears almost paradoxical [7].

This has opened the way to alternative business models, characterized by a different graduation in the use of DRM measures to keep control over users' use of digital content: (1) subscription/rental/pay per download (assisted by DRM measures, e.g., Rhapsody), relatively successful, mainly because of DRM-triggered issues (see above); (2) "superdistribution" models based on a P2P subscription model, which allows a limited exchange of content controlled through DRM (e.g., Wippit); and (3) distribution models where the remuneration is based on advertising revenues (rather than a single-user license fee) or by the economic value ascribed to the collection of users' personal data, whether or not associated with other services with fee, sponsorships, and e-commerce services (this is the model that is usually adopted by user-generated content social network platforms to guarantee the availability of free content, e.g., Mjuice and We7).

14.4 UPCOMING SCENARIOS AND RECOMMENDATIONS

The development of new technologies is significantly changing the existing business models and legal frameworks, and puts in contact the fundamental rights that were not in conflict before (e.g., privacy and copyright). This is pushing toward finding new balances between the concurrent interests of the right holders, users, CSPs, and other service providers and intermediaries.

Although streaming and on-demand supply of digital contents have exacerbated the incidence of copyright piracy, they have also contributed to opening the way to new distribution models via legal digital platforms. Yet distribution of digital content still remains too segmented due to limitations imposed through territorial, linguistic, platform, and or technical boundaries.

---

* The European Commissioner for Digital Agenda Nelli Kroes highlights the inadequacy of the current legal copyright framework, http://blogs.ec.europa.eu/neelie-kroes/digital-copyright-way-forward/.

Cloud computing services can further contribute to the achievement of a digital single market. To this extent, it is necessary to keep building a diffused trust in the digital environment at technical, business, legislative, and contractual levels.

From a technical standpoint, a digital single market requires a definition of common interoperability standards between products and services. It remains necessary to keep developing new alternative business models and to adjust the existing legal scenario to simplify copyright clearance, management, and cross-border licensing, and increase the relevance of users' interests.

In this respect, in July 2012, the Commission published a proposal for a directive of the European Parliament and the Council on collective management of copyright and related rights and multiterritorial licensing of rights in musical works for online uses in the internal market. The proposal pursues two complementary objectives: (1) to promote greater transparency and improved governance of collecting societies through strengthened reporting obligations and the right holders' control over their activities, so as to create incentives for more innovative and better quality services, and (2) to encourage and facilitate multiterritorial and multirepertoire licensing of authors' rights in musical works for online uses in the EU/EEA.* And, indeed, some musical rights collecting societies have already taken initiatives toward the actual implementation of a pan-European licensing of online rights; on April 29, 2012, the Italian, French, and Spanish copyright collecting societies [Società Italiana degli Autori ed Editori (SIAE), SACEM, and SGAE, respectively] announced the launch of Armonia, the first licensing hub operating at a EU level for the granting of digital licenses on a multiterritorial basis.†

The achievement of an "enabling framework" requires *inter alia* a review of the regime of copyright exceptions set forth in the Information Society Directive, so as to possibly extend the range of exceptions to which

---

* EC media release on the proposal.
† Armonia is the single effective European online music licensing hub, formalized as an European Economic Interest Group (EEIG). It aims at serving its members' interests by providing the best conditions for the exploitation of their digital rights. Founded by SACEM, SIAE, and SGAE, Armonia is open to other collective management societies sharing its vision. Built on international standards, Armonia is a licensing hub that offers rights owners, through their societies, an integrated one-stop shop solution: (1) Armonia aggregates both international and local repertoires, representing today 5.5 million works and growing; (2) it facilitates licensing of music with DSPs in terms of negotiation and rights settlement; and (3) it uses streamlined licensing and negotiation processes, derived from the founding members' successful track record in licensing DSPs. This licensing hub is therefore willing to operate for the benefits of both rights owners and digital service providers (information gathered on www.armoniaonline.eu).

Article 6(4) applies and a standardization of agreements between users and right holders.

As regards the active role of right holders, they shall have to keep the pace of the swift developments in technological field to remain competitive and adapt their business offer in an effective way to meet users' expectations. The offer of digital content via cloud services to users shall have to be planned in due advance so as to (1) identify ahead of time the opportunities arising from technological developments in the short–mid term (e.g., a new unexplored platform or device for making available digital content, such as streaming or progressive download to set-top boxes and/ or portable devices); (2) adapt existing business models or identify new ones in respect of the said technological opportunities arising (e.g., extending the offer of pay-TV movies on portable devices to current residential customers and/or new customers, whether on a subscription or free trial basis); and (3) assess ahead of time whether current licensing agreements in place enable the right holders to make available to the public the digital content according to their envisaged business models or otherwise require further investments. In a nutshell, the competitiveness of right holders shall require a continuous internal dialog and cohesion between internal departments (technology, legal, marketing, CRM, finance, etc.) to ensure an effective and timely business plan is adequately devised.

REFERENCES

1. Opinion 05/2012 on Cloud Computing, available at http://ec.europa.eu/justice/data-protection/article-29/documentation/opinion-recommendation/files/2012/wp196_en.pdf
2. Recommendations for companies planning to use cloud computing services, available at http://www.cnil.fr/fileadmin/documents/en/Recommendations_for_companies_planning_to_use_Cloud_computing_services.pdf
3. Cloud Computing, available at http://www.garanteprivacy.it/documents/10160/2052659/CLOUD+COMPUTING+-+Proteggere+i+dati+per+non+cadere+dalle+nuvole-+doppia.pdf
4. Data protection "in the cloud", available at http://dataprotection.ie/viewdoc.asp?DocID=1221&m=f
5. Guidance on the use of cloud computing, available at http://www.ico.org.uk/for_organisations/data_protection/topic_guides/online/~/media/documents/library/Data_Protection/Practical_application/cloud_computing_guidance_for_organisations.ashx
6. EUR-Lex, available at http://eur-lex.europa.eu/LexUriServ/LexUriServ.do?uri=CELEX:52012PC0011:en:NOT.
7. Gervais, D.J. and Hyndman, D.J., Cloud control: Copyright, global memes and privacy, available at http://works.bepress.com/daniel_gervais/37, pp. 57, 75, 91.

CHAPTER 15

Cloud Computing and Adult Literacy

How Cloud Computing Can Sustain the Promise of Adult Learning?

Griff Richards, Rory McGreal, and Brian Stewart

Athabasca University
Athabasca, Alberta, Canada

Matthias Sturm

AlphaPlus
Toronto, Ontario, Canada

CONTENTS

15.1 INTRODUCTION

Adult literacy in Canada consists of a patchwork of large and small adult education providers: many of them are autonomous community societies, some are school boards, and others are community college based, as well as a range of independent community-based groups. Funding for adult literacy comes from several pockets: from different provincial and/or federal government departments and from charitable organizations. Much of the federal funding is short term in response to shifting government priorities. Indeed, Crooks et al. [1] suggest that the ongoing funding search, with the attendant application and reporting activities, detracts from the ability to provide more effectively planned and sustainable adult education programs. A major challenge for adult literacy providers is that while their client base has significant human and economic potential, low-literacy adults are not perceived as large contributors to the economy, and thus, much of the funding is intermittent—from project to project. Without sustained and sustainable resources to exploit technologies, nor exposure to the use of technologies for teaching, adult literacy providers will remain very traditional in their use of face-to-face pedagogy and remain relatively unexposed to the potential benefits of technology-enhanced learning and cloud computing.

The structures of adult learning and adult education organizations and learners in Canada make the use of cloud computing particularly appropriate. Informal learning and semiformal community-based learning are the dominant modes of adult learning within small businesses, trade unions, cooperatives, industrial and commercial enterprises, hospitals, prisons, and religious and cultural organizations. There are no statistics on the amount of informal learning that is occurring, but according to Cross [2], there is general agreement that it is growing rapidly. Cloud computing can be used to address the increasing cost and complexity of providing the state-of-the-art e-learning services that are beginning to outstrip the abilities and resources of adult education institutions and organizations. The degree of integration and interoperability required to provide seamless service is becoming too complex for smaller entities to manage efficiently. In addition, higher level functions such as sophisticated data analytics that could be valuable tools in understanding adult education processes cannot be developed as quickly as necessary, if at all. Computer service departments struggle to keep up with the growing demand for information technology (IT) services on campuses. New approaches are required if adult education institutions and organizations are to effectively meet the demands of learners and other stakeholders for ever more sophisticated services, while still working within the growing budgetary constraints of both the organization and the adult learning sector as a whole. Cloud computing could form a major part of an effective solution to this problem.

Many institutions and companies are moving rapidly to adopt *cloud computing*, a term that refers to accessing Information and Communications Technology (ICT) services across the Internet. The computers and software applications are housed on Web servers in large industrial-scale computing centers rather than provided locally. The first benefit of these commercial "computing utilities" is that they can harvest the economies of scale and offer services at a fee that is far lower than most organizations would require to implement and maintain their own computing infrastructure. To lower energy costs, cloud providers locate their data centers near power generation facilities; to lower staff costs per machine, cloud providers install vast numbers of computers in each server farm. Many institutions already benefit from these economies of scale by outsourcing e-mail to Google or Microsoft.

The second benefit of cloud computing is in having large-scale data processing resources available "on demand." Scientists with analyses that might take hours or days to execute on a single computer can speed the processing by tasking the cloud to provide the equivalent of hundreds of computers for a few minutes. Lower costs and flexible computing on demand are the two key advantages of cloud computing. The impact is already being felt in some institutions and businesses; cloud computing will soon spread to other areas of the economy and to adult literacy organizations that become aware of its benefits.

Cloud computing can be an industrial-scale replacement of the "cottage industry" approach to institutional computing that now exists within institutions and organizations. Much of the capital costs of institutional computing can be converted to lower operating costs. With the cloud, the physical space and the energy ICT consumes are reduced in-house, yet the available computing power is greatly increased. In addition, elastic scalability allows users to increase or decrease computing capacity as needed [3].

At first blush, cloud computing seems to be an entirely technical issue since adult literacy educators, like most consumers, are blissfully unaware of the technologies they access. They search the Web or book airplane tickets with little thought to the layers of hardware and software that provides these services. However, a major paradigm shift will lead those using technology to rethink the services they offer and how they are offered. For example, the emergence of the World Wide Web Mosaic Browser in 1995 made it possible to both publish and retrieve information without having an intermediary, while also reducing the difficulty in publishing information quickly and at a much reduced cost. This had a huge impact on the world of distance learning that until then leaned toward "big mass media" paper publications and television. The "anyone-can-publish" environment brought on by the World Wide Web meant that almost any institution could offer distance education, a capability they are now adopting in ever increasing numbers. By 2005, the integration of mobile telephones with the Internet literally meant that anyone, almost anywhere, could connect to the world's information systems. This has been particularly beneficial to democratizing information access in developing countries, and mobile phones have become the main consumer channel for both voice and data services. The ability to "leapfrog" the millions of miles of copper wire and boxes that plug into electrical outlets has enabled emerging and developing countries to partake in the knowledge economy at a faster rate and to partially close the digital divide.

Piña and Rao [4] argue that cloud computing is creating "new IT [Information Technology]-enabled market constructs" and it will have a profound effect on IT management. The cloud will challenge everyday business models from which the educational and economic sectors cannot escape. The shift to cloud computing provides an opportunity for adult literacy providers to implement and/or restructure their online operations and decide what services to offer and how they will be provided. However, this will not happen automatically. The adult literacy sector in Canada faces endemic regionalization and programming challenges that have little to do with computing, and everything to do with politics, funding, community leadership, and professional collaboration.

A symposium of e-learning experts sponsored by Contact North [5] identified a number of specific operational and technical challenges, all of which could be viably addressed using cloud computing. These include addressing content quality, learner support, the e-learning compatibility of administrative systems, ongoing IT management infrastructures, tools, broadband availability, support services (helpdesk), and the evergreening of IT.

de Broucker and Myers [6] recommended the implementation of a public policy framework for adults that acknowledges the "right to learn." This includes financial support, incentives for employers, and more government investment using a "coordinated approach to respond to adult learners' needs." Support for cloud computing would go a long way in addressing these recommendations.

While cloud computing can be used to lower the costs of providing a technological infrastructure for adult literacy, there will still be real costs—the economics of cloud provision have yet to be fully defined and understood. The cloud investment can reasonably only be realized with sufficient stable funding. Building a collaborative community around cloud computing might be a way to bring a large number of educational resources together to develop and sustain a coherent and cost-effective delivery model for adult literacy training that would benefit many. It may also provide the cross-fertilization of ideas and talents to see a new range of literacy services that will help low-literacy Canadians cope with our text-laden society.

This chapter is organized around four questions:

1. What is cloud computing and why is it important for adult literacy?

2. What is the current state of adult literacy education in Canada and is a cohesive online adult literacy community feasible?

3. What is the current use of IT to support adult literacy?

4. What might a cloud computing strategy for adult literacy look like and what are the challenges to realize such a vision?

Technical issues aside, the changes cloud computing brings may provide an unprecedented opportunity to revolutionize the way in which we offer adult literacy training and new literacy services that can hasten the integration of low-literacy adults into society. The cloud could facilitate the alignment of institutional processes, and therefore enable the reduction of system complexity. There are legitimate reasons for institutional or organizational differences: size, programming, structure, and operational mandate, all of which provide significant reasons for differentiation. The initial benefits from an adult education cloud are in outsourcing the infrastructure costs. However, the areas of significant gain can still be realized at the application level with, for example, e-mail and shared learning management systems, content management systems, automated assessment systems, and Web conferencing systems. These would represent the initial applications that would formulate a common cloud provision.

15.2 QUESTION 1: WHAT IS CLOUD COMPUTING AND WHY IS IT IMPORTANT FOR ADULT LITERACY?

Cloud computing is a nebulous term

—Anonymous

Wikipedia notes that the cloud concept originated among telephone networks and that "The first scholarly use of the term *cloud computing* was in a 1997 lecture by Ramnath Chellappa."

According to Pingdom [7], the term "cloud computing" was launched into the mainstream in 2006 when Eric Schmidt, CEO of Google, used the term when describing Google's own services during a search engine conference: "It starts with the premise that the data services and architecture should be on servers. **We call it cloud computing**—they should be in a 'cloud' somewhere."

The National Institute of Standards and Technology (NIST) defines the term as follows:

> Cloud computing is a model for enabling convenient, on-demand network access to a shared pool of configurable computing resources (e.g., networks, servers, storage, applications, and services) that can be rapidly provisioned and released with minimal management effort or service provider interaction [8].

In common usage, cloud computing has grown to mean Internet access to large-scale computing facilities provided by others. There are a few key concepts which are described below.

15.2.1 Economies of Scale

The cost of providing ICT has become a growing concern for many organizations. For example, a large university with tens of thousands of students might be expending over $500,000 each year just for the infrastructure (servers, software, storage, staff, and communications) to provide e-mail. However, Google has a huge e-mail facility that currently provides millions of Gmail accounts for no fee (thus the cost to Google of adding a few thousand or a few hundred thousand academic mail users is nearly negligible). Cloud providers have located their computing facilities near power generation facilities (so the electricity is "greener" and cheaper since less power is wasted in transmission) and their large facilities are more robust and require fewer staff per e-mail account to maintain than small facilities. Google Apps for Education is currently providing free e-mail and other applications such as document sharing to entice universities to make the switch to greener and cheaper cloud computing services. Microsoft and Amazon (and others) are also offering cloud services on a large scale.

In traditional ICT organizations, increasing computing capacity requires additional capital investment followed by increased operating and maintenance costs. As with erecting a new building, the infrastructure needs to be maintained regardless of usage. In contrast, cloud computing is like renting space in a building—you only pay for the space and services as long as you need them. The cost of the building is amortized over a large number of tenants. Since cloud tenants connect via the Internet, their number can be very great and their share of costs can be very small compared with traditional ICT costs.

The economies of scale can also apply to the adult literacy community. The development of an adult literacy cloud could help reduce this funding sustainability gap by allowing more effective planning and provision of services. This is not a simple task, however, and will require significant and involved collaboration across the adult education sectors, yet institutions and organizations appear to have few viable alternatives. A freely accessible adult learning cloud computing environment or medley of environments could provide significant financial savings for learners, employers, and adult learning organizations and institutions while at the same time forming the basis for coordinated approaches to learning delivery provincially or even nationally. A long-term investment in a cloud for adult learning in Canada would not only reduce the cost and increase the scope of technology services but also enable institutions to create more meaningful and realistic technology plans that address the short- and long-term technology needs of their program delivery. Of course, this coordination of services would also have to address issues such as data and personal privacy. Low-literacy refugees from war-torn countries may be reluctant to use free services if there is the slightest chance that their identities are not protected.

15.2.2 Virtualization

Today's computers are both fast and powerful and are capable of serving several users at the same time. Each user is given a share of the computer's time and resources, and several "virtual" computing sessions can be run at the same time; the typical user does not even notice that they are sharing a computer. Every job that accesses the cloud through the Internet is assigned to the next available virtual space—often on a different physical computer than the last virtual session. The cloud management software looks after the job allocations, constantly shifting usage to optimize the use of several computers connected together in the cloud. Fewer computers are needed in the workplace than in the current desktop environment where each user has his/her own personal computer.

15.2.3 Device Independence

Since the data processing is done "in the cloud," the user no longer needs a powerful (nor expensive) desktop computer. Smaller and cheaper workstations, "notebook" computers, and mobile devices such as tablet computers or even "smart phones" can connect to the cloud via the Internet. The cloud will be able to reformat the output to suit the user's device—perhaps reading out loud to a mobile phone rather than sending text to its small screen [9]. Moreover, users can alternate devices and access their applications and content independently from wherever they are located using any Internet-capable device. For adult learning institutions, device independence may result in using the scarce financial resources for software and hardware purchases and maintenance more effectively as they do not need to provide and support physical computers. The lower cost may also enable greater access to computers by learners, as they can find less-expensive alternative access devices. Technical support could be provided from more aggregated central units, and therefore lower cost, addressing a current need especially in community-based agencies where practitioners commonly provide their own support.

15.2.4 Elasticity

With desktop computing, each user is limited to the resources (processing, memory, and data storage) available in his/her personal computer. With cloud computing, users can request as much computing power as they need. For example, Roth [10] discusses how he recently used a cloud computing facility to find a missing security code by testing every possible combination until he found the one that fit. With a desktop computer, this might have taken years, but by programming a cloud to run hundreds of virtual copies of his/her program at the same time, the missing code was found in minutes, at a cost of about $2. Cloud resources are said to be "elastic"—they can expand or contract to the amount of computing power needed at any given time. This means that very powerful analyses can be conducted more readily than would be feasible on a desktop computer. Keahey et al. [11] note how several scientists can schedule the use of a shared cloud and that open-source cloud software makes it possible to quickly create new cloud installations. Of course, licensing approaches will need to be more flexible for this to be advantageous. A more flexible, "pay-as-you-go" approach will need to be integrated into licensing structures.

15.2.5 Cloud Service Models

Cloud services typically fall into one of three technical/marketing categories: infrastructure as a service (IaaS) in which the expert user implements his/her own software to optimize the use of the computing facility; platform as a service (PaaS) in which the client customizes his/her application to run inside the cloud management software; and software as a service (SaaS), such as Gmail, in which the user simply uses the software provided. This flexible approach means that an organization with special needs and appropriate technical skills can build their own computing solution, while customization and the use of generic software can meet most users' requirements. As a rough analogy, if IaaS were renting a car at the airport, then PaaS would be hailing a taxi and SaaS would be taking the public bus. The service models provide options to suit user independence, expertise, budget, and technical needs. Different services will have different benefits; the uptake rate will be influenced by the applicability within organizations. The models will need to evolve with requirements of the adult literacy providers and their needs for the cloud; executing working cloud models and ensuring satisfactory quality of service are essential.

15.2.6 Concerns over Cloud Computing

The major concern is about *security*. Since it is difficult to know where a virtual job will be processed (i.e., where the computer is physically located), data may easily cross international boundaries and suddenly be open to legal inspection in other countries—this would be a concern, for example, should Canadian data that are supposed to be protected under Protection of Privacy Laws cross over to the United States and be subject to the Patriot Act. Haigh [12] notes that Microsoft located its European e-mail server farm in Dublin to avoid client concerns that their data would be open to the US government. Private, secure, or mission critical data should not be processed in third-party public cloud computing environments. Secure data could be processed in private clouds—for example, Danek [13] notes that the Canadian government forecasts to set up its own secure cloud computing environment to rationalize the use and cost of government ICT infrastructure across several departments. A cloud run on systems based in Canada would be an essential investment for the adult education and adult literacy sector across the country. Clients of these programs often belong to marginalized groups that share concerns about the protection of their privacy and use of personal information. In jurisdictions where adult literacy programs are publicly funded, client data include information about government services that needs to be secure and protected.

The second concern is the need for a *fast and reliable Internet connection*. Cloud computing involves rapidly moving the data to be processed elsewhere, and then quickly returning the results. A slow or intermittent Internet connection can interrupt the data flow and separate the user from the virtual machine. (One author of this report had to retype several paragraphs when a communications interruption disconnected him from the word processing application in a cloud environment.) As more and more

Internet traffic flows through fiber optic cables, bandwidth will increase and communications costs will decrease. However, cloud computing may not be a successful strategy for users in rural and remote communities until they can be assured continuous robust connectivity.

The third concern is about *switching costs*. Many legacy software applications will need to be moved to the cloud environment and incompatibility in design standards can pose significant hurdles and be quite costly when porting them to a cloud platform. Fortunately, as has been mentioned earlier, very few adult literacy organizations have investments in ICT. However, the costs of "lock-in" cannot be avoided. The "Monkey and the Coconut" tale suggests that you can catch a monkey by chaining a coconut to a tree and boring a hole just large enough for a monkey to reach its hand in and grab a fistful of honeyed rice. The closed fist is too large to go backward through the hole. For the monkey to be free of the trap, it has to let go of its investment in the bait. The costs of "letting go" from a cloud service or an internal ICT infrastructure may be insurmountable— just as it is difficult for a homeowner to dispute the rate hikes by the local electricity provider by threatening to get energy from another source. It is conceivable that in future, the "free" Google and Microsoft academic and e-mail services might be charged for as the providers will eventually need to recover their investment and operating costs. At that point, institutions may be "locked in" to those services.

The fourth concern is *hype*. Katz et al. [14] note that cloud computing seems to have caught the attention of almost every technology journalist to the point where it might be oversold. While the cloud has arrived for common services such as e-mail, for many other services the transition may take much longer. Much technical and policy work remains to be done by the adult literacy community to determine which applications can go to the cloud and which require a more conservative approach. Expectations will need to be adjusted to reflect realistic and achievable applications. Figure 15.1 shows the exponential growth in the number of Google searches using the term "Cloud computing." As typical of new technologies, the "hype cycle" peaked early in 2011 followed by a leveling-off period as understanding became widespread and pilot implementations took place. This may hit a further inflection point as another wave of cloud-based services is created, which may include education. The graph does not necessarily show a lessening interest, rather a lessening novelty. As cloud computing becomes mainstream, there is less need to discuss what it means anymore, just how to do it, from envisioning to engineering.

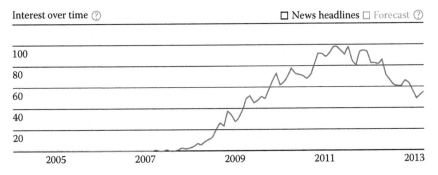

FIGURE 15.1   Google trend plot of the term "cloud computing" taken on February 22, 2013. The number 100 represents the peak search volume.

15.2.7 Summary

Cloud computing changes the efficiencies and economics of providing ICT services. Large cloud "utilities" are being developed that will make it cost-effective to move many if not most ICT services "to the cloud"; the nature of the services provided can be negotiated with the cloud provider. Virtualization will enable several computing jobs (such as word processing or e-mail users) to run on a single computer, while elasticity makes it possible to have huge amounts of computing resources instantly available to meet the demands for intensive data processing. Cloud computing is evolving rapidly and new methods to ensure effective management and security will emerge. Currently, most applications of cloud computing are in administration and research, but the ability to build and share powerful new processes will rapidly expand the variety of services available. This is where the greatest potential might lie for adult learning and literacy training.

Katz et al. [14] provide the following list of the benefits of a cloud computing approach:

- Driving down the capital costs of IT in higher education

- Facilitating the transparent matching of IT demand, costs, and funding

- Scaling IT

- Fostering further IT standardization

- Accelerating time to market by reducing IT supply bottlenecks

- Countering or channeling the *ad hoc* consumerization of enterprise IT services

- Increasing access to scarce IT talent

- Countering a pathway to a five nines (99.999% system availability) and 24 × 7 × 365 environment

- Enabling the sourcing of cycles and storage powered by renewable energy

- Increasing interoperability between disjointed technologies and within institutions

These benefits can explain the growing interest in cloud computing among a wide variety of organizations, institutions, and businesses around the world. Figure 15.1 reflects a typical "Gartner hype cycle" for a new technology that is moving from hype to implementation.

The line shows the exponential growth in the relative number of Google searches followed by a decline in searches as the term becomes a part of mainstream computer understanding.

15.3 QUESTION 2: WHAT IS THE CURRENT STATE OF ADULT LITERACY EDUCATION IN CANADA AND IS A COHESIVE COMMUNITY APPROACH POSSIBLE?

It is beyond the scope of this chapter to completely portray the current state of adult literacy in Canada. There exist a number of excellent studies and literature reviews already published on this topic by researchers and government organizations [15–23]. Their portrayal is consistent with the Organisation for Economic Co-operation and Development (OECD) [24] thematic report on adult learning: Canada is a vast country, and despite a wide variety of regional and federal programs that contribute to adult literacy, there remains a shortage of programs especially in rural and remote areas. There is a general need for additional programming for adults, particularly for Aboriginal peoples and for the working poor. The thematic report also expresses concern that the lack of a coordinated federal–provincial policy on adult literacy makes it difficult to resolve many issues such that

- The special needs of adults are generally neglected.

- There is no sense of a coherent system of adult education.

- Adult education is vulnerable to instability in government [24, pp. 42–43].

Adult education and literacy in Canada is also divided by different approaches and organizational types. In some regions, it is community groups that deliver the bulk of adult literacy education, whereas in other areas, this is left to community colleges or partnerships of both. Funding comes from a mix of federal employment initiatives and provincial education programs. The funding is usually short term, and literacy providers spend a good deal of their time applying for the next grant or writing reports. The Movement for Canadian Literacy [20] claims that the lack of a long-term funding strategy makes it difficult to sustain programs and staff. Horsman and Woodrow [19] describe adult basic education as "the poorest cousin of the education system."

There are three main target audiences for adult literacy education:

1. Canadians from rural and remote areas where access to education is limited. (This includes a large number of people with Aboriginal ancestry, some of who have been educated in English and others in their native Aboriginal language.)

2. School leavers who fail to complete high school due to a complex array of reasons and become trapped in the "working poor" layer of the economy and may require to upgrade their skills to retain their job or to search for alternate employment.

3. "Newcomers to Canada," that is, recent immigrants from around the world who are generally (but not always) literate in their own language. [In some jurisdictions, adult literacy and English as a Second Language (ESL) programs are funded and delivered separately.]

Federal funding is generally targeted to assist newcomers to Canada to become functional in one or the other of the official languages, and there is a pattern of successful economic integration particularly by the second-family generation in urban areas. The OECD [23] identifies Aboriginals and the working poor as the two populations least served by adult education programs. Many Aboriginals grow up in isolated areas and learn English from parents for whom English was an imperfectly learned second language. Many of the current generation also often fail to master their own native language and are caught between two cultures. The increasing urbanization of the Aboriginal population brings many within reach of targeted literacy programs, and there are a number of e-learning approaches that are being initiated to reach those in remote areas. However, low-literacy adults in isolated communities are among those with the least access to Internet connectivity and computers.

Some 20% of Canadians form "the working poor" earning less than one-third of the median wage [23]. Many of them are also in rural and remote areas and traditionally earned their living in the primary resources and agriculture sectors. With the decline of the resource economy, many lack sufficient education to access retraining for other jobs. Others simply cannot access the existing daytime literacy programs because of commitments to work or family care.

While there are a lot of people falling through the cracks, some adult literacy practices are making significant inroads. Prior Learning Assessment and Recognition enables individuals to get recognition for life experiences and skills, and the resulting academic credits make academic credentials accessible. In British Columbia (BC), considerable work has also taken place in "laddering" or transferring credits earned in college or trades as entry paths into higher education. In Alberta and the Northwest Territories, the Alberta–North collaboration of higher education institutions and community organizations that provide technology access and educational support in 87 remote communities enables a large number of learners to become the first in their family to earn a degree.

Despite the low level of federal–provincial coordination in adult literacy, the community is organizing itself into regional and national networks to exchange information and educational resources. Of particular note is the National Adult Literacy Database (www.nald.com) that maintains a repository of up-to-date research and AlphaPlus (www.alphaplus .com), which also shares learning resources. When the Canada Council on Learning ended its mandate in 2009, the Adult Learning Center spun out the Movement for Adult Literacy, which is now the National Literacy Learning Network, a forum for all of the regional literacy networks across Canada.

Adult literacy deficits are not unique to Canada, but are also found in Australia, the United States, and other industrialized countries, some of which are large developed countries with remote areas populated by resource workers and Indigenous peoples, and others have large urban populations. Literature from these countries reveals many of the same issues and offers relevant approaches to provide adult literacy education. Ideally, it would seem that the place to prevent adult literacy problems is in primary school education. However, literacy education starts in the home and the influences of early community literacy are well documented [16].

Life-long learning has become ever more important as adults have to readapt to ever-increasing demands of their skills and knowledge. As the OECD states in the introduction of the Programme for the International Assessment of Adult Competencies (PIAAC) survey that has been undertaken in many countries in 2012,

> Governments and other stakeholders are increasingly interested in assessing the skills of their adult populations in order to monitor how well prepared they are for the challenges of the modern knowledge-based society. Adults are expected to use information in complex ways and to maintain and enhance their literacy skills to adopt to ever changing technologies. Literacy is important not only for personal development, but also for positive educational, social and economic outcomes. [23]

PIAAC assesses the current state of the skills in the new information age and in that builds upon earlier conceptions of literacy from International Adult Literacy Survey (IALS) in the 1990s and the Adult Literacy and Lifeskills (ALL) Survey in 2003 and 2006. In the process, the definition of literacy has changed from reading and writing to including skills essential to successful participation in work, family, and community environments in the information age. This reconception of literacy has not only driven the need of governments in industrialized countries to assess and better prepare their population for the workforce but also put the importance of technology-based learning and sharing of resources on the fast approaching horizon.

15.4 QUESTION 3: WHAT IS THE CURRENT USE OF INFORMATION TECHNOLOGIES TO SUPPORT ADULT LITERACY?

Although technology rapidly evolves, there are four basic patterns of using technology for literacy education:

1. Learners receive *individualized computer-based lessons* from physical disks or via Web sites. The Web delivery is becoming more practical as it resolves the software distribution issues and learners can maintain records of their progress; however, in areas with poor Internet access, it may be more practical to transfer the lessons by CD-ROM or DVD. Drill and practice sessions are particularly effective for initial

skills and knowledge including phonetics, building vocabulary, and improving spelling and learning grammar. Audio–video materials such as podcasts can also help create a contextual awareness of language conventions. Literacy might borrow techniques from a number of very effective second language learning Web sites such as japanesepod101.com that match services to the motivation and budget of the learner. Free materials are very useful, but study texts, drills, and maintenance of a vocabulary portfolio require a subscription. Tutor-mediated online conversation sessions are available for an additional fee. An unexpected boon has been the wealth of free informal learning materials available in the video format on Web sites such as youtube.com.

2. *Online course or workshops* can be used to offer higher order learning activities such as reading and discussing articles from newspapers with other learners in a text or voice chat. Cohort-paced online courses enroll learners in a group so they move through the learning activities about the same time and speed. The cohort reduces the feeling of isolation; learners can interact to discuss the course content and to give each other support. A course facilitator or instructor or tutor helps the group move through the materials in a timely fashion and provides answers to questions that may arise. Cohort-paced courses typically have lower dropout rates than independent courses or self-study materials. In some instances, cohorts may involve synchronous computer conferencing; however, the scheduling of such events can be complicated and they can make it difficult for learners who have other obligations such as child care, shift work, or travel. Some community learning centers also equipped with broadband video-conferencing facilities that make it possible to bring small groups of learners together for work or study sessions, although the main use to date appears to be for the professional development of the tutors rather than for literacy instruction [25].

3. *Web searches, e-mail, conferencing, writing, blogging, and digital media projects* are authentic everyday communications activities that provide rich opportunities for literacy instruction. This type of support is best provided in (or from) a learning center where a staff member can be available to assist learners with the technology and with their literacy tasks. The completed artifacts can be copied into an e-portfolio to promote reflection on progress over time. There is

no reason why the instructional support could not be given at a distance. This would benefit transient literacy learners, especially if they could access their personal files from any Internet connection.

4. Another area is the use of *assistive technologies*, for example, software that can help the learner by reading electronic text files out loud, or providing online dictionaries and other reference materials. Some assistive software that patches onto Office software and reads text as it is composed has been particularly useful for English language learners and learners with dyslexia [26]. Assistive software will become portable and personal as the number of smartphones that link to the Internet increases and a wide variety of assistive applications emerge for that platform.

Despite this enormous potential, the usage of technology by literacy providers is not strong. Holun and Gahala [18] note that technology has a reputation as a "moving target"—by the time a serious intervention can be developed and evaluated, the technology has moved along. Another reason is the lack of technology accessible to literacy learners and the relatively low number of studies examining the use of technology for literacy training. Finally, Fahy and Twiss [15] note that while adult literacy educators are beginning to use technology for their personal communications and professional development, few have adopted technology to their teaching practices.

However, there are also many literacy programs that have embraced the use of technology in their program provision. In Ontario, adults can learn online through the Web-based literacy training provided by five e-Channel programs using a variety of synchronous and asynchronous delivery methods [27]. Several classroom-based programs across Canada use technology-based resources as an integrated part of literacy training or to supplement in-class learning providing opportunities for reinforcement and scheduling flexibility for their clients. The following describes a few of these programs.

As one of the e-Channel providers, Sioux-Hudson Literacy Council's Good Learning Anywhere program has used technology-based resources to reach clients in remote communities since 2003. The program employs six to seven instructors and five mentors who work remotely to meet the literacy needs of 300 adults across Ontario. For the last 3 years, various cloud services have been used to facilitate program delivery and administrative activities, such as Google Apps, Gmail (organizational), Google Docs, and Google Drive. Instructors collaborate on learner plans from a distance, which are shared with the mentors and learners to work on goal achievement and career selection. A wiki is used to store PowerPoint slides used for courses delivered in a live online classroom through Saba Centra, which is provided free adult literacy programs in the province. The wiki is also used to house internal working documents such as expense reports and client registrations and assessments, and records of attendance and goal completion. Staff training is provided online and technical support is provided using online tutorials. Last but not least, an online chat client provides on-demand support directly from the program's Web site. One of the program managers reports that it took a year for the staff to get comfortable with the technology and that there is a varying level of comfort with them as well as some frustration with the constant change of technology applications. Overall, however, providing their services online has enabled the agency to grow and provide literacy training to their clients more effectively.

Across the country, there have also been some well-documented uses of technologies in class-based programs. At the Saskatchewan Institute for Applied Science and Technology (SIAST) in Saskatoon, a range of trades, technology, and educational upgrading programs are offered. The Basic Education Program uses SMART Boards or BrightLink with a digital projector as well as adaptive technologies to read text aloud. At the Antigonish County Adult Learning Association (ACALA) and People's Place Library in Antigonish, Nova Scotia, USTREAM (www.ustream.tv) is used to stream documentaries created by program participants, who also work on developing and maintaining the television channel. At the Northwest Territories Literacy Council, a project was launched which offered adult basic educators workshops in Inuvik about how to incorporate blogging and digital storytelling into their practice [28]. In Winsor and Oshawa, adult literacy learners worked with Glogster to create an interactive poster and PhotoStory to make a "How to" video at the Adult Literacy Program of the John Howard Society of Durham Region and the Adult Literacy Program at the Windsor Public Library [29].

In 2011–2012, AlphaPlus, an adult literacy support organization specializing in the use of technologies, used a case study approach to "generate a better sense of how staff, volunteers and students in literacy agencies are working with digital technologies, and to better understand the opportunities and challenges presented by digital technologies in adult literacy teaching and learning." Among the key points of the short-term study [30] were as follows:

- There is no one-size-fits-all model of digital technology integration.

- Maintenance of technology infrastructure is an issue.

- Sufficient financial resources to cover basic costs of developing and maintaining a robust technology infrastructure is crucial to success.

- Sufficient financial resources to enable programs to provide practitioners with time to explore and develop their own digital technology skills, and to incorporate and integrate digital technologies in instruction are crucial to overall success. Release time for professional development and the resources to cover release time to learn are critical issues.

- Organizational culture is important—a culture that fosters and enables professional learning and that values and promotes the use of digital technologies for teaching and learning is key to effectively integrating digital technology with adult literacy practice.

- Strategic planning and prioritization are key drivers for successful use and integration of digital technologies.

- Even students at the most basic levels of literacy can learn using digital technologies.

In these and other programs working on integrating technology-based resources, challenges are many and varied. Raising issues about their use and a critical analysis of their appropriateness for adult literacy learners is also important. Chovanec and Meckelborg [31] argue, based on research with adult literacy learners and practitioners in Edmonton, that using social media, a cloud-based service, does not necessarily bring about text-based literacy development and that ways to bridge the rich informal learning at social networking sites with nonformal and formal adult education settings need to be found. A greater use of technology-based resources is the benefit of adult literacy programs and their clients if issues that hinder their integration are addressed. Even more benefit of instructional technology can be achieved if technology-enhanced learning is made accessible in a cloud computing environment that encourages localization and sharing across the wider community.

15.5 QUESTION 4: WHAT MIGHT A CLOUD COMPUTING STRATEGY FOR ADULT LITERACY LOOK LIKE AND WHAT ARE THE CHALLENGES TO REALIZE SUCH A VISION?

Whenever a new technology is implemented, there is a tendency to first think of it and use it in terms of whatever it replaced, similar to the way automobiles were first thought of as horseless carriages. Gradually, as technology improves, it finds acceptance and stimulates new ideas and new ways of using it—much the way mobile phones merged with personal digital assistants (PDAs) to become smartphones that can access the Internet. Cloud computing is not simply an extension of the Internet; it represents a convergence of Web service provision with high-performance computing, delivered on demand over broadband networks.

Although the initial entry point of cloud computing into the education sector is the outsourcing of e-mail and collaboration software, we are beginning to see ubiquitous access to an unprecedented variety of on-demand computing services—services that require tremendous processing power for short instances—enough power to instantly convert a tourist's digital snapshot of a street sign into text, to translate the text to the target language, and to return an audio message to the user, perhaps with an accompanying map and directions back to the hotel. Such appliances are already being used and can be adapted for a wide variety of literacy applications.

However, augmenting knowledge is not the same as amplifying human learning—while we still do not fully understand how people learn best, we do know many useful ways in which technology can support learning and support the performance of daily tasks. Unfortunately, such promising practices are currently scattered and not collected together into a cohesive framework. For this, we need community building and agreements to make it possible to cut and paste instructional ideas and resources from one computing environment into another. Cloud computing can serve to provide a ubiquitous platform to make such techniques coalesce into a common infrastructure for adult literacy.

The following sections imagine a progression of cloud computing applications from simple (what we are doing now) to complex (what we might do in the future). We pass through our current state of online applications or "Apps" that provide personal computing support and community collaborations to the power that comes from being able to track language acquisition and analyze one's performance in order to prescribe appropriate learning methods and appropriate instructional resources for literacy training. As we may also see the rise of contextualized reading devices that will help everyone decipher the text back into the spoken words it represents, the latest level are applications that make illiteracy no more an impairment than an astigmatism is for those wearing corrective eyeglasses. There are two paths to end illiteracy, and while educators might persevere in efforts to train low-literacy adults, perhaps the real power of cloud computing will be in developing methods and devices that make the stigma irrelevant.

15.5.1 Provision of Personal Computing

No fee provision of application services means anyone who can get on the Internet can have basic word processing, spreadsheets, and e-mail. Gmail, for example, also provides personal file storage and some collaboration tools. No fee cloud access is important for literacy learners as it provides an easier computing environment to learn in, and low-literacy rates go hand in hand with low computer literacy. No fee access provides an Internet identity and a continuing address for the homeless and low-income earners forced to move on a frequent basis.

Moreover, with the appearance of more inexpensive notebook computers, tablets, and smartphones, the cost of each access point is lowered, and thus, the cost of setting up public service and education Internet access facilities is decreasing rapidly. Everyone can afford these cheap devices, and with an expansion of no charge public WiFi, they will have continuing access to the Internet and the cloud.

The mobile phone market has grown to the point where there are now more mobile phones than any other computing device. Each year, more of these are smartphones capable of higher order computing tasks, displaying text, images, and video, and accessing the Internet. These devices are capable of connecting to and through the cloud computing systems. With widespread coverage and a growing installed base of users, wireless networks have the potential for supporting a variety of new on-demand data processing services. Mobile technology providers are quick to encourage growth in the number of applications (apps) by providing efficient online marketplaces such as the Apple Store or Android Market for developers to sell their products or provide them free of charge. Unfortunately, Canada still has one of the most expensive bandwidth costs for wireless access over the cellular telephone networks, so market growth of smart phones will likely be slower for lower income individuals and for those in rural areas where many low-literacy adults reside and where free WiFi service is uncommon [32].

The hardware/software paradigm suggests that anything that could be done in hardware should be replicable by software. This is becoming true for low-cost assistive technologies such as screen readers and talking typewriters that can now be configured on the small touch screen of the smartphone. Wearable and implantable technologies are also emerging, with the potential of being connected to an omnipresent cloud that monitors one's personal health and safety. The matrix of possibilities is so vast that it might be harder to guess when these trends will appear than what will appear. Cloud computing makes it possible to augment the processing power of personal technologies in unprecedented ways.

The practitioners of adult literacy are not rushing to adopt new emerging technology practices. Best et al. [33] provide a recent compilation of promising practices for online literacy training, most of which are text-laden and involve human rather than machine facilitation. While literacy is important for scholarly activity, smart devices may soon help discretely accommodate limited language users by reading aloud or prompting contextually appropriate actions.

15.5.2 Shared Community Resources

Google Docs was originally conceived as a shared space for collaboration in creating and revising documents. This application has potential for supporting shared professional development and educational resources (computer teaching, coaching). Miller [34] suggests that the shared cloud platform also offers greater opportunities for community and work collaborations. An advantage of cloud computing in education noted by the Seattle Project [35] is that students learning programming were no longer disadvantaged by differences in their workstations (although they might be affected by differences in bandwidth). Each student was provided a virtual computer to configure and program, and shared resources were available to all the educators involved.

Since clouds have a potentially unbounded elasticity, it is possible that millions of users can be interacting at once, giving rise to spontaneous communities and interactions. In a social networking environment, there is potential for communities of literacy learners to grow and for literacy providers to develop and test shared resources and enable volunteers working from home. The resulting analytics can also greatly facilitate the ability to evaluate the usage and effectiveness of any materials provided. This is possible now under Web services models, but with a cloud there is potential for having more interchanges of experiences, techniques, content, and learning applications. This amplifies the need for policy directions supporting openness in terms of intellectual exchanges among professionals, release of information using open licenses as open educational resources (OERs), or learning application development as open source. If millions of computer users are connected to the same cloud, essentially they could all access services using the shared network. (Facebook.com already operates a large monolithic cloud that has millions of concurrent users.) This common platform increases the potential for new types of resources that might be cooperatively developed and shared including localized lexicons, information overlays to provide directions or assist adult learning, and employer-specific job training materials.

Programmers in a cloud's user population could contribute in developing or customizing the software and services, much as they do in creating open-source software. Sharing of applications will accelerate the development and spread of new functions the way creative common licensing has accelerated the spread of content and lessons as OERs.

Another possibility is the "crowdsourcing" of volunteer literacy coaches and translators. In the real world, online crowdsourcing is used to recruit volunteer language "role models" to help Spaniards learn English. Diverbo (http://www.diverbo.com) is a Spanish language training organization that offers one to two weeks free room and board to hundreds of anglophone teens and adults each summer to create an English town ambiance, a "Pueblo Inglés" where Spaniards can be tutored in the English language. Lucifer Chu has also demonstrated crowdsourcing of 20,000 volunteers for the translation of MIT OpenCourseWare into Chinese [36]. Using the cloud to build a social network for the adult literacy community, providers can similarly harness the power of volunteers across Canada to support learners and build a useful collection of artifacts and exercises. The United Nations has created an international network of online volunteers who aid in course development, translation, programming, advice, and support (http://www.onlinevolunteering.org/). This type of service for developing countries can be duplicated in Canada to take advantage of the growing number of educated retirees who wish to volunteer their time to support adult literacy initiatives.

A pan-Canadian literacy cloud, combined with accessible and inclusive repositories of OERs that can be used, reused, mixed and mashed, and localized for specific populations would also be of immense help in augmenting the capacity of the diverse adult literacy organizations across the country. The beginnings of such a community of practice can be seen in Tutela.ca, a repository of learning resources for teachers serving newcomers to Canada.

15.5.3 Persistent Personal Storage: Augmenting Cognition

In addition to massive computing power, cloud computer farms also offer rapidly accessible and massive file storage. Cloud-based personal portfolios could readily be used to track the acquisition and use of learning content by learners, and allow the storage of learning artifacts captured on pocket cameras or mobile phones. These ideas exist in some custom server applications, but the reality is that the cloud will make them faster, with more memory, and more accessible from almost anywhere that bandwidth is sufficient and affordable. Local organization or employers could create verbal lexicons. Today, GPS*-equipped smartphones can serve as just-in-time training aids—for example, Øhrstrøm [37] has demonstrated the use of smartphones in Norway as procedural aids for autistic teenagers. Routine tasks such as taking a bus are presented as a series of location-triggered action prompts that the child can refer to as required. This allows the autistic child freedom to travel in a relatively large geographic area while having the security of a smartphone equipped with a repertoire of situational procedures. A personalized "my guide to my community" could help newcomers understand and access services available in their Canadian location.

15.5.4 Analytics and Personalization

Analytics refers to a wide range of data processing methods that use data from a wide range of sources to make inferential decisions about a situation and recommend a path of action. At the low end are a wide variety of computer-based learning tutorials, some of which have been linked to course management systems to keep track of student progress. Performance tracking involves the collection of data about an individual's progress through a set of online learning activities. By tracking the speed and outcomes of learning activities, an individual's performance can be compared to aggregate histories of a large numbers of learners moving through the same courses. The resulting analysis can lead to pattern matching and identification of persistent learner errors and personal

---

* GPS—Geographic Positioning System.

characteristics (such as speed of cognitive processing) that could forecast learner outcomes or be used to prescribe remedial exercises.

These computational methods are used to track credit card purchases and identify activities that are uncharacteristic of the cardholder's previous purchasing patterns, potentially indicating inappropriate use. The emerging research in this area involves tracking data and providing analytics to suggest optimal learning paths based on learners' preferences and observed performance.

The elasticity of cloud computing is ideal for this kind of large-scale instantaneous analysis. Not all the data need to be gathered automatically—teachers at the Open High School of Utah track student progress by making notes in a constituent relationship management (CRM) system. As teachers interact with students, they make notes of progress and problems, and the system prompts the teacher whenever a student falls behind or fails to keep in touch [38]. If installed in a cloud computer, such a tracking system could help teachers everywhere monitor the progress of learners and provide the social contact and personalization that is so important for learner engagement and retention.

Cloud computing already supports a wide range of virtual worlds and online multiplayer games; teenagers spend innumerable hours on their XBOXes, Playstations, and other gaming systems, using avatars to form teams for virtual assaults on military targets in cyberspace. Today's games are highly collaborative and interactive. Players can communicate with each other using headsets or text and they learn how to form groups to cooperatively develop strategies and solutions in team-based game environments. While much learning takes place with these games, it has little intentional learning related to the skills of reading, writing, and arithmetic. Educational games come across as being rather dull in comparison—imagine the gains that could be made if content and applications enabling literacy learning were embedded in such massively subscribed cloud-based edutainment systems.

15.5.5 Policy Issues

Policy and control issues are crucial. The provincial/federal disputes are a major cause of fragmentation across the country. This and other issues such as regulatory compliance to ensure security, enable audits, and preserve privacy represent significant barriers to the adoption of cloud computing in adult literacy circles. Although a common platform affords easier collaboration, it also increases security risks. In particular, the areas of identification and authentication will require new schemes to preserve privacy and gain the trust of the users, while developing measures to boost the security of publicly accessible systems that may come under attack. Much work has been done in these areas with the creation of federations that act as local authentication agents for individuals to access broader cloud assets. However, the continuous parade of lost identity cases serves to both remind and undermine the degree of confidence that should be afforded service providers.

15.5.6 Beyond Text: Is Literacy Obsolete?

Early digital computers had to be programmed using binary code, and only in the 1970s, did we see higher level computer languages that allowed programmers to specify directions in English-like text commands. Today many computers (like those used in a car's navigation system) can be directly interfaced by voice commands. Indeed, smartphones equipped with cameras can easily read quick response (QR) codes and retrieve related messages from the Internet—including short video clips or other situation-relevant material. With enhanced processing, text analysis can be made available to scan and interpret text—not just into English, but through other Web services such as Google Translate, into the target language of choice. For the large number of new Canadians who struggle in adult basic education classes, this form of literacy appliance can be an excellent assistive technology. Voice to text, text to voice, French to English, or Chinese or any other language, we are approaching the era where the universal translator once the stuff of science fiction (like the Babel Fish translator in *The Hitchhiker's Guide to the Galaxy*) is becoming a reality.

Universal literacy is a fairly modern concept that came along with the industrial revolution and the need to have a literate population to work and communicate in the era of the post office. Before literacy, specialists called "scribes" were called upon to write and read letters dictated by the illiterate members of their community. Perhaps, voice and video over Internet and mobile phones have flattened the need for this type of training, and with electronic book readers, the illiterate have gained access to copious amounts of text information. In parts of Africa, the tribal drums have given way to solar-powered FM radio transmitters and mobile phones—neither of which rely on the heavy burden of text that extracts so many years of anguish on the dyslexic population and others that have the misfortune of reading difficulties. In the near future, speech-to-text and text-to-speech applications will help to level the playing field for those with learning difficulties or who have not had the advantage of a good school in early life. While text literacy might not become obsolete, it may, like Latin, become less significant as an element to a person's immediate and direct participation in society. Nonetheless, computer literacy and access to computing resources will continue to increase in importance and will grow as a critical component in the curriculum of adult education.

15.5.7 Conclusion: The Impact of Cloud Computing on Adult Education and Literacy

This chapter provides a glance at rapidly emerging technology and attempts to grasp its potential impact for the world of adult learning and literacy. Let us recap some basic notions.

First of all, "cloud computing" is a movement toward utility computing where large "server farms" located next to "green energy" sources and connected by low-power high-bandwidth fiber optics will provide the computing infrastructure for many small, medium, and large organizations that can no longer cost-effectively provision their own in-house IT systems. The first of these commercial systems have already been launched by companies such as Amazon, Google, and Microsoft, and many more are being planned. Cloud computing facilities are also being used for research and for government services. Some clouds are public and can be used by anyone; others are private and tightly secured to protect the privacy of the information contained. Both Microsoft and Google are giving away cloud computing capacity to educational organizations to run custom e-mail and other documentation sharing services. This appeals to universities because student e-mail alone is costing them hundreds of thousands of dollars each year.

Thus, a *first step* toward the use of cloud computing by an adult literacy community could be to recommend to learners the *use of the free services* available or make a special e-mail arrangement with one of these providers if a branded e-mail address is preferred. It would be ideal if a significant number of adult literacy providers in Canada could collaborate on this approach, because then the same cloud provider could host a portfolio of specialized services of benefit to learners with literacy difficulties. It would also make it easier to codevelop and share other services in the future. Every adult literacy learner would benefit by having free e-mail and free access to these services, and the adult literacy community could benefit by using the data collected to refine software and determine new services that might be useful. This could all be achieved without losing traditional organizational or institutional e-mail identities or "logo brands." Probably, it would take more time to negotiate the collaboration agreement among the literacy providers than to implement the technical service, so this would require vision and leadership to pave the way. The emergence of a consolidated collaborative cyber community for adult literacy would show the way to future collaborations in literacy training software, literacy appliance software, instructor professional development, and research. It would also be possible for an adult literacy learner to have continuity of e-mail and literacy support if they moved from one community to another.

The *second* important notion is that cloud computing is "elastic" and provides computing power on demand. Just as cyber security codes can be quickly hacked by tasking a thousand virtual machines to work for 2 minutes, powerful analysis routines could help track and coach literacy learners in a just-in-time analysis of their needs. This is not "ready to go," but it is within the realm of current knowledge and systems; however, the knowledge and routines are scattered in pockets. Identifying requirements and unifying the system to do, this should be the *second step*. This can only be done effectively by organizing the community of practice to become involved.

Assuming that a community can be coordinated, once the basic parameters are known, many of the lessons can be assembled from OER repositories and documented; others might be created or mixed through the "wisdom of crowds" wherein tasks are distributed among the many community literacy volunteers and researchers. Collaborative research projects could be sought after creating the analytic software to track and coach individuals as individuals who are working toward common goals. (Richards proposed this concept in 2009 as the Community Open Literacy Toolkit.)

Third, in order to take full advantage of the affordances enabled by cloud computing, the adult learning community needs to support the development, adaptation, assembly, and dissemination of OER. With proprietary content and applications, the burden of requesting permission and/or having to pay again and again as the materials are used and reused in different formats significantly negates the advantages of the cloud. Users need to have free reign to mix and remix the content and adapt it for voice and video as appropriate for their learners. The cloud can provide learners and their organizations with access to the growing number of free open education resources as well as open-source applications supporting social interaction, publishing, collaborating, editing, content creation, computing, and so on [39].

The *fourth* notion is that literacy training can be augmented with literacy appliance software that provides just-in-time assistance to low-literacy adults.

The range of software could include text-to-voice, voice-to-text, bar code and QR code reading, and language translation. Much of these services exist as *Web services*, but they need to be harnessed and brought together in a suite of applications accessible and usable by the low-literacy population. Cloud computing can both provide a collaborative portal for these services as well as the high-power computing necessary to extract the text or shapes from pictures and generate the appropriate response including related information that might be available. While literacy would be ideal, such applications may make it possible for low-literacy adults to participate more inclusively in everyday life. The benefits of such adaptations can be expected to benefit other populations such as seniors or tourists.

Fifth, cloud computing and the Internet are available through an increasing number of mobile devices—in fact, more adult literacy learners have mobile phones rather than personal computers and mobile tablets are becoming increasingly more popular and are beginning to augment and even replace laptops and netbooks. Thus, mobile devices as a delivery platform should be given priority for research and technical development— over printed texts and personal computers. These mobile devices represent the state of the art, and they are always with the adult learners go, and are becoming the platform of choice for accessing a wide range of services including training through *mobile learning*.

Sixth, finally and most significantly, is the reality that it is becoming impossible to conceive a modern definition of literacy that excludes ICT literacy. The growing importance of the Internet and networking skills for adults must be recognized. A 21st century literacy is not possible without the skills for accessing and using the Internet. The cloud can be the doorway to these skills.

Cloud computing is at the adult literacy doorstep, but it will take time to implement the above ideas. Some of these ideas face technical barriers, others face cultural and political barriers, and some have distant ideas in need of more research. However, they do provide a unified vision of what is possible, if the adult literacy community can collaborate together for mutual benefit. Then all literacy providers and the adult literacy learners will surely benefit from the synergies that emerge. Canada is large and vast—the literacy movement needs to coordinate its efforts in a way that retains and reinforces the local roots and human face. Cloud computing provides an affordable opportunity to plan a new future together.

ACKNOWLEDGMENTS

This chapter originated from a study funded by AlphaPlus, a nonprofit adult literacy agency in Toronto, Ontario.

REFERENCES

1. Crooks, S., Davies, P., Gardner, A., Grieve, K., Mollins, T., Niks, M., Tannenbaum, J., and Wright, B. (2008). Connecting the dots. Accountability in adult literacy: Voices from the field. The Centre for Litercy [SIC] of Quebec. Retrieved from http://www.literacyandaccountability.ca/File/03_CTD_Field_report_Oct_2008.pdf
2. Cross, J. (2007). *Informal Learning: Rediscovering the Natural Pathways That Inspire Innovation and Performance.* San Francisco, CA: Wiley & Sons.
3. Powell, J. (2009). Cloud computing—What is it and what does it mean for education? Retrieved from http://erevolution.jiscinvolve.org/files/2009/07/clouds-johnpowell.pdf
4. Piña, R. A. and Rao, B. (2010). The emergence and promise of cloud computing for under-developed societies. Paper presented at the *Proceedings of PICMET 2010 Technology Management for Global Economic Growth.* Phuket, Thailand, July 18–22. Retrieved from http://faculty.poly.edu/~brao/2010.Cloud.PICMET.pdf
5. Contact North. (2010). The future of e-learning: Realities, myths, challenges and opportunities. Retrieved from http://contactnorth.ca/sites/default/files/contactNorth/files/pdf/discussion-papers/the_future_of_e-learning_-_realities__myths__challenges__and_opportunities.pdf
6. de Broucker, P. and Myers, K. (2006). Too many left behind: Canada's adult education and training system: Research report W|34. Retrieved from http://www.cprn.org/doc.cfm?doc=1479
7. Pingdom. (2009). The origin of 9 popular Web buzzwords. Retrieved from http://royal.pingdom.com/2009/04/07/the-origin-of-9-popular-web-buzzwords/
8. Mell, P. and Grance, T. (2009). The NIST definition of cloud computing. National Institute of Standards and Technology. Information Technology Laboratory, Version, 15(10.07), Retrieved from http://www.csrc.nist.gov/groups/SNS/cloud-computing/index.html
9. Chen, X., Liu, J., Han, J., and Xu, H. (2010). Primary exploration of mobile learning mode under a cloud computing environment. Paper presented at the *International Conference on E-Health Networking, Digital Ecosystems and Technologies,* Shenzhen, China. Retrieved from http://ieeexplore.ieee.org/xpl/freeabs_all.jsp?arnumber=5496435
10. Roth, T. (2010). Cracking passwords in the cloud: Amazon's new EC2 CPU instances (Web blog of November 15, 2010). Retrieved from http://stacksmashing.net/2010/11/15/cracking-in-the-cloud-amazons-new-ec2-gpu-instances/.
11. Keahey, K., Figueiredo, R., Fortes, J., Freeman, T., and Tsugawa, R. (2008). Science clouds: Early experiences in cloud computing for scientific applications. *Conference on Cloud Computing and Its Applications,* Chicago, IL, October. Retrieved from http://www.nimbusproject.org/files/Science-Clouds-CCA08.pdf

12. Haigh, G. (2010). Baby steps into the cloud: ICT as a service for education. (Corporate brochure). Reading: Microsoft Corporation. Retrieved from http://blogs.msdn.com/b/ukschools/archive/2010/12/07/microsoft-education-white-paper-baby-steps-into-the-cloud.aspx

13. Danek, J. (2010). Government of Canada cloud computing: Information technology shared services (ITSS) roadmap. (Powerpoint presentation). Retrieved from http://isacc.ca/isacc/_doc/ArchivedPlenary/ISACC-10-43305.pdf

14. Katz, R., Goldstein, P. J., and Yanosky, R. (2009). Demystifying cloud computing for higher education. *ECAR Research Bulletin*. Retrieved from http://www.educause.edu/ecar

15. Fahy, P. J. and Twiss, D. (2010). Adult literacy practitioners' uses of and experiences with online technologies for professional development. *Journal of Applied Research on Learning* 3(2): 1–18. Retrieved from www.ccl-cca.ca/pdfs/JARL/Jarl-Vol3Article2.pdf

16. Fleer, M. and Raban, B. (2005). Literacy and numeracy that counts from birth to five years: A review of the literature.

17. Folinsbee, S. (2008). Online learning for adults: Factors that contribute to success (A literature review). Sudbury, ON: College Sector Committee for Adult Upgrading. Retrieved from http://www.nald.ca/library/research/csc/litreview/litreview.pdf

18. Holum, A. and Gahala, J. (2001). Critical issue: Using technology to enhance literacy instruction. (Web posting). North Central Regional Educational Laboratory. Retrieved from http://www.ncrel.org/sdrs/areas/issues/content/cntareas/reading/li300.htm

19. Horsman, J. and Woodrow, H. (Eds.) (2006). *Focused on Practice: A Framework for Adult Literacy Research in Canada*. St. John's, NL: Harrish Press. Retrieved from http://decoda.ca/wp-content/uploads/FocusedOnPractice.pdf

20. Movement for Canadian Literacy (2007). Environmental scan: Literacy work in Canada. Retrieved from http://www.literacy.ca/content/uploads/2012/03/Environmental-Scan-of-Literacy-Work-in-Canada-2007-2.pdf

21. Myers, K. and de Broucker, P. (2006). Too many left behind: Canada's adult education and training system. (Report for the Canadian Policy Research Network). Retrieved from http://www.cprn.org/documents/43977_en.pdf

22. Nadin, M. (2001). *The Civilization of Illiteracy.* [Project Gutenberg electronic text #2481. Originally published Dresden University Press, 1997]. Retrieved from http://digital.library.upenn.edu/webbin/gutbook/lookup?num=2481

23. Organization of Economic Cooperation and Development. (2002). Thematic review on adult learning: Canada country note. Retrieved from http://www.oecd.org/dataoecd/51/31/1940299.pdf

24. Organization of Economic Cooperation and Development. (2013). Education, economy and society: Adult literacy. Retrieved from http://www.oecd.org/edu/educationeconomyandsociety/adultliteracy.htm

25. Innovative Communities Connecting and Networking [iCCAN] (2010). Literacy tutor training pilot program a first in Alberta. *iCCAN Connected.* Winter. Retrieved from http://www.iccan.ca/newsletters/119-winter-2010-newsletter

26. Kurzweil Educational Systems (2005). Scientifically-based research validating Kurzweil 3000—An annotated review of research supporting the use of Kurzweil 3000 in English language learner classrooms. (Monograph). Retrieved from https://www.kurzweiledu.com/files/K3000%20ELL%20Research.pdf

27. Ontario Ministry of Training, Colleges, and Universities (2013). *Literacy and Basic Skills: Learning Online.* Retrieved from http://www.tcu.gov.on.ca/eng/training/literacy/online.html

28. Smythe, S. (2012). Incorporating digital technologies in Adult Basic Education: Concepts, practices and recommendations. AlphaPlus. Retrieved from http://incorporatingtechnologies.alphaplus.ca

29. Greig, C. and Hughes, J. (2012). Adult learners and digital media: Exploring the usage of digital media with adult literacy learners. AlphaPlus. Retrieved from http://digitalmedia.alphaplus.ca

30. AlphaPlus (2012). Learning together with technologies: Illustrative case studies. Retrieved from http://learningtogether.alphaplus.ca

31. Chovanec, D. and Meckelborg, A. (2011). Social networking sites and Adult Literacy: Raising the issues. AlphaPlus. Retrieved from http://socialnetworking.alphaplus.ca

32. Kaminer, A. and Anghel, B. (2010). Death grip: Caught in a contract and cannot quit? Toronto Sun. Retrieved from http://www.seaboardgroup.com/main/index.php?option=content&task=view&id=825&Itemid=212

33. Best, L., Kaattari, J., Morgan, D., Trottier, V., and Twis, D. (2009). Bridging distance: Promising practices in online learning in the Canadian literacy community. (Monograph). Retrieved from http://www.nald.ca/gettingonline/goresources/bridgingdistance/bridgingdistance.pdf

34. Miller, M. (2008). *Cloud Computing: Web-Based Applications That Change the Way You Work and Collaborate Online.* New York: Pearson.

35. Cappos, J., Beschastnikh, I., Krishnamurthy, A., and Anderson, T. (2009). Seattle: A platform for educational cloud computing [Electronic Version]. *ACM SIGCSE Bulletin*, 41(1). Retrieved from http://portal.acm.org/citation.cfm?id=1508905

36. Radio Taiwan International [RTI]. (2010). Newsmakers: Lucifer Chu. RTI+Plus blogpost http://blog.rti.org.tw/english/2010/10/03/newsmakers-lucifer-chu/.

37. Øhrstrøm, P. (2010). Helping autism-diagnosed teenagers navigate and develop socially using e-learning: Some reflections on design and ethics. Paper presented at the *Arctic Frontiers 2010—Thematic Conference on Distance Learning.* February, Tromso, Norway.

38. Wiley, D. (2011). Presentation at the *Open Education Conference.* Barcelona, Spain.

39. Bittman, T. (2008). Cloud computing and K-12 education. (Blog posting). Retrieved from http://blogs.gartner.com/thomas_bittman/2008/11/26/cloud-computing-and-k-12-education/.

Index